Neurotoxicology of Insecticides and Pheromones

Neurotoxicology of Insecticides and Pheromones

Edited by

Toshio Narahashi

Northwestern University
Chicago, Illinois

PLENUM PRESS • NEW YORK AND LONDON

Library of Congress Cataloging in Publication Data

Symposium on Chemistry of Neurohormones and Neurotransmission, Anaheim, Calif., 1978.
Neurotoxicology of insecticides and pheromones.

"Proceedings of the Symposium on Chemistry of Neurohormones and Neurotransmission held in conjunction with the 175th anniversary of the American Chemical Society, at Anaheim, California, March 14–15, 1978."
Includes index.
1. Pesticides – Physiological effect – Congresses. 2. Neurotoxic agents – Physiological effect – Congresses. 3. Pheromones – Physiological effect – Congresses. 4. Insects – Physiology – Congresses. I. Narahashi, Toshio. II. American Chemical Society. III. Title.
SB951.5.S95 1978 595.7'02'4 78-10913

ISBN-13:978-1-4684-0972-7 e-ISBN-13:978-1-4684-0970-3

DOI: 10.1007/978-1-4684-0970-3

Proceedings of the Symposium on Chemistry of Neurohormones and Neurotransmission held in conjunction with the 175th Anniversary of the American Chemical Society, at Anaheim, California, March 14–15, 1978

Softcover reprint of the hardcover 1st edition 1979

A Division of Plenum Publishing Corporation
227 West 17th Street, New York, N.Y. 10011

Preface

Substantial progress has been made during the past three decades in the development of a variety of chemical means to control insect pests. A large number of highly effective insecticides have been developed, and many of them have contributed to a great extent to increasing agricultural productivity and eradicating diseases transmitted by vectors. However, contamination of our environment caused by various pesticides has become a serious problem, and various attempts have been made to develop newer methods of controlling insect pests. One of the approaches is to develop newer insecticides which cause less contamination of the environment without losing their insecticidal potencies. Another approach, among other things, is to utilize pheromones to control certain species of insects.

No matter what approach one may choose, it is imperative to understand the mechanism of action of pesticides for efficient development and utilization of these chemicals. One important feature common to most insecticides and pheromones is their interaction with the nervous system which is the main target site. Thus the mechanisms of action of these chemicals on the nervous system represent one of the most critical aspects in pesticide toxicology. In comparison to the chemical and biochemical studies dealing with the metabolism of various pesticides, the study of neurotoxicology has lagged behind to a considerable extent for various reasons. Technical complications involved in neurotoxicological experiments appear to be one of the contributing factors. Unlike the two-dimensional scheme of the pesticide metabolism study, neurotoxicological study involved multi-dimensional factors such as the physico-chemical nature of membrane permeability, enzymatic reactions, and the integrated nervous function as an organ.

Recognizing the critical importance of neurotoxicology in further development of effective chemical means of pest control, the Division of Pesticide Chemistry of the American Chemical Society held a Symposium on Chemistry of Neurohormones and Neurotransmission on March 14 and 15, 1978, in conjunction with the 175th National Meeting of the American Chemical Society at Anaheim, California.

The purpose of this Symposium was to provide a forum where a variety of scientists who were interested in the interactions of insecticides and pheromones with the nervous system got together to exchange their views. Thus chemists, biochemists, electrophysiologists and neurochemists gathered, and thirteen papers were presented. Three of them dealt with pheromones and chemical perception, seven papers were concerned with the electrophysiological modes of action of insecticides on the nervous system, and three papers were related to the neurochemical modes of action of insecticides. This book represents the publication of all these papers. It is my hope that this publication will be a useful reference in the field of pesticide neurotoxicology and that it will provide the basis on which further advancement in nerurotoxicology is made.

Toshio Narahashi

Contents

INTRODUCTION: NERVOUS SYSTEM AS A TARGET SITE OF PHEROMONES AND INSECTICIDES

Toshio Narahashi

Department of Pharmacology, Northwestern University

Medical School, Chicago, Illinois 60611

The chemistry of neurohormones and insecticides has achieved remarkable progress during the past decade or two. These chemicals act primarily on the nervous system of insects thereby exerting unique effects. Neurotoxicological studies have also made substantial progress recently. However, there has been virtually no chance for both chemists and neurotoxicologists to get together to exchange views. The time is ripe and this symposium is going to provide them with a unique forum to bring their current data together.

Behavioral responses induced in insects by pheromones are undoubtedly the result of stimulation of the sensory nervous system. Electrophysiological techniques have clearly demonstrated such responses in the insect antennae. Specificity of the pheromone-induced behavior is due to the specific configuration of the sensory receptor and its specific affinity for the pheromone, but the detailed molecular mechanism of such interaction between the pheromone and the receptor remains largely to be explored.

For most insecticides currently in use, it has clearly been established that the nervous system is the major target site. However, the mechanism of action on the nerve is considerably different among those insecticides acting on the nerve. Cholinesterases in the nervous system are indeed the prime target for certain insecticides such as organophosphates and carbamates, whereas no specific enzyme system has been clearly established as the target for some other insecticides. It should be emphasized that excitation of the nerve takes place as a result of changes in nerve membrane permeabilities to ions such as sodium and potassium. These permeability changes are physico-chemical phenomena of the membrane not dependent upon the metabolic energy, and the sites of ion permeation are called "ionic channels" which have been shown to be holes, not carrier mechanisms. It has indeed been demonstrated that permeabilities of ionic channels are drastically altered by certain chlorinated hydrocarbon and pyrethroid insecticides thereby causing severe symptoms of poisoning in the nervous system.

Studies of the insecticidal action on the nervous system are of importance in various respects. It is quite obvious that the interactions between insecticides and nerve membranes must be explored in order to understand the molecular mechanism of action of insecticides. Besides this, the study of insecticides with the nervous system has a great impact on our understanding of the structure-activity relationship of insecticides.

In order to establish the structure-activity relationship for any group of insecticides acting primarily on the nervous system, we have to understand the processes of toxic action of insecticides. There are at least three major steps for an insecticide to exert its killing action on insects, i.e. penetration through the cuticle or other routes, activation and/or detoxication, and action on the target nerve. In addition, a variety of reactions may be interposed during these processes. Thus the potency of an insecticide to kill insects is a final manifestation of a series of complex reactions, and it could be misleading to relate the molecular structure of insecticides to the killing potency. The structure must be compared with the action on the target site. This is relatively easy if we can isolate the target component *in vitro*, and much progress has indeed been made along this line using organophosphate and carbamate insecticides which inhibit cholinesterases (Fukuto, 1971; Metcalf, 1971).

However, experiments are more complicated for many other insecticides because their effects must be compared on the nerve. We have recently developed a technique whereby a large number of insecticide derivatives can be compared for their direct effect on the isolated crayfish abdominal nerve cord which is fairly sensitive to many insecticides (Takeno et al., 1977). Through the study of a variety of synthetic pyrethroids it has been found that the potency on the nerve and the potency to kill insects do not necessarily run parallel with each other. Some pyrethroids are potent neuropoisons but lack strong insecticidal activity, whereas some others are relatively weak on the nerve yet can kill insects effectively. One such example is shown in Table 1. The potency to stimulate the nerve is measured as the concentration to stimulate the frequency of spontaneous discharges of the crayfish abdominal nerve cord to 200% of the control level (nerve stimulation 200 or NS_{200}), and the insecticide potency is measured as LC_{50}. Although (-)-*trans* phenothrin is only about half as potent as (+)-*trans* phenothrin in stimulating the nerve, the former virtually

Table 1. Potencies of phenothrin in stimulating the crayfish abdominal nerve cord and in killing houseflies.

Compounds	NS_{200} ($\times 10^{-8}$M)	LC_{50} ($\times 10^{-4}$M)
(+)-*trans* phenothrin	5.2	4.85
(-)-*trans* phenothrin	11.5	>143
(+)-*cis* phenothrin	6.8	5.71
(-)-*cis* phenothrin	102	114

lacks insecticidal potency while the latter is a potent insecticide. On the other hand, the correlation between the nerve and insecticidal potencies is good for (+)- and (-)-*cis* phenothrin. This example clearly indicates the need for comparing the direct action on the target site for the purpose of elucidating the true structure-activity relationship. There are a number of such interesting examples (Narahashi, 1978; Nishimura and Narahashi, 1978; Narahashi et al., 1977).

The topics covered by the present symposium may be divided into three large categories, i.e. pheromone perception and receptors (W.L. Roelofs, T.L. Payne and D.M. Norris), neurophysiology of insecticide action (T.A. Miller, D.W. Gammon, A.E. Lund, D.L. Shankland, D.E. Woolley, J. van den Bercken and T. Narahashi), and neurochemistry of insecticide action (F. Matsumura, R.D. O'Brien and T.R. Fukuto). The aim of this symposium is not only to provide a forum to present current developments in this field but also to facilitate communications and collaborations between chemists and neurotoxicologists for further advancement of newer pest control techniques.

REFERENCES

Fukuto, T.R., 1971, Relationships between the structure of organophosphorus compounds and their activity as acetylcholinesterase inhibitors, Bull. W.H.O. 44: 31-42.

Metcalf, R.L., 1971, Structure-activity relationships for insecticidal carbamates, Bull. W.H.O. 44: 43-78.

Narahashi, T., 1978, Neurophysiological study of the structure-activity relations of insecticides, in: Pesticides and Venom Neurotoxicity, ed. D.L. Shankland, R.M. Hollingworth and T. Smyth, Jr., Plenum, New York. p. 119-131.

Narahashi, T., K. Nishimura, J.L. Parmentier, K. Takeno and M. Elliott, 1977, Neurophysiological study of the structure-activity relation of pyrethroids, in: Synthetic Pyrethroids, ACS Symposium Series No. 42, ed. M. Elliott, Amer. Chem. Soc. p. 85-97.

Nishimura, K. and T. Narahashi, 1978, Structure-activity relationships of pyrethroids based on direct action on nerve. Pesticide Biochem. Physiol. 8: 53-64.

Takeno, K., K. Nishimura, J. Parmentier and T. Narahashi, 1977, Insecticide screening with isolated nerve preparations for structure-activity relationships, Pesticide Biochem. Physiol. 7: 486-499.

PHEROMONE PERCEPTION IN LEPIDOPTERA

Wendell L. Roelofs

New York State Agricultural Experiment Station

Cornell University, Geneva, New York 14456

The sex pheromone of many lepidopteran species has been identified as a mixture of 2-4 components. Even the first identified pheromone, bombykol, in *Bombyx mori* silkworms (Butenandt et al. 1959) has recently been found to involve a second component, bombykal (Kasang et al. 1978). If a single pheromone component is used, it is assumed that the sequence of evoked behavioral responses is elicited by it, along with any visual, tactile, or auditory stimuli that may operate at close range. With multichemical pheromone systems, however, it has been found (Cardé et al. 1975; Baker et al. 1976) that some of the components, named *primary components* (Roelofs and Cardé 1976), elicit long-distance upwind anemotaxis, whereas other components named *secondary components*, are not essential for eliciting upwind anemotaxis, but are used in combination with the primary components to evoke close-range responses, such as landing, wing fanning, hair pencilling, and copulatory attempts. For example, the redbanded leafroller moth, *Argyrotaenia velutinana*, has a 3-component sex pheromone system consisting of (Z)- and (E)-11-tetradecenyl acetate (91:9) (Z11- and E11-14:Ac) and dodecyl acetate (Roelofs et al. 1975). The long-range upwind anemotactic responses of the males are elicited by the geometrical isomers, whereas, the presence of dodecyl acetate in combination with the isomers elicits an increase in the frequency of landing, walking to the source while wing fanning, and finally copulatory attempts (Baker et al. 1976).

We know little about the ethological functions of individual pheromone components and we know little about the perception processes involved in pheromone blends. At the peripheral level, studies have shown that species utilizing a mixture of two primary components can possess two functionally different acceptor sites. Recordings (Kochansky et al. 1975) from *Sensilla trichodea* on male wild silk moth, *Antheraea polyphemus*, antennae showed that there were 2 sex pheromone receptor cells. The cell with the larger spike amplitude was responsive to the major component of the blend (*E*,*Z*)-6,11-hexadecadienyl acetate, whereas the other cell was responsive to the corresponding aldehyde pheromone component. Two pheromone receptor cells also were found (Den Otter 1977) with *Sensilla trichodea* olfactory hairs of male summerfruit tortrix moth, *Adoxophyes orana*, antennae. The cell with the larger spike amplitude responded to the major pheromone component of the blend, (*Z*)-9-tetradecenyl acetate (Z9-14:Ac), whereas, the other cell responded to the minor component, Z11-14:Ac, and to a lesser extent to Z9-14:Ac. Studies with component mixtures indicated that the 2 cells are not tuned to specific mixtures, and thus refutes the presence of a "labeled line" system of quality coding. It was concluded that the specific bahavioral responses to a 9:1 ratio are determined in the central nervous system by input from the 2 cells. Den Otter (1977) suggests that the pheromone information is a function of the activities of each individual olfactory hair, and that the afferent information from many hairs serves to improve the signal-to-noise ratio. O'Connell previously (1975) had recorded from 2 scent cells of *Sensilla trichodea* on male redbanded leafroller antennae. Both cells responded to the major component, Z11-14:Ac, but the cells with the highest spike amplitude had a greater response for this compound, whereas the other cells had a greater response for the geometrical isomer pheromone component, E11-14:Ac. O'Connell found that the response frequencies to the pheromone components and to 6 other related compounds varied between the 2 cells, and concluded that these differences were due to intrinsic factors within the receptor neurons. He suggests that the pheromone perception process involves a pattern of activity arising across an ensemble of receptor neurons, each having its own distribution of sensitivities. Blend specificity would be determined by the central nervous system with input required from numerous olfactory hairs.

Secondary components can further complex the peripheral input by increasing the impulse frequency of one of the scent cells. With the redbanded leafroller moth (O'Connell 1972), dodecyl acetate elicits small responses by itself, but greatly increases the response of the high spike amplitude cell when mixed with Z11-14:Ac compared to the response to Z11-14:Ac alone. The same phenomenon was found with the tobacco cutworm moth, Spodoptera litura (Aihara and Shibuya 1976). In this case, recordings from single olfactory cells showed that the secondary pheromone component, (Z,E)-9,12-tetradecadienyl acetate, elicited no response by itself, but when mixed with the primary pheromone component, (Z,E)-9,11-tetradecadienyl acetate, it increased the impulse frequency at least 4-fold over the response to the primary component used alone.

The exact processes for perceiving specific pheromone blends are not known, but the use of geometrical and positional isomers in species specific pheromone component mixtures has been found to be quite common. Examples of some specific blends used by leafroller moths is given in Table 1. The component ratio for each species can be determined by analyzing female gland extract or effluvium, and it is generally found that the best ratio for luring males in field trapping studies closely approximates that found to be produced by the corresponding female moth. There are notable exceptions to this, however, and sometimes males are captured by a

Table 1.-Ratios of positional and geometrical isomers in attractant systems for various tortricid species (some additional compounds are used with some of the species).

Species	14:Ac's		
	Z-11	E-11	Z-9
Adoxophyes orana[a]	100	-	900
Adoxophyes fasciata[b]	100	-	200
Clepsis spectrana[c]	100	-	10
Pandemis limitata[d]	100	-	10
Choristoneura rosaceana[e]	97	3	-
Argyrotaenia velutinana[f]	91	9	-
Archips mortuanus[g]	90	10	1
Archips argyrospilus[g,h]	60	40	4
Archips podana[i]	60	40	-
Archips semiferanus[j]	33	67	-
Archippus breviplicanus[k]	30	70	-
Archips cerasivoranus[g]	15	85	-
Platynota stultana[l]	11	89	-

[a]Meijer et al. 1972; Tamaki et al. 1971b
[b]Tamaki et al. 1971a
[c]Minks et al. 1973
[d]Roelofs et al. 1976a
[e]Roelofs and Tette, 1970; Hill and Roelofs, 1978
[f]Roelofs et al. 1975
[g]Carde et al. 1977
[h]Roelofs et al. 1974
[i]Persoons et al. 1974
[j]Miller et al. 1976
[k]Sugie et al. 1977
[l]Hill and Roelofs 1975

broad range of ratios. These findings suggest that the male perception processes are not as finely tuned and specific as previously thought. However, a reevaluation of the data indicates that the apparent lack of specificity might be caused by the testing methods. In this paper I will discuss the specificity of male moths as related to their behavioral responses to precise pheromone component blends.

BLEND SPECIFICITY FOR MOTH ACTIVATION

As stated above, the redbanded leafroller moth uses a 91:9 blend of primary pheromone components, Z11- and E11-14:Ac. Laboratory bioassays (Baker et al. 1976) were conducted to determine if the males would respond specifically to this blend, or if there was "broad tuning" for responses to a range of ratios. At low concentrations (2 ng pheromone on filter paper), the 91:9 mixture elicited the greatest moth activation response (Fig. 1). It would

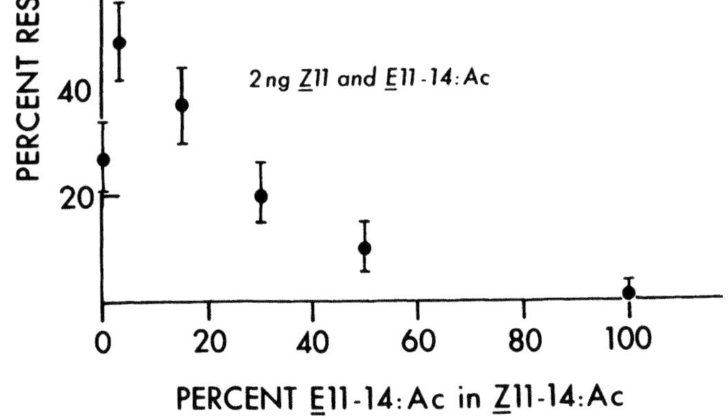

Fig. 1.-Activation responses of redbanded leafroller males (from Baker et al. 1976).

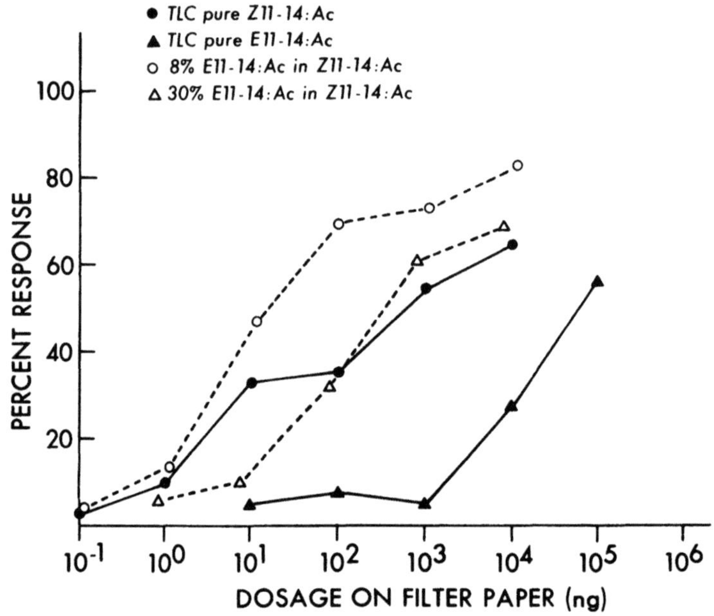

Fig. 2.-Wing-fanning responses of redbanded leafroller males (from Baker et al. 1976).

appear from this figure taken alone that increasing amounts of the E-isomer cause a decrease in the moth responses. Further studies, however, showed that activation responses could be obtained from the whole range of mixtures if the concentrations were substantially increased (Fig. 2). For example, pure Z11-14:Ac or the 70:30 blend of Z-11/E-11 elicited responses similarly to those obtained with the 91:9 blend if the concentrations were increased by 10-fold, and similar responses were obtained with pure E11-14:Ac if its dosage was increased 1000-fold. These data indicate that specificity for the perception of the blends is related to the

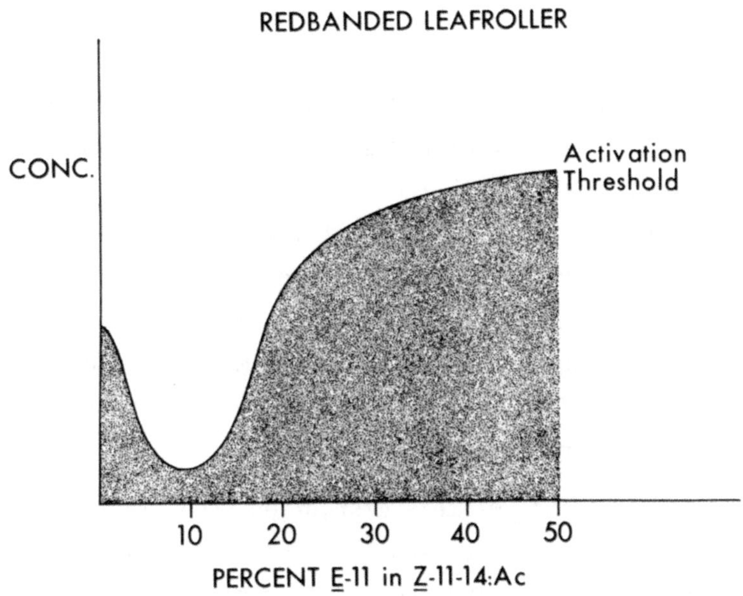

Fig. 3.-Postulated activation threshold to various blends of E- and Z11-14:Ac (from Roelofs 1978).

various threshold levels for behavioral responses. A graph for the moth activation thresholds for redbanded leafroller males could be similar to that postulated in Fig. 3.

The peripheral receptors can send their messages at any time, but the thresholds undergo large daily changes as affected by light cycles, temperature, and other environmental inputs in the central nervous system. Thus the actual "shape" of the threshold curve most likely is quite changeable.

EFFECTS OF EXCESS INPUT

In 1964, Wright (1964) predicted that antennal receptors would fatigue or saturate if subjected to pheromone concentrations up to a hundred thousand times higher than the threshold concentration of detection. The exact mechanism is not known, but a disruptive effect on male moth attraction does occur at concentrations of 10^2-10^3 times higher than the male response threshold level (Shorey et al. 1967; Marks 1976). With the red bollworm, Diparopsis castanea, the male activation threshold was estimated as 9.2 molecules of pheromone/mm^3/sec over a 90-sec exposure period (5000 molecular impacts on the male antennae) (Marks 1976), whereas, 95% disruption of Diparopsis mating in the field was estimated to require a pheromone release rate of 1.5×10^4 molecules/mm^3/sec, which is 1.6×10^3 greater than the threshold concentration.

The modus operandi of this disruption probably involves both sensory adaptation of the antennal receptors and habituation of the central nervous system. Involved in this disruption process is an alteration of the male's upwind anemotactic responses to the odor source. The males are lured upwind at some optimum release rate, but the flight responses become disoriented when the release rate is increased 3 to 100 times (Roelofs and Cardé, 1977). This was first demonstrated (Shorey et al. 1968) with the cabbage looper moth, Trichoplusia ni, which was trapped maximally with a release rate of ca. 60 μg/h of pheromone, (Z)-7-dodecenyl acetate, but few males were trapped when the release rate was increased 3-fold.

Alteration of the male's flight responses was studied further in a sustained-flight tunnel (Miller and Roelofs, 1978a; 1978b). The gypsy moth, Lymantria dispar, was studied initially to determine the effects of various optical isomer mixtures of pheromone (cis-7,8-epoxy-2-methyloctadecane, disparlure) on the male's flight responses. Field trapping studies (Miller et al. 1977; Cardé et al. 1977) showed that the purest (+)-disparlure captured the most males and that racemic (±)-disparlure was not significantly more attractive than unbaited traps. In the flight tunnel, however, traps placed at the end of the tunnel (2.44 m long) captured male gypsy moths equally well when baited with either (+)- or racemic-disparlure. This apparent discrepancy between the laboratory and the field tests was resolved when the moths were sustained in flight by taking advantage of the moth's optomotor responses to a floor moving beneath them (Kennedy and Marsh, 1974). The males could be suspended in the odor plume for long periods of time by adjusting the speed of the floor belt passing backwards beneath the flying moth. These studies showed that the flight duration was directly related to the optical purity of pheromone. Males responding to 1 μg of (+)-disparlure on a filter paper disc flew 10 times longer and 20 times farther than males responding to 2 μg of racemic disparlure. There was no difference in the percentage of activated males and the males did fly long enough to racemic-disparlure to traverse the 2.44 m long tunnel and get captured in the traps, but the important difference was an alteration in their flight duration when (-)-isomer was present. This showed that the disorientation effects of flight behavior were separate from effects on moth activation and were much more discriminating in terms of determining active pheromone mixtures in the laboratory.

The same type of flight alteration can occur with binary pheromone mixtures as the concentration of one or the other pheromone component is increased. One of the components could exert this effect at a lower threshold level than the other, and so the disorientation threshold curves with increasing concentrations of each component would be different. The flight disorientation threshold for various pheromone component ratios puts an upper limit on the concentration range over which males can be attracted to odor sources or traps. The attractant range, then, consists of pheromone release rates that are above the activation threshold level for various mixtures and below the release rate that effects an alteration in the flight responses for the various mixtures. Fig. 4

shows a postulated diagram for the redbanded leafroller moth. The activation threshold curve shows a minimum at the natural pheromone blend of 91:9. An increase in the release rate of this blend eventually reaches a level (upper shaded area) that causes disorientation in the male's flight. If an unnatural blend is used, for example an 85:15 mixture, a higher release rate is required to get moth activation, but disorientation would occur at a higher release rate than when using the natural blend. This would indicate that male moths could possibly be trapped with high release rates of an unnatural blend, whereas the same release rate of the natural pheromone could already be too high for drawing males into the trap.

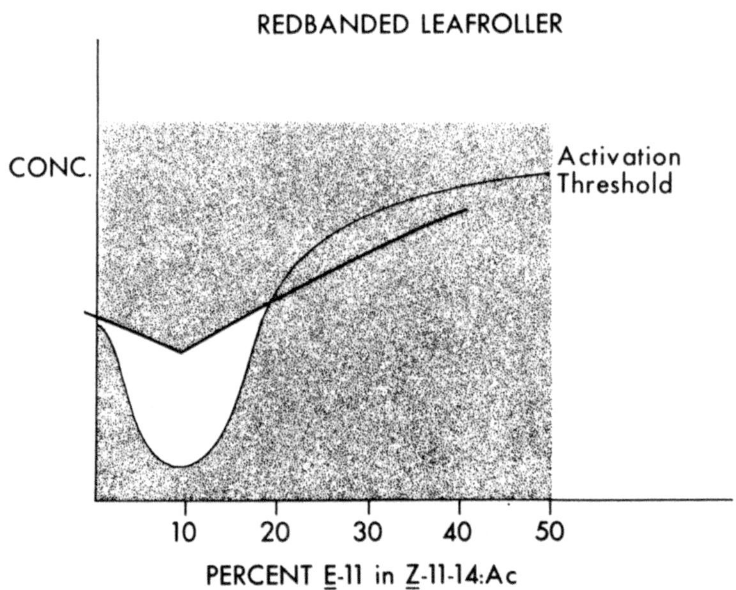

Fig. 4.-Threshold diagram with the "attraction" area bounded by the flight disorientation threshold (upper lines) and the activation threshold (lower curve) (from Roelofs 1978).

OMNIVOROUS LEAFROLLER MOTH

The threshold hypothesis presented above can be applied to some previously puzzling field trapping data obtained with the omnivorous leafroller moth, Platynota stultana, in California. In late fall and spring, the males were trapped mainly with pheromone component ratios of 94:6 to 88:12 (E/Z11-14:Ac) with essentially no males being trapped with traps containing a 74:26 ratio (Baker et al. 1978). In the summer months, however, the 74:26 ratio trap was catching a mean of 335 males/trap compared to only 49 males/trap with the 94:6 blend trap. This shift to a lower ratio of E-isomer in the summer months was obtained in 1973 and in 1974. The isomer ratio actually produced by female omnivorous leafroller moths was checked throughout the year by collecting pupae from the field at different times. The ratio consistently was found to be ca. 88:12 at all times.

It would be strange to find the males changing their sensitivity to the natural blend at various times of the season, so one logical explanation is one based on the threshold hypothesis. In the fall and spring periods, the temperatures were between 0-21°C and the pheromone component mixture release rates from the rubber septa were probably fairly low. In the summer months, the temperature range was 15-36°C and the release would have been much higher. It is possible that the high release rate was too high for the natural pheromone mixture, but was still within the attractant range for some of the unnatural mixtures, such as 74:26. The diagram in Fig. 5 shows a postulated threshold diagram for the omnivorous leafroller and the mixtures that possibly would be effective in trapping males at low and high release rates. In the spring (low release rates), only ratios containing between 5-20% Z11-14:Ac would be attractive, whereas in the summer only ratios containing between 12-30% Z11-14:Ac would be attractive if release rates are high. Presumably mixtures containing 5-20% Z11-14:Ac mixtures would be attractive in the summer if released at low re-

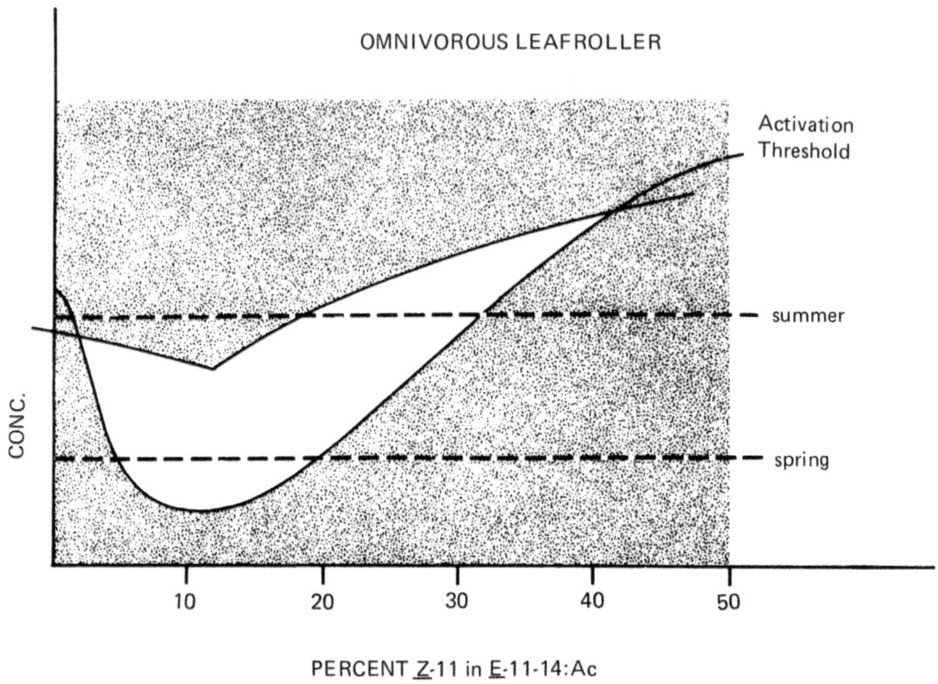

Fig. 5.-A postulated "attraction area" for male omnivorous leafroller moths. The dashed lines indicate the release rates possible in the spring and summer and show the ratios expected to trap males at those release rates (from Roelofs 1978).

lease rates. According to the hypothesis, the males are not changing their sensitivity to the natural pheromone blend throughout the year, but rather male catches with changing component ratios is a result of different release rates produced throughout the year by temperature changes.

PINK BOLLWORM

Field trapping results also have been variable with the pink bollworm moth, <u>Pectinophora gossypiella</u>. The female moths were found to produce a 66:34 ratio of pheromone components, (Z,Z)-/(Z,E)-7,11-hexadecadienyl acetates. In early season testing, the males were captured best with a 67:33 component ratio, but in the later flights they were trapped by a range of ratios between 40:60 and 70:30 (Flint et al. 1977). In other studies (Bierl et al. 1974) using high concentrations of pheromone components on filter paper, the males were trapped with ratios of 30:70 and 50:50, but hardly at all by a 70:30 ratio, which is the closest to the ratio used by the females. These data also could be explained by a threshold diagram. In Fig. 6 it can be seen that low release rates, as produced in the cooler months, would result in catches with a narrow

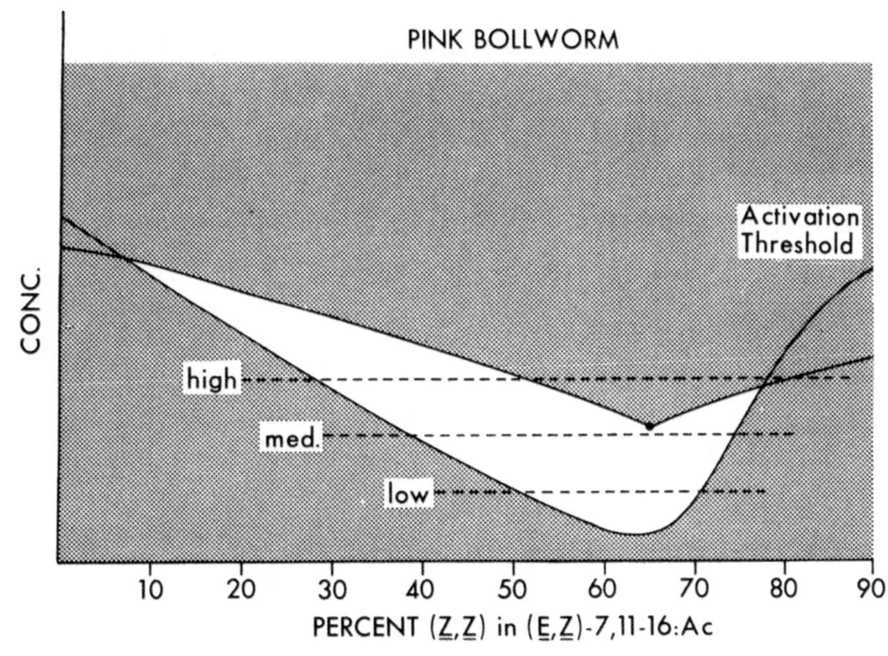

Fig. 6.-A postulated threshold diagram for the male pink bollworm moths. The various release rates indicate the component ratios expected to attract males at each rate (from Roelofs 1978).

range of ratios close to the natural blend. Higher release rates would expand that range of ratios to include those between 40:70 to 75:25, whereas very high release rates would be above the disorientation level for the natural blend, but would result in trap catches with unnatural blends, such as 30:70 to 50:50. Again, the catches would be an artifact of the system, rather than a change in the male's perception of the various blends.

REDBANDED LEAFROLLER MOTH

The redbanded leafroller moth was used in a study designed to test the threshold hypothesis. It was first determined that the female moths produced a fairly constant mixture of pheromone components. Analyses of the pheromone titer in pupae and adult moths showed that the ratio of components was approximately the same when the titer first appeared in the pupal stage as it was throughout the entire adult life, even though the pheromone titer increased sharply in the first 4 days following moth emergence (Miller and Roelofs, 1977). Analyses of over 300 individual female glands from field-collected pupae showed that the mixtures were fairly precisely controlled at 9.1±1.8 (S.D.) E11-14:Ac in Z11-14:Ac.

The male moth responses to various component ratios were then tested in the sustained-flight tunnel and in the field. Preliminary tests in the flight tunnel (Miller and Roelofs, unpublished) showed that a 3-fold increase in the release rate of a 92:8 Z11/E11 mixture dramatically affected flight behavior. Moth activation was still the same, but moths responding to the high release rate exhibited abnormal anemotactic flights, with the forward movement reduced at least by half and the net upwind distance travelled before leaving the plume was decreased to at most 1/10 that of flights to a more optimum release rate. The anemotactic zig-zag pattern of the moth to high release rates seemed to account for the reduction in upwind speed and net upwind distance travelled. This disorientation effect is then typical of the behavior expected in the shaded areas of the threshold diagrams above the disorientation threshold lines.

The same type of erratic flying was obtained with release rates of a 70:30 ratio that were high enough to get the moths activated. It appeared that high release rates were required to get the moths activated, and that these release rates were already above the disorientation threshold level.

Field tests were conducted in 1977 to test the threshold hypothesis, but were not conclusive. A test involving a range of release rates from 1, 3, 10 and 30 caps/trap of mixtures containing either 3, 9, 20 or 50% E-isomer was supportive of the hypothesis, but the number of males trapped was low. Only the 3% and 9% mixtures attracted significant numbers of males, and male catches with the 2 lowest release rates of the natural blend, 9%, were significantly better than the 3 lowest release rates of 3%. Interestingly, the male catches of the highest release rate of 3% was not significantly different from the catches with the lowest release rate of 9%. These data are in line with the threshold diagram for the redbanded leafroller moth (Fig. 4). It is seen that low release rates of 9% E11-14:Ac would attract males, while low release rates of 3% could be below the threshold of moth activation. Higher release rates of the 3% mixture would eventually exceed the activation threshold and be expected to attract male moths.

SIGNIFICANCE

Binary pheromone blends are detected peripherally by male moths with scent cells located on the Sensilla trichodea olfactory hairs. The specific blend apparently is perceived by the central nervous system from the input pattern of many hairs. Behavioral responses are elicited if the input exceeds a certain threshold level, which is controlled in the central nervous system and must be an integral part of the overall perception mechanism. It appears that the threshold is the lowest for perceiving the natural blend of pheromone components, but that behavioral responses can be elicited with other blends if higher concentrations are used. Responses to increasing concentrations are finally limited by a flight disorientation effect that occurs when the optimum release rate for attraction to the natural blend is increased about 3-10 times. The flight disorientation threshold can be below the activation threshold for blends that are quite different from the natural blend and moths would not be trapped by any release rate of these mixtures.

The threshold hypothesis underscores the importance of defining pheromone blends at a release rate comparable to that used by the female moth. Field trapping with a range of ratios does not necessarily mean a lack of specificity if abnormally high release rates are used. It is conceivable to trap 2 species with a high release rate of an 80:20 blend, when low release rates of 90:10 could be attractive to species A, and low release rates of 70:30 attractive to species B. These two species could be reproductively isolated in nature and low release rates of 80:20 would not attract either species. Only the abnormally high release rate of an unnatural blend, 80:20, attracts both species and erroneously gives the impression that there is no specificity in their pheromone systems.

The threshold diagrams could be useful in designing insect control programs using the disruption method. In this method, mating is disrupted by permeating the air with pheromone or other communication disruptant. The threshold diagrams indicate that the disorientation effect would be achieved at the lowest level with the natural pheromone blend, and that higher release rates of each individual component would be needed to reach the disorientation threshold level. In practice, this has been shown with several species. With redbanded leafroller moths, the male moth catches in monitoring traps were reduced 98% in apple blocks sprayed with microencapsulated pheromone (90:10), but only 89% reduced with a 50:50 mixture, and only 67% reduced with a 0:100 mixture (pure

E11-14:Ac) (Roelofs et al. 1976). Additional tests with defined release rates from 8 mil hollow fibers showed that the natural blend (91:9) was 99% disruptive at a release rate of 5 mg/h/ha, whereas a rate of 15 mg/h/ha was required to obtain that level of disruption with a 50:50 mixture (Reissig and Roelofs, unpublished; Taschenberg and Roelofs, 1978). With pink bollworm moths, the 50:50 mixture of pheromone components was ca. 100-fold more disruptive at low release rates than either component used alone (Shorey et al. 1976).

The threshold diagrams could be used to develop disruptive blends that would not be attractive at low concentrations - for example, at some distance outside the treated area. With red-banded leafroller moths, it could be better to permeate the atmosphere with an 80:20 mixture, rather than with a 91:9 natural blend, which would be attractive at some concentration level away from the test area.

In summary, the threshold hypothesis presents some ideas that could be useful in interpreting field trapping data, in setting up insect control programs, in defining species specificity, and in describing the overall pheromone perception process.

REFERENCES

Aihara, Y. and Shibuya, T. 1976. Responses of single olfactory cells to the sex pheromone in the tobacco cutworm moth, Spodoptera litura F. Proc. Symp. Insect Pheromones and Their Applications, Nagaoka and Tokyo, Japan, Dec. 8-11, 1976, pp. 41-8.

Baker, J. L., Hill, A. S. and Roelofs, W. L. 1978. Seasonal variations in male omnivorous leafroller moth trap catches with pheromone component mixtures. Environ. Entomol. 7: In press.

Baker, T. C., Cardé, R. T. and Roelofs, W. L. 1976. Behavioral responses of male Argyrotaenia velutinana (Lepidoptera: Tortricidae) to components of its sex pheromone. J. Chem. Ecol. 2: 333-52.

Bierl, B. A., Beroza, M., Staten, R. T., Sonnet, P. E. and Adler, V. E. 1974. The pink bollworm sex attractant. J. Econ. Entomol. 67: 211-16.

Butenandt, A., Beckmann, R., Stamm, D. and Hecker, E. 1959. Uber den Sexuallockstoff des Seidenspinners Bombyx mori. Reindarstellung und Konstitution. 2. Naturforsch. 146: 283-4.

Cardé, R. T., Baker, T. C. and Roelofs, W. L. 1975. Ethological function of components of a sex attractant system for Oriental fruit moth males, Grapholitha molesta (Lepidoptera: Tortricidae). J. Chem. Ecol. 1: 475-91.

Cardé, R. T., Cardé, A. M., Hill, A. S. and Roelofs, W. L. 1977. Sex pheromone specificity as a reproductive isolating mechanism among the sibling species Archips argyrospilus and A. mortuanus and other sympatric tortricine moths (Lepidoptera: Tortricidae). J. Chem. Ecol. 3: 71-84.

Cardé, R. T., Doane, C. C., Baker, T. C., Iwaki, S. and Marumo, S. 1977. Attractancy of optically active pheromone for male gypsy moths. Environ. Entomol. 6: 768-72.

Den Otter, C. J. 1977. Single sensillum responses in the male moth Adoxophyes orana (F.v.R.) to female sex pheromone components and their geometrical isomers. J. Comp. Physiol. 121: 205-22.

Flint, H. M., Smith, R. L., Forey, D. E. and Horn, B. R. 1977. Pink bollworm: response of males to (Z,Z-) and (Z,E-) isomers of gossyplure. Environ. Entomol. 6: 274-5.

Hill, A. S. and Roelofs, W. L. 1975. Sex pheromone components of the omnivorous leafroller moth, Platynota stultana. J. Chem. Ecol. 1: 91-9.

Hill, A. S. and Roelofs, W. L. 1978. Sex pheromone components of the obliquebanded leafroller moth, Choristoneura rosaceana. J. Chem. Ecol. 4: In press.

Kasang, G., Kaissling, K.-E., Vostrows, O. and Bestmann, H. J. 1978. Bombykal, a 2nd pheromone component of silkworm moth Bombyx mori L. Angew. Chem. 17: 60.

Kennedy, J. S. and Marsh, D. 1974. Pheromone-regulated anemotaxis in flying moths. Science 184: 999-1001.

Kochansky, J., Tette, J., Taschenberg, E. F., Cardé, R. T., Kaissling, K.-E., and Roelofs, W. L. 1975. Sex pheromone of the moth, Antheraea polyphemus. J. Insect Physiol. 21: 1977-83.

Marks, R. J. 1976. Laboratory evaluation of the sex pheromone and mating inhibitor of the red bollworm, Diparopsis castanea Hampson (Lepidoptera, Noctuidae). Bull. ent. Res. 66: 427-35.

Meijer, G., Ritter, F., Persoons, C., Minks, A. and Voerman, S. 1972. Sex pheromones of summer fruit tortrix moth, Adoxophyes orana: two synergistic isomers. Science 175: 1469-70.

Miller, J. R., Baker, T. C., Cardé, R. T. and Roelofs, W. L. 1976. Reinvestigation of oak leaf roller sex pheromone components and the hypothesis that they vary with diet. Science 192: 140-3.

Miller, J. R., Mori, K. and Roelofs, W. L. 1977. Gypsy moth field trapping and electroantennogram studies with pheromone enantiomers. J. Insect Physiol. 23: 1447-53.

Miller, J. R. and Roelofs, W. L. 1977. Sex pheromone titer correlated with pheromone gland development and age in the redbanded leafroller moth, Argyrotaenia velutinana. Ann. Entomol. Soc. 70: 136-9.

Miller, J. R. and Roelofs, W. L. 1978a. Gypsy moth responses to pheromone enantiomers as evaluted in a sustained-flight tunnel. Environ. Entomol. 7: 42-4.

Miller, J. R. and Roelofs, W. L. 1978b. Sustained-flight tunnel for measuring insect responses to wind-borne sex pheromones. J. Chem. Ecol. 4: In press.

Minks, A. K., Roelofs, W. L., Ritter, F. J. and Persoons, C. J. 1973. Reproductive isolation of two tortricid moth species by different ratios of a two-component sex attractant. Science 180: 1073-4.

O'Connell, R. J. 1972. Responses of olfactory receptors to the sex attractant, its synergist and inhibitor in the redbanded leafroller, Argyrotaenia velutinana. In D. Schneider [ed.], "Olfaction and Taste, IV", Stuttgart: Wissenschaftliche GmbH, pp. 180-6.

Persoons, C., Minks, A., Voerman, S., Roelofs, W. and Ritter, F. 1974. Sex pheromone of the moth, Archips podana: isolation, identification, and field evaluation of two synergistic geometrical isomers. J. Insect Physiol. 20: 1181-8.

Roelofs, W. L. 1978. Threshold hypothesis for pheromone perception. J. Chem. Ecol. 4: In press.

Roelofs, W. L. and Cardé, R. T. 1977. Responses of lepidoptera to synthetic sex pheromone chemicals and their analogues. Ann. Rev. Entomol. 22: 377-405.

Roelofs, W., Cardé, A., Hill, A. and Cardé, R. 1976a. Sex pheromones of the threelined leafroller, Pandemis limitata. Environ. Entomol. 5: 649-52.

Roelofs, W. L., Cardé, R. T., Taschenberg, E. F. and Weires, R. W. 1976b. Pheromone research for the control of Lepidopterous pests in New York. Advan. Chem. Ser. 23: 75-87.

Roelofs, W., Hill, A., and Cardé, R. 1975. Sex pheromone components of the redbanded leafroller, Argyrotaenia velutinana (Lepidoptera: Tortricidae). J. Chem. Ecol. 1: 83-9.

Roelofs, W., Hill, A., Cardé, R., Tette, J., Madsen, H. and Vakenti, J. 1974. Sex pheromone of the fruittree leafroller moth, Archips argyrospilus. Environ. Entomol. 3: 747-51.

Roelofs, W. L. and Tette, J. P. 1970. Sex pheromone of the oblique-banded leaf roller moth. Nature 226: 1172.

Shorey, H. H., Gaston, L. K. and Jefferson, R. N. 1968. Insect sex pheromones, p. 57-126, in R. L. Metcalf (ed.), Advances in Pest Control Research. Interscience, N. Y.

Shorey, H. H., Gaston, L. K. ad Kaae, R. S. 1976. Air-permeation with gossyplure for control of the pink bollworm. Advan. Chem. Ser. 23: 67-74.

Shorey, H. H., Gaston, L. K. and Saario, C. A. 1967. Sex pheromones of noctuid moths. XIV. Feasibility of behavioral control by disrupting pheromone communication in cabbage loopers. J. Econ. Entomol. 60: 1541-5.

Sugie, H., Yaginuma, K. and Tamaki, Y. 1977. Sex pheromone of the Asiatic leafroller, Archippus breviplicanus Walsingham (Lepidoptera: Tortricidae): isolation and identification. Appl. Ent. Zool. 12: 69-74.

Tamaki, Y., Noguchi, H., Yushima, T. and Hirano, C. 1971a. Two sex pheromones of the smaller tea tortrix: isolation, identification, and synthesis. Appl. Entomol. Zool. 6: 139-41.

Tamaki, Y., Noguchi, H., Yushima, T., Hirano, C. and Sugawara, H. 1976b. Sex pheromone of the summerfruit tortrix: isolation and identification. Kontyu 39: 338-40.

Taschenberg, E. F. and Roelofs, W. L. 1978. Male redbanded leafroller orientation disruption in vineyards. Environ. Entomol. 7: 103-6.

Wright, R. H. 1964. Insect control by non-toxic means. Science 144: 487.

PHEROMONE AND HOST ODOR PERCEPTION IN BARK BEETLES

Thomas L. Payne

Department of Entomology, Texas Agricultural Experiment Station, Texas A&M University, College Station, TX 77843

INTRODUCTION

In recent years, there has been considerable interest in olfactory perception in insects. This has primarily been because of the identification of pheromones. However, more recently the interest in olfactory perception results from investigations on the potential use of behavioral chemicals in pest management. In this light, bark beetles (Coleoptera: Scolytidae) are unique and challenging subjects for study.

BEHAVIORAL ASPECTS

The perception of behavioral chemicals, pheromones and host odors, is intricately woven into the behavior of bark beetles and is critical to their survival. Primarily, pheromones and host odors are important in host selection and colonization. However, they are also critical in communication leading to species propagation.

A knowledge of the basic behavior of bark beetles and how and where behavioral chemicals are involved in that behavior is essential in developing an understanding of the perception of these materials and their potential for use in pest management. Recent reviews have dealt with details of the various aspects of the behavior of bark beetles and the roles of behavioral chemicals (Borden, 1974, 1977; Vité and Francke, 1976; Wood, 1972). In general, bark beetles go through a sequence of behavioral phases which has to do with host selection and colonization. The sequence involves emergence, dispersal, host selection, concentration or aggregation, and colonization (Wood, 1972; Vité and Francke, 1976). Borden

(1974) elaborated on the sequence and indicated where behavioral chemicals are involved. The first beetles to emerge from the host tree or overwintering site ('pioneer' beetles) are involved in host selection. The 'pioneer' beetles may be either sex depending upon the genus. In *Dendroctonus*, females are generally the 'pioneer' beetles, whereas, in *Ips* they are the males. During host selection the 'pioneer' beetles orient to and land on the host tree. For some species olfactory perception of primary attractants may be involved. In two species of ambrosia beetle, for example, primary attraction was shown to result in response to ethanol (Cade *et al.*, 1970; Moeck, 1970, 1971), an apparent product of anaerobic metabolism in the host (Graham, 1968). These findings support an early theory that primary attraction is stimulated by olfactory cues resulting from respiratory fermentation in susceptible host trees (Person, 1931). In other bark beetles, primary attraction has been attributed to host odors such as terpene hydrocarbons (Rudinsky, 1966; Oksanen *et al.*, 1970; Perttunen *et al.*, 1970; Kangas *et al.*, 1971). For still other species host selection is believed to result after the 'pioneer' beetles land on the host tree and that landing occurs at random (Callaham, unpublished 1953 in Wood, 1972; Brown and Franklin, 1977).

Once the 'pioneer' beetles have selected a host tree they bore through the bark and initiate gallery construction in the conductive tissue of the tree. Just prior to or at the onset of the boring activity pheromones (secondary attractants) are released. Perception of the pheromones by members of the species in the area, as well as host odors for some species, initiates secondary attraction and the concentration or aggregation phase of the sequence. Responding beetles, generally of both sexes, orient to the selected host tree and initiate boring and gallery construction. These 'homestead' beetles also produce pheromone(s), which, along with host odors for some species, continue to aggregate beetles on the host tree. In the southern pine beetle, *D. frontalis* Zimmerman, for example, aggregation often occurs *en masse* over a span of a few days (Figure 1) (Coster *et al.*, 1977). As a result of this aggregation and subsequent boring activity, the tree becomes colonized by the beetles and progeny are produced.

Following mass attack, fewer beetles orient to attacked tree. Instead, orientation is to neighboring trees. This 'switching' behavior (Gara and Coster, 1968) may result from secondary attraction established in the neighboring trees by additional 'pioneer' beetles or from the influence of pheromones still being released from beetles in the mass attacked tree. In the latter case, those beetles in the mass attacked tree may release inhibitor or antiattractant pheromones which could stimulate flying beetles to orient to neighboring trees.

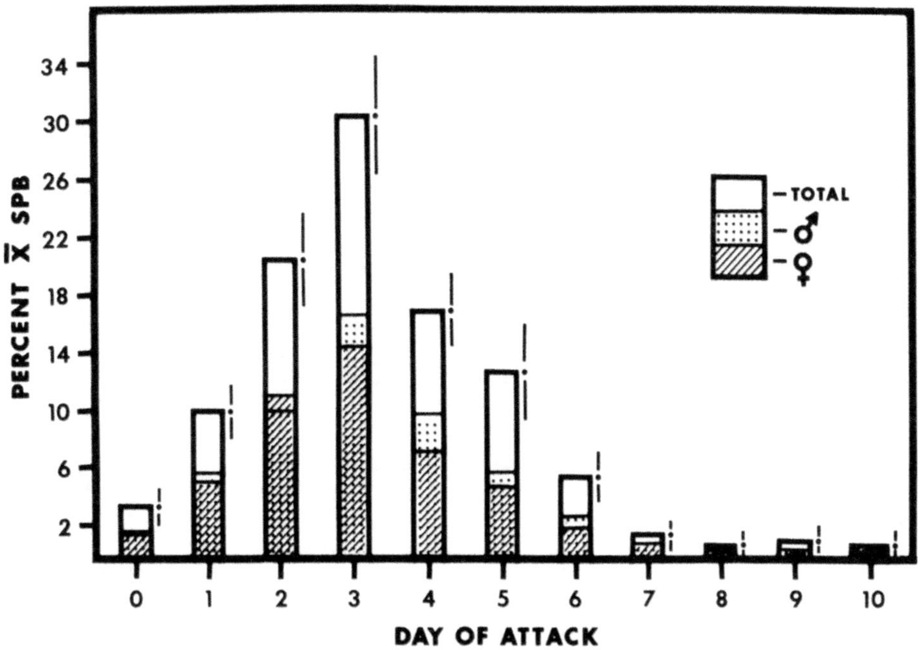

Fig. 1. Distribution of southern pine beetles trapped at pine trees during 11 days of mass attack. Based on 11 trees, n = 32,077 beetles, (after Coster *et al.*, 1977).

Vité and Francke (1976) proposed as simple model by which pheromones and host odors may regulate colonization of host trees by the southern pine beetle (Figure 2). The model consists of three phases of beetle attack: 1) initial attack, 2) mass attack and 3) shifting of the attack. During the initial phase females attack the host tree and release the attractant pheromone frontalin (1,5-dimethyl-6,8-dioxabicyclo [3.2.1]octane; Kinzer *et al.*, 1969). In conjunction with host odor, eg. α-pinene, both sexes aggregate on the attacked tree with males responding predominantly over females. In the mass attack phase both sexes release pheromones from the attacked tree. Females continue to release frontalin, which, with host odor, attracts more males and females. *Trans*-verbenol (Renwick, 1967) is also released by the attacking beetles and functions as a synergist with frontalin in the absence of host odor (Pitman *et al.*, 1969; Payne *et al.*, unpublished). Males in the tree release verbenone (Renwick, 1967) which functions to balance to sex ratio of responding beetles by inhibiting the attraction of males somewhat over females. During the shifting phase, a third pheromone, *endo*-brevicomin (*endo*-7-ethyl-5-methyl-6,8-dioxabicyclo [3.2.1]octane; Silverstein *et al.*, 1968) is released and flying beetles are deterred from aggregating on the attacked tree.

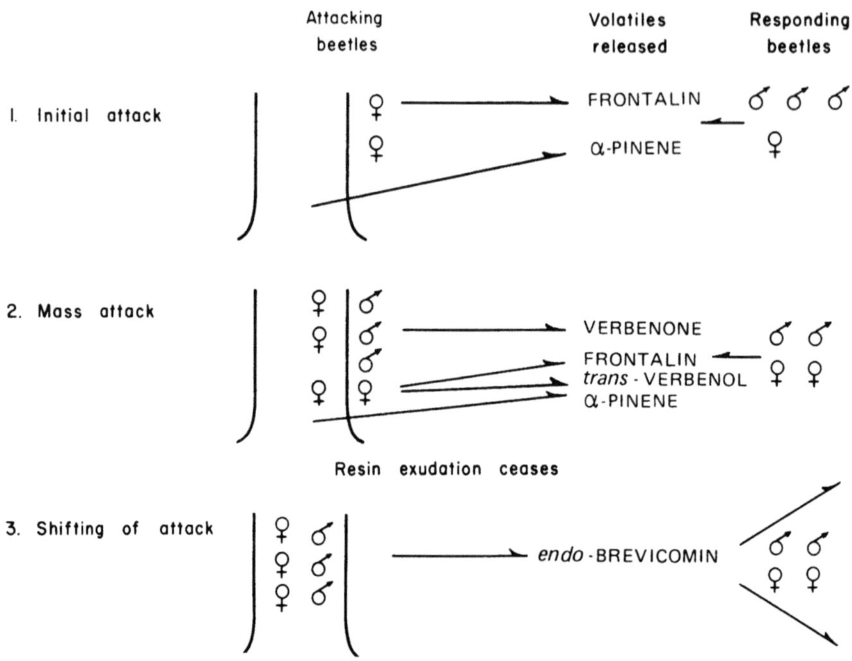

Fig. 2. Proposed mechanism by which pheromones regulate colonization of pine trees by southern pine beetles (revised after Vité and Franke, 1976).

The discussion presented here on the general role of pheromones and host odors in the behavior of bark beetles is indeed simplified. The full story, most of which is unknown, is undoubtedly quite complex. For most species it is likely that behavioral chemicals are intricately linked with other factors in their roles. For example, in some species vision has been implicated in host selection (Pitman and Vité, 1969), in others sound appears to play a role in aggregation and colonization (Rudinsky and Michael, 1973) and in still other species compounds produced by beetle associated microorganisms may be involved in the sequence of behaviors (Brand *et al.*, 1975, 1976, 1977). For the present, however, the level of understanding of behavior is sufficient for comparison at the antennal olfactory response level in order to begin to develop a general understanding of the perception of the pheromones and host odors.

METHODS OF INVESTIGATION

Extrapation experiments and electrophysiological investigations have shown that olfactory perception is a function of the antennae and more specifically, certain sensilla on the antennae (Borden and

Wood, 1966; Payne *et al.*, 1973; Angst and Lanier, 1978; Dickens and Payne, 1978; Mustaparta *et al.*, 1978).

Insect olfactory sensilla contain primary neurons which consist of dendrites, cell bodies and axons. Upon stimulation of an acceptor (thought to be located on the dendritic membrane: Schneider, 1970; Steinbrecht and Kasang, 1972) by a pheromone or host odor molecule, a series of electrophysiological events take place. Initially a slow, graded potential occurs at the acceptor. This is followed by the generator potential (also a slow, graded potential) which occurs in the dendrite. Action potentials or spikes (fast rising, short duration impulses) then follow from the cell body, when the generator potential is of sufficient intensity to overcome the threshold of the cell body. The strength of the olfactory stimulus is reflected directly in the intensity or height of the slow potentials (Boeckh *et al.*, 1965; Stürckow, 1965; Payne, 1971, 1975; Angst and Lanier, 1978) which determine the frequency of the subsequent action potentials (Priesner, unpublished in Boeckh *et al.*, 1965; Kaissling and Priesner, 1970; Mustaparta *et al.*, 1978).

The electroantennogram (EAG) technique (Schneider, 1955, 1957a, b; Schneider and Hecker, 1956) and the single cell technique (Boeckh, 1962) can be used to measure these electrophysiological events as they relate to the perception of pheromones and host odors. The EAG consists of a measurement of the summed slow potentials from several olfactory sensilla responding simultaneously to a stimulus (Boeckh *et al.*, 1965). Electroantennograms are recorded from bark beetles by inserting a recording electrode in the club of the antenna and an indifferent electrode in the base or in the insect's head (Figure 3a). There are limitations with the use of the EAG technique, however, and care must be taken to ensure that sufficient controls are included (Schneider, 1957a, b; Stürckow, 1970; Adler, 1971; Birch, 1971).

The more intricate single cell technique involves recording electrophysiological responses (action potentials) from individual cells associated with single olfactory sensilla. In this technique, the recording electrode is inserted in the base of the sensillum and the indifferent electrode is situated in the haemolymph of the antenna (Figure 3b).

Both techniques have been used successfully to elucidate various aspects of olfactory perception in bark beetles. The EAG has been used qualitatively to determine the presence or absence of olfactory sensilla for a given material on the antenna as a whole. In addition, the EAG has been used to determine the influences of biotic and abiotic factors on olfactory perception. A differential adaptation technique was recently developed in which the EAG is used to determine the relative specificity of the antennae as a whole, as well as an olfactory system for various

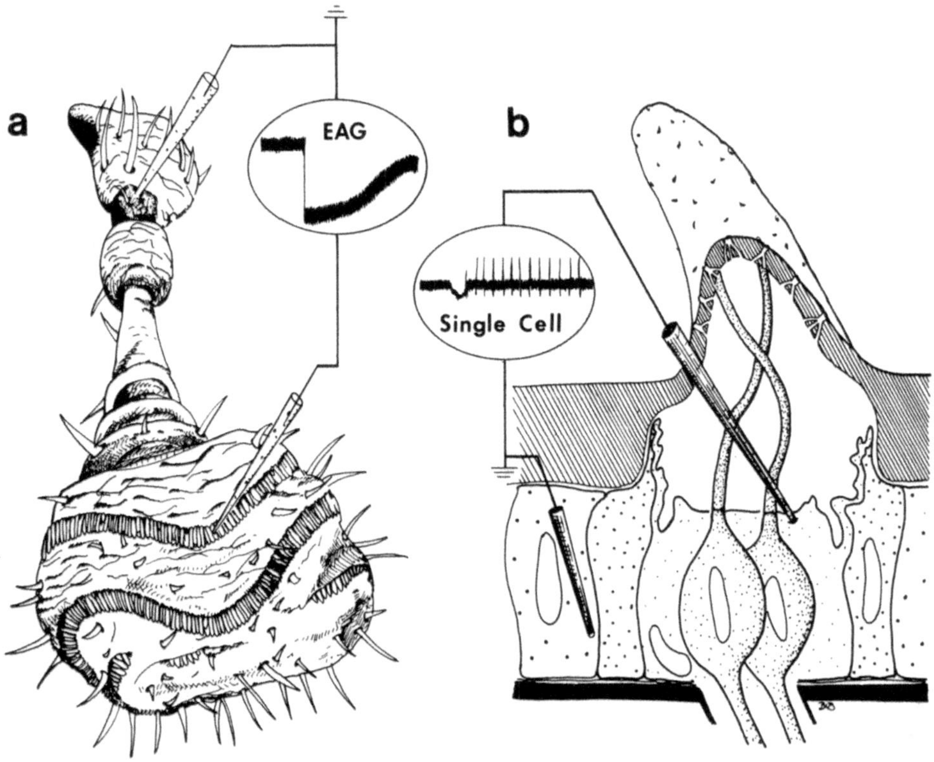

Fig. 3. Schematic diagram of the (a) electroantennogram (EAG) and (b) single cell recording techniques.

pheromones and host odors (Payne and Dickens, 1976; Dickens and Payne, 1977). In short, the technique involves stimulating the antennal preparation with adaptation-stimulation pairs of test compounds. For each pair the preparation is exposed to the adaptation compound until completely adapted; that is, no response results upon a second stimulation by the same material. Within msec., the preparation is then exposed to the stimulation compound. The presence or absence of response, or the degree of response to the stimulation compound, is a measure of the relative interaction of the compounds with the same acceptors. For example, if two compounds stimulate the same acceptors, adaptation to one will prevent response to the other. If, however, two compounds share some, but not all of the same acceptors, adaptation to one will reduce but not prevent response to the other.

The single cell technique is also used to elucidate antennal olfactory specificity, but from the level of the individual sensillum. With the technique it is possible to determine whether or not two compounds stimulate the same cell. And with the differential adaptation technique it is possible to determine whether or not they stimulate the same acceptors on a single cell (Seabrook, 1977; Dickens and Payne, 1977). In conjunction with morphological and histological information, the single cell technique makes it possible to develop detailed information on antennal sensilla structure and function in relation to olfactory perception (Kaissling and Priesner, 1970; Dickens and Payne, 1978).

ORGANS OF PERCEPTION

The antennae of bark beetles are club-shaped and consist of from seven to nine segments. With the exception of a few mechanosensilla, all sensilla are located on the distal club (Payne *et al.*, 1973). The general antennal form and arrangement of sensilla are well adapted for the habitats of the beetles. The localization of sensilla on the enlarged terminal club provides the beetles with a concentration of olfactory sensilla for perception of host odors and/or pheromone during the host selection and/or aggregation phases of their behavior. In addition, the antennal form enables the beetles to hold their antennae close to their heads to prevent damage from abrasions or resin coating as they tunnel in the host tree (Figure 4).

Antennal olfactory sensilla are generally located in sensory bands which encircle the club in some species or sensory fields which only cover one face of the club in others (Payne *et al.*, 1973). The sensilla take the form of pegs and hairs of various sizes and have been referred to as sensilla basiconica and sensilla trichodea, respectively (Payne *et al.*, 1973; Angst and Lanier, 1978; Dickens and Payne, 1978).

In *Dendroctonus* beetles, for example, there are at least two complete sensory bands around the club (Figure 5a). In some species a partial or complete third sensory band may occur distally. The bands consist primarily of long pegs, (sensilla basiconica long type) with short pegs (sensilla basiconica short type and fluted sensilla) present only in relatively small numbers. Sensory hairs (sensilla trichodea II) occur primarily along the margins of the club and between the sensory bands.

In contrast, the sensilla on the antennal club of *Ips* beetles are located primarily on one surface (Figure 5b). The club is oval in shape and slightly convex on the anterior surface. The majority of the sensilla are located on the anterior surface with a few occuring sporatically on the posterior surface. Most of the sensilla on the anterior surface are arranged in three serpentine-shaped sensory fields which do not encircle the antenna, but only extend

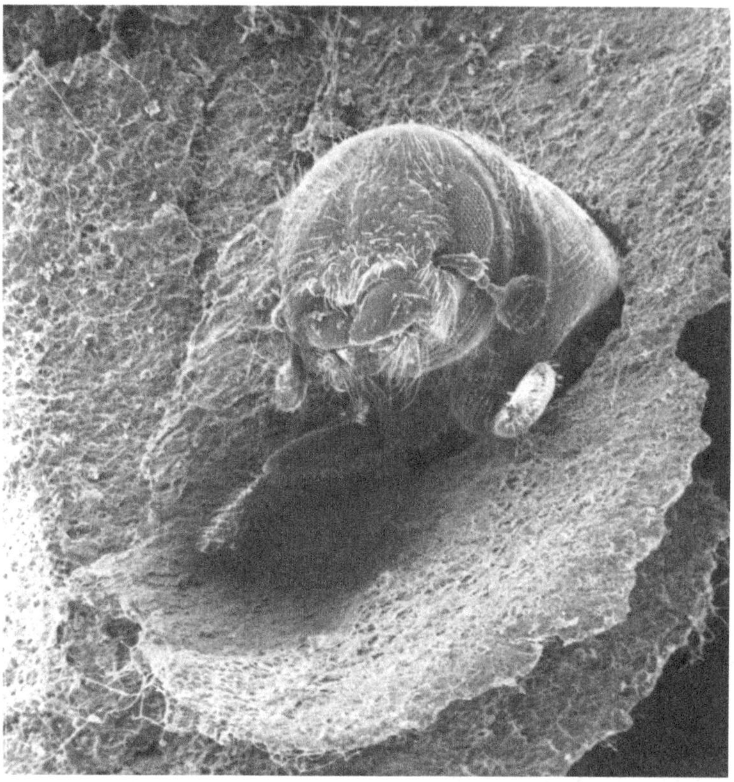

Fig. 4. Scanning electron micrograph of an adult southern pine beetle. 50x.

to the margins of the anterior surface. As in *Dendroctonus* long and short pegs and hairs make up the sensilla on the club. The sensory fields are made up primarily of the long pegs with the occurence of a few short pegs.

The antennal sensilla on a few bark beetles have been investigated for internal morphology (Borden, 1968; Moeck, 1968; Borg and Norris, 1971; Dickens and Payne, 1978). In general, the peg-like sensilla (those which make up the sensory bands and fields) were found to have thin walls perforated with numerous pores (Figure 6a). The shorter hair like sensilla were also found to have thin, porous walls. Such porous walled sensilla are generally classified as chemosensilla of one type or another (Slifer *et al.*, 1959; Boeckh, 1962; Schneider *et al.*, 1966; Slifer, 1967; Schneider and Steinbrecht 1968; Ernst, 1969; Steinbrecht, 1969; Mustaparta, 1973, 1974, 1975; Dickens and Payne, 1978).

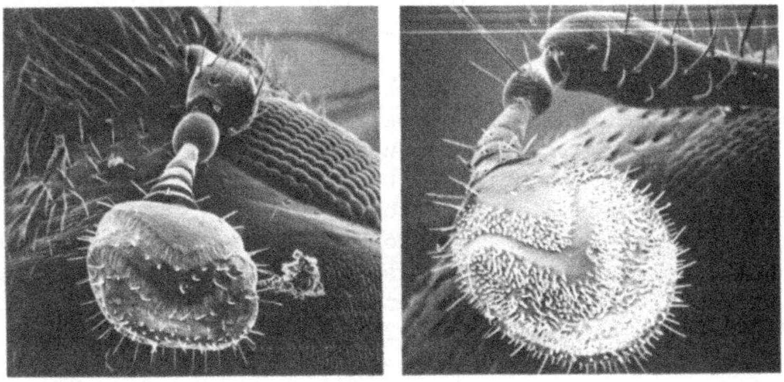

Fig. 5. Scanning electron micrograph of a *Dendroctonus* (a) and *Ips* (b) antenna. 200x. (after Payne *et al.*, 1973).

The size and number of the sensilla pores are important to the perceptual ability of the beetle by providing it with the ability to collect the air-borne molecules of pheromones and host odors that stream across the antennae. The relatively small size of olfactory sensilla necessitates that the pores be comparatively small; although their diameters may vary from 100Å to over 1000Å (Moeck, 1968; Schneider and Steinbrecht, 1968; Dickens and Payne, 1978). Also, depending upon the insect species, olfactory sensilla may have from 3 to 100 pores/μm^2 of external sensillum surface area (Schneider and Steinbrecht, 1968; Kaissling, 1971; Dickens and Payne, 1978).

In the southern pine beetle the pores are approximately 140Å in diameter on the long pegs of the sensory bands (Dickens and Payne, 1978). There are approximately 70 pores/μm^2 of sensillum surface area, which amounts to approximately 2100 pores/sensillum. Since the beetle has somewhere in the neighborhood of 700 long peg-like sensilla per antennal club, it is equipped with approximately 2.9 x 10^6 pores available to conduct pheromone and host odor molecules from the air and surface of the sensilla to the sensory mechanisms within the sensilla. However, the complexity of the olfactory sensillum does not reside exclusively in its external morphology. Quite the opposite is the case. Associated with each cuticular pore is a pore funnel approximately 430-Å in diameter which tapers to a pore canal approximately 140-Å in diameter (Figure 6b). The pore canal opens to a broad pore kettle 650-700-Å wide and 500-Å deep from which 10-11 pore tubules approximately 100-Å in diameter radiate (Figure 6c). There are from 10500-21000 tubules/sense cell depending upon the presence of one or two cells respectively. In some cases the tubules appear to contact the membrane of the sense cell dendrite (Figure 6d), which has been found in other insects to be surrounded by a sensillum liquor (Ernst, 1969).

The short thin walled porous hairs (sensilla trichodea II) were found to respond to both pheromones and host odors (Dickens and Payne, 1978). Stimulation by the pheromones frontalin and *endo*-brevicomin and the host tree terpene hydrocarbon α-pinene elicited phasic-tonic responses from the associated neuron (Figure 7). The long, thin walled porous pegs (sensilla basiconica long) which primarily constitute the sensory bands, were also found to be chemosensilla for the behavioral chemicals. The pheromones frontalin, *exo*- (*exo*-7-ethyl-5-methyl-6,8-dioxabicyclo[3.2.1]octane; Silverstein *et al.*, 1968) and *endo*-brevicomin, trans-verbenol and verbenone, and the host tree terpene hydrocarbons α-pinene and 3-carene elicited characteristic phasic-tonic responses from neurons associated with the sensilla (Figure 8).

In the pine engraver, *I. pini* (Say), cells associated with pegs and hairs (sensilla basiconica and trichodea) of the sensory fields were found to respond to major volatiles from frass (boring dust which includes fecal pellets and host fragments) of the species

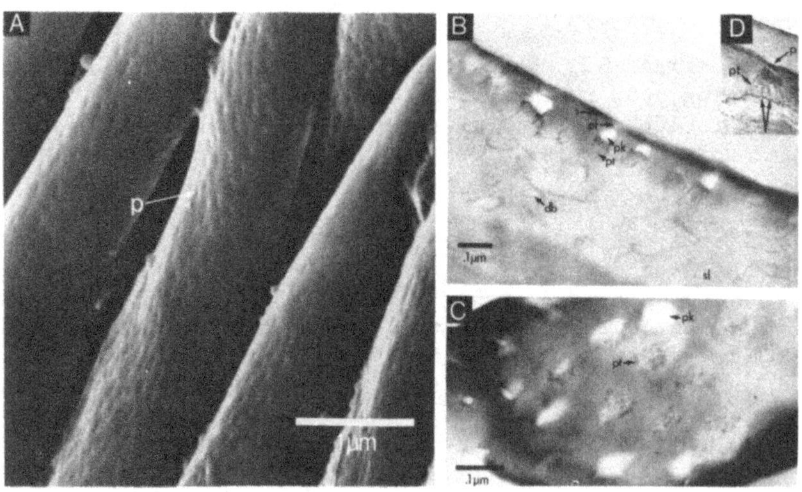

Fig. 6 a. Scanning electron micrograph of peg sensilla (s. basiconica long) of a southern pine beetle.
b. Transmission electron micrograph of longitudinal section of a peg sensillum.
c. The lower tangential section through sensillum.
d. Transmission electron micrograph showing pore tubules contacting dendritic branch in a peg sensillum. Forked arrows indicate points of apparent contact. db = dendritic branch; p = pore tubule; s = surface layer; sl = sensillum liquor, (Dickens and Payne, 1978).

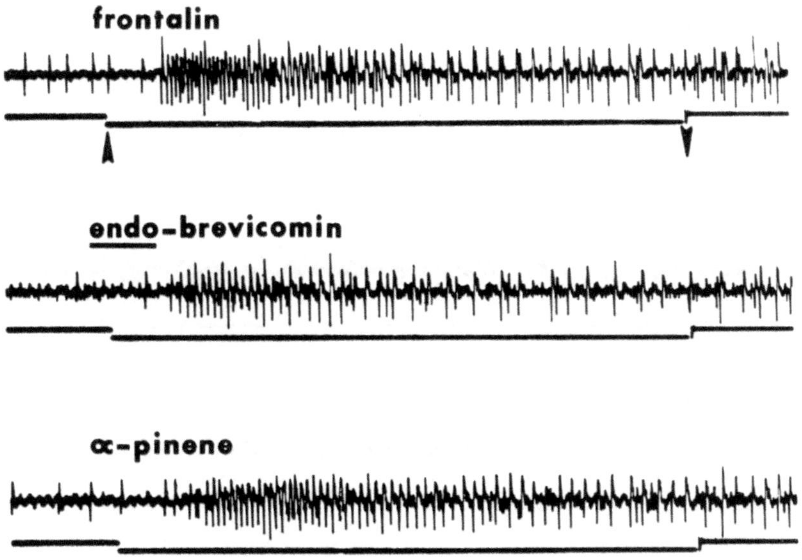

Fig. 7. Single cell responses from a hair sensillum (s. trichodeum type II) of a southern pine beetle to pheromones and a host tree terpene. Displaced horizontal bar (indicated by arrows) represents 1 sec. stimulation. Delay from stimulus onset to spike initiation was due to an artifact in the delivery system, (Dickens and Payne, 1978).

(Mustaparta *et al.*, 1978). These included ipsdienol [(+)-2-methyl-6-methylene-2,7-octadien-4-ol; Silverstein *et al.*, 1966], *trans*-verbenol, verbenone, linalool (3,7-dimethyl-1,6-octadien-3-ol; Young *et al.*, 1973) and 1-octanol. In addition, the pheromones ipsenol [(-)-2-methyl-6-methylene-7-octen-4-ol; Silverstein *et al.*, 1966] and *cis*-verbenol (Silverstein *et al.*, 1966) and the host tree terpene hydrocarbons α-pinene and myrcene elicited responses. Olfactory cells were also found in the California five-spinned *Ips*, *I. paraconfusus* (Lanier), for ipsdienol, ipsenol, *cis*- and *trans*-verbenol and host terpenes (Mustaparta, personal communication).

ASPECTS OF PERCEPTION

Transduction

In order for molecule-bound information to be transferred to the bark beetles, transduction must occur. That is, pheromone and host odor molecules must come into contact with the olfactory acceptors, thereby initiating the transformation of the stimulus into a bioelectrical potential. Transduction is not well understood;

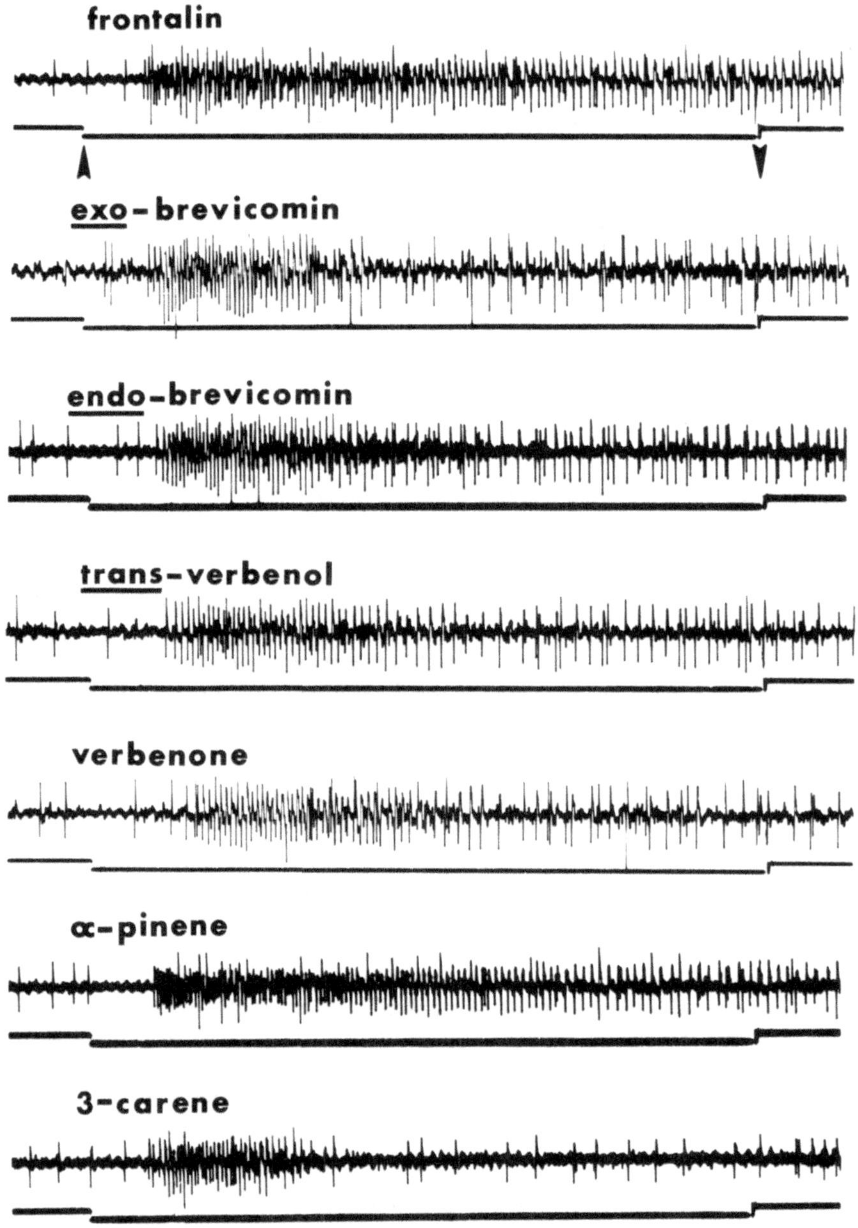

Fig. 8. Single cell response from a peg sensillum (s. basiconicum long) of a southern pine beetle to pheromones and host tree terpenes. Displaced horizontal bar (indicated by arrows) represents 1 sec. stimulation. Delay from stimulus onset to spike initiation was due to an artifact in the delivery system, (Dickens and Payne, 1978).

which, in part, may account for the several interesting hypotheses (see reviews Dravnieks, 1966; Moulton and Beidler, 1967; Davies, 1971; Kafka, 1970). There is evidence, however, to suggest that for some insects an acceptor protein on the dendrite is involved (Norris *et al.*, 1971; Ferkovich *et al.*, 1973 a, b).

Thresholds

The degree of efficiency of olfactory sensilla can best be realized through information on thresholds of response at the peripheral and whole organism levels. Data are available for bark beetles on the thresholds of response at both levels, but only in relation to the amount of behavioral chemical placed on a filter paper substrate. Calculations are not available on the various parameters needed to determine the number of odor molecules which are required to elicit an electrophysiological and/or behavioral response. Such calculations have been made for the male silk moth *Bombyx mori* L. which may provide a general model with which other insect olfactory systems can be compared (Kaissling and Priesner, 1970). Using tritium-labelled pheromone and considerations of numerous parameters such as antennal size, number of olfactory sensilla, surface diffusion of molecules, physics of the stimulus system and other factors, they determined that the threshold of response of the pheromone acceptor is 'one molecule'. The behavioral response threshold was calculated to be 200 pheromone-induced spikes per second elicited by 200 pheromone molecules. The olfactory cell was found to have a spontaneous activity of 160 spikes per second. The signal to noise ratio was 200 to 1600, which is just above the theoretical minimum (120 to 1600) required for communication of information.

It seems probable that a single molecule - single acceptor response threshold exists for bark beetles as well as other insects. However, the efficiency of the olfactory sensilla is also determined by the number of pores and associated acceptors for a given stimulus.

Antennal Olfactory Response

Bark beetles must deal with an array of pheromones and host odors in their environment. The compounds may affect the behavior of the beetles as attractants, synergists or inhibitors (Table 1). The fact that beetles possess olfactory sensilla for such compounds and that the compounds and their mixtures have various effects on behavior has been demonstrated in laboratory and field experiments (Borden, 1974; Payne, 1975; Vité and Franke, 1976; Angst and Lanier, 1978; Mustaparta *et al.*, 1977, 1978; Payne *et al.*, unpublished).

Table 1. Bark Beetles species and compounds for which behavioral and electrophysiological data are available

	Compound		Species	Behavioral Effect		Electrophysiological Data	
Pheromones							
	*endo*-brevicomin (racemic)	1/	*Dendroctonus frontalis*	inhibitor	13,19/	EAG, Single cell	20,21/
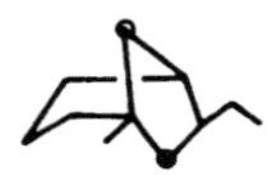	*exo*-brevicomin (racemic)	1/	*D. brevicomis*	attractant	10/	EAG	19/
	"	3/	*D. frontalis*	no effect	13,14/	EAG, Single cell	19-21/
	(+)		"	--		Single cell	21/
	(-)	"	"	--		"	
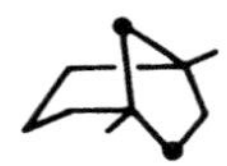	frontalin (racemic)	2/	*D. brevicomis*	attractant	11/	EAG	19/
	"	4/	*D. frontalis*	"	2,14/	EAG, Single cell	19-21/
	(+)		"	weak attractant	14/	EAG	14/
	(-)	"	"	attractant	"	"	"
	verbenone	5/	*D. frontalis*	weak inhibitor	14,15/	EAG, Single cell	21/
	*trans*-verbenol	5/	*D. frontalis*	synergist	14/	EAG, Single cell	21/
	*cis*-verbenol	6/	*Ips paraconfusus*	attractant	6/	Single cell	22/
	(racemic)	8/	*I. typographus*	weak attractant	17/	EAG	26/
	(R)	"	"	no effect	"	"	
	(S)		"	attractant	"	"	

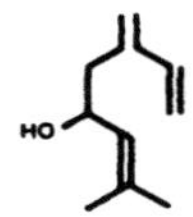	ipsdienol	6/					
	(racemic)		*I. pini* (Idaho)	weak attractant	16/	EAG, Single cell	23-25/
	"		" (N.Y.)	attractant	"	"	
	"	7/	*I. typographus*	"	18/	EAG	26/
	(+)		*I. paraconfusus*	--		Single cell	22/
	"		*I. pini* (Idaho)	weak attractant	16/	EAG, Single cell	23-25/
	"	6/	" (N.Y.)	"		"	
	(-)		*I. paraconfusus*	attractant	6/	Single cell	22/
	"		*I. pini* (Idaho)	attractant	16/	EAG, Single cell	23-25/
	"		" (N.Y.)	weak attractant	"	"	
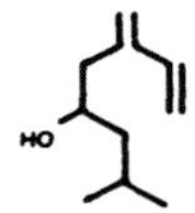	ipsenol	6/	*I. paraconfusus*	attractant	6/	Single cell	22/
	"		*I. pini* (Idaho)	inhibitor	16/	EAG, Single cell	23-25/
	"		" (N.Y.)	"		"	
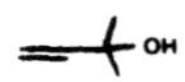	methylbutenol	9/	*I. typographus*	attractant	9/	EAG	26/
HOST TERPENES							
	α-pinene		*D. frontalis*	synergist	14,15/	EAG, Single cell	19-21/
	3-carene		*D. frontalis*	no effect	15/	EAG, Single cell	19-21/

1/ Silverstein *et al.*, 1968.
2/ Kinzer *et al.*, 1969.
3/ Mori, 1974.
4/ Mori, 1975.
5/ Renwick, 1967.
6/ Silverstein *et al.*, 1966.
7/ Silverstein, personal communication.
8/ Mori *et al.*, 1976.
9/ Bakke *et al.*, 1977.
10/ Bedard *et al.*, 1969.
11/ Vité and Pitman, 1969.
12/ Browne, personal communication.
13/ Vité and Renwick, 1971.
14/ Payne *et al.*, unpublished.
15/ Renwick and Vité, 1969.
16/ Birch and Lanier, personal communication.
17/ Krawielitzki, *et al.*, unpublished in Vité *et al.*, 1976b.
18/ Bakke, 1976.
19/ Payne, 1975.
20/ Dickens and Payne, 1977.
21/ Dickens and Payne, 1978.
22/ Mustaparta, personal communication.
23/ Angst and Lanier, 1978.
24/ Mustaparta *et al.*, 1977.
25/ Mustaparta *et al.*, 1978.
26/ Dickens, personal communication.

In the forest environment it is unlikely that bark beetles make contact with molecules of a single compound in the absence of all other compounds. Both pheromones and host odors are likely to be present in varying combinations and concentrations on a continuous basis. As a result, it is probably safe to assume that bark beetles depend upon the input from a multiple of antennal olfactory sensilla for any given behavioral response. There is little known; however, on how beetles might perceive the compounds such that a given behavioral response results. At this point one can only speculate on how they are able to code the information of the odor molecule such that the information reaches the central nervous system and a behavioral response results.

Characteristics associated with antennal olfactory response to various behavioral chemicals lead to some insights on how bark beetles might code the chemical message and indicate that acceptor number and specificity are probably involved.

Pheromone and Host Odor Acceptors. In general, bark beetles have more acceptors for pheromones than for host odors. This can be concluded from both EAG and single cell data where increasing concentrations of pheromones and host odors elicited increased response (Payne, 1975; Angst and Lanier, 1978; Mustaparta *et al.*, 1978). In EAGs, depolorizations to host odors reach an upper plateau at concentrations significantly below the same concentrations for pheromones which continue to elicit increased responses (Figure 9). Since the intensity of the response provides a relative measure of the number of acceptors for one given stimulus as compared to another, the results show the presence of different numbers of acceptors for pheromones over host odors (Payne, 1975). Stimulation of single cells by increasing concentrations of behavioral chemicals results in an increased spike frequency (Mustaparta *et al.*, 1978). And significantly more spikes result to a given concentration of compound in cells responsive to pheromone compared to those responsive to host terpene (Figure 10). This further supports the presence of more acceptors for pheromones than host odors since spike frequency is believed to be directly related to the number of stimulus-acceptor interactions (Kaissling and Priesner, 1970).

Specificity. There appears to be various degrees of specificity in the olfactory cells for pheromones and host odors. For some species it appears that pheromones and host odors alike can interact on the same cell and in some cases via some of the same acceptors. In other species there appears to be separate cells with varying degrees of specificity for the behavioral chemicals.

In the southern pine beetle, there appears to be an olfactory system for pheromones and host odors in which all acceptors have a high degree of specificity for frontalin (Dickens and Payne, 1977), and in which both classes of compounds can stimulate the same cell

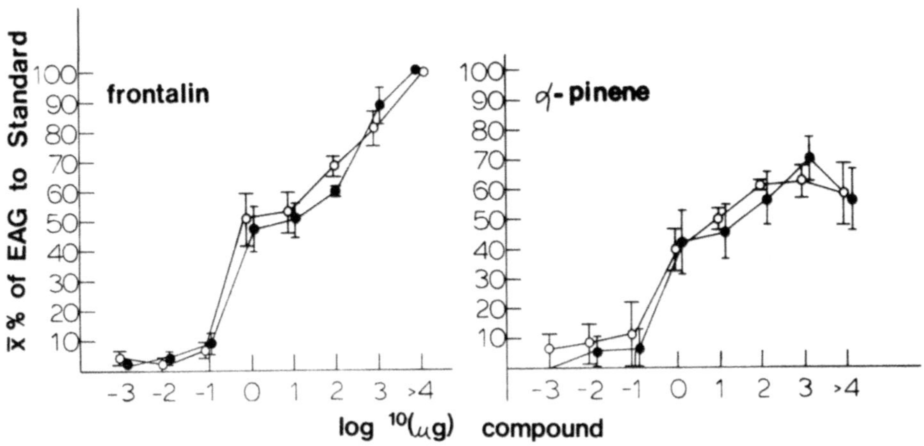

Fig. 9. Mean percent response of southern pine beetles to a pheromone and a terpene. Responses presented as percent of response to the standard (ca. 10^4μg frontalin). Mean response to standard = ca. 2.09 mV. Filled circles = male, open circles = female, and vertical bars = standard errors (Payne, 1975).

(Figure 8) (Dickens and Payne, 1978). Using the differential adaptation of acceptors, frontalin, *endo*-brevicomin, *trans*-verbenol, verbenone, α-pinene and 3-carene were compared for their relative interactions with the antennal olfactory acceptors (Figure 11). Adaptation of the acceptors for frontalin was found to block response of the acceptors for all other compounds tested for both male and female beetles and showed that frontalin was capable of interacting with the acceptors for the compounds. However, the reverse was not found to be true. Adaptation to any of the other compounds did not prevent response to frontalin. Percent interactions of the compounds with the frontalin acceptors further support the conclusion that there are generally greater numbers of acceptors for pheromones than for host odors.

The findings were substantiated at the single cell level with frontalin and α-pinene odors (Dickens and Payne, 1977). A cell adapted to either frontalin or α-pinene did not respond with a change in spike frequency from its tonic response level when restimulated within msec. with the same compound (Figures 12a and b). Adaptation to frontalin followed by stimulation with α-pinene failed to elicit a response (change in spike frequency) (Figure 12c). However, when the cell was adapted to α-pinene, stimulation by frontalin did elicit an increase in spike frequency (Figure 12d), indicating that although all acceptors on the cell for α-pinene were adapted, acceptors were available with which frontalin could interact.

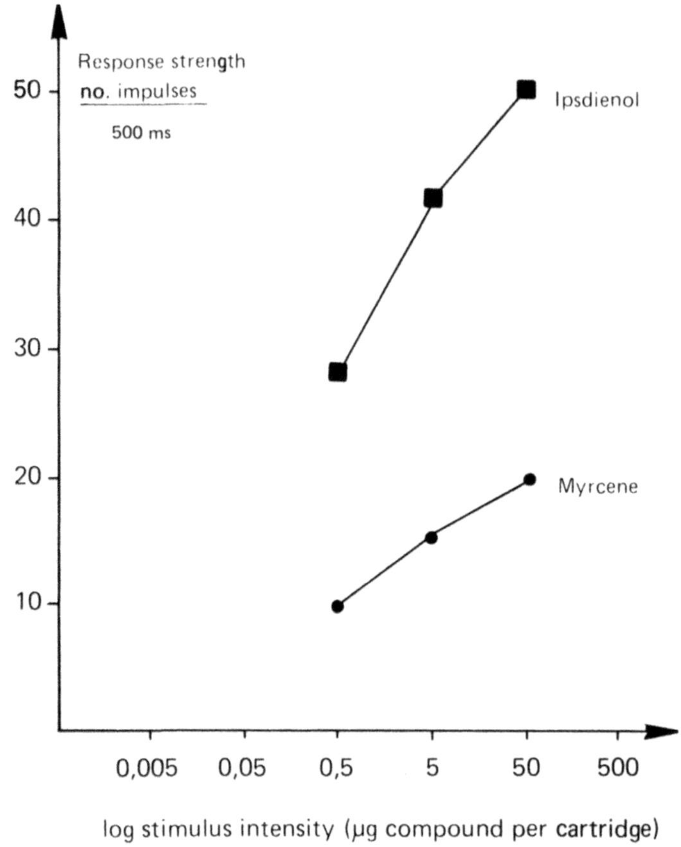

Fig. 10. Stimulus response curves of separate, single olfactory cells of *I. pini* to a pheromone (ipsdienol) and a host terpene (myrcene) (after Mustaparta *et al.*, 1978).

The results indicate that the beetles have added flexibility in coding information in that a single cell can possess more than one acceptor type and furthermore, that the cell is capable of transferring information from one acceptor type while another type is occupied. Similar results were found for the spruce budworm moth, *Choristoneura fumiferana* (Clem.), in response to an attractant, an inhibitor and a host terpene (Seabrook, 1977).

A slightly different situation appears to exist for *I. pini* (Mustarpta *et al.*, 1978). Single cell recordings from over 90 cells revealed the presence of a predominence of cells (over 40) essentially specialized for an attractant pheromone of the species, ipsdienol. Response was elicited to other compounds; however, the levels were low and only to amounts which likely do not occur at the perceptual level in nature. Fifteen cells were found primarily specific for ipsenol even though the pheromone is not found produced in the species. However, ipsenol is known to be an inhibitor of aggregation in *I. pini* (Birch and Wood, 1975; Birch and Light, 1977) and is an attractant pheromone produced by *I. paraconfusus* (Silverstein *et al.*, 1966) a species often occuring commonly with *I. pini* in the same host. Eight of those cells responded to ipsdienol but weakly and at higher concentrations. Two cells were found which gave significant response to both pheromones. Stimulation of one of the cells by a mixture of ipsdienol and ipsenol elicited a response significantly greater than to either pheromone alone or the sum of the two responses (Mustaparta *et al.*, 1977). Other cells were found which responded with varying degrees of specificity to other pheromones (some found in *I. pini*) and to host tree terpenes, but for which no behavioral effects are known in *I. pini*. A few cells were found which responded to camphor and linalool found present in the frass *I. pini*. Twenty-three cells responded at various levels to *trans*-verbenol, *cis*-verbenol and verbenone. Six cells were found which responded exclusively to the host tree terpene myrcene. Very few cells responded to α-pinene, and none selectively.

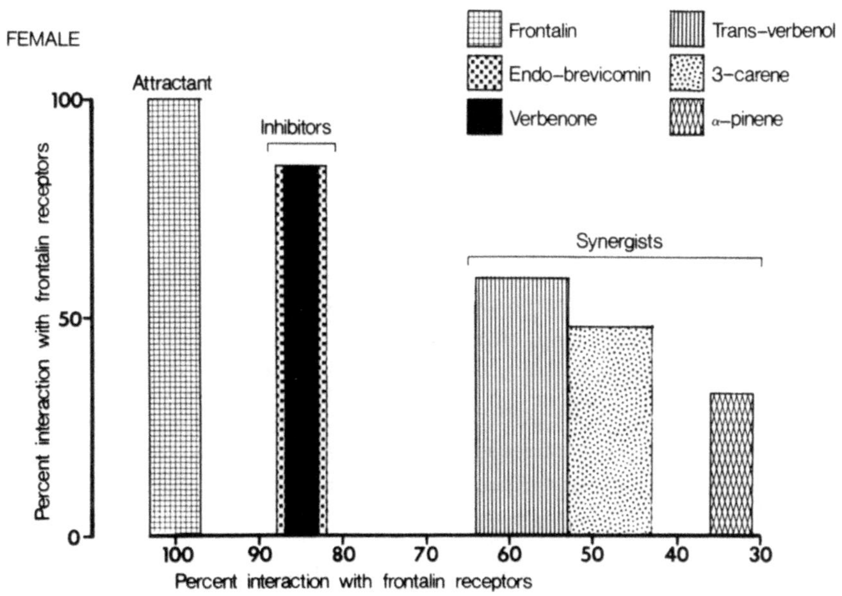

Fig. 11. Mean percent interaction of pheromones and host terpenes with frontalin acceptors of female southern pine beetles. Width of columns represents $\bar{X} \pm SE$ for each compound with the exception of the frontalin column which has an arbitrary width (Dickens and Payne, 1977).

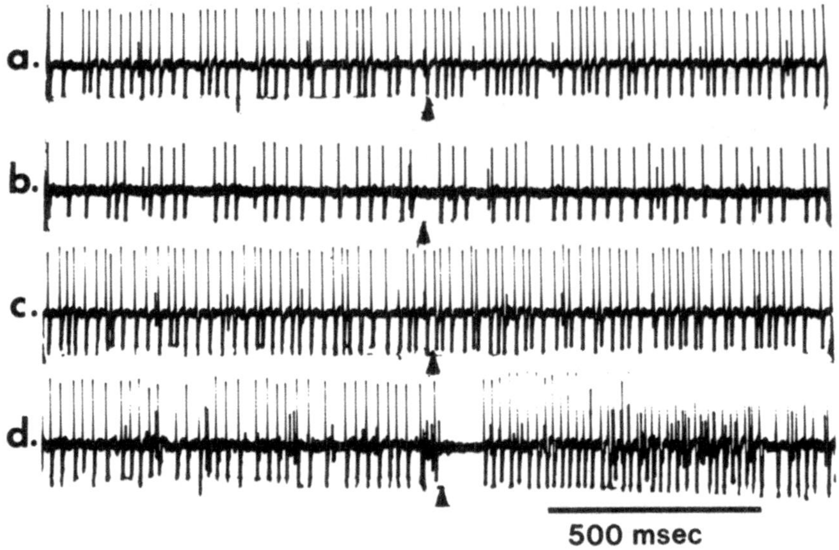

Fig. 12. Differential adaptation of cell associated with peg sensillum in a southern pine beetle. Arrow indicates onset of stimulus compound. When the preparation was adapted to frontalin (a.: c.) stimulation with either frontalin (a) or α-pinene (c) elicited no significant change in spike frequency from tonic response level. Likewise, no change in spike frequency occurred upon restimulation with α-pinene (b). Cell adapted to α-pinene (d) did respond with an increase in spike frequency when stimulated by frontalin. All four traces recorded from the same preparation. Gain was increased in (c) and (d), (Dickens and Payne, 1977).

Pheromones and host tree terpenes, even though they may differ significantly in structure, can interact with the same cells and even the same acceptors and yet elicit quite different behaviors. In the southern pine beetle, for example, an attractant, a synergist and an inhibitor (frontalin, α-pinene and *endo*-brevicomin, respectively) share some of the same acceptors. Each compound alone has essentially no effect on natural populations, in the behavioral tests used. However, when frontalin and α-pinene are present at the same time, significant numbers of beetles are attracted (Kinzer *et al.*, 1969; Payne *et al.*, unpublished). *Endo*-brevicomin, which interacts with a large percentage of the frontalin acceptors, significantly reduces the attractiveness of frontalin and α-pinene to flying beetles.

Behavioral data indicate that specificity for enantiomeric forms of aggregation pheromones may be widespread in bark beetles (Borden *et al.*, 1976; Hedden *et al.*, 1976; Pitman, personal communication in Borden *et al.*, 1976; Vité *et al.*, 1976; Harring and Mori, 1977;

Payne *et al.*, unpublished; Birch and Lanier, personal communication). In species where data on the production of enantiomers are available, the beetles respond primarily or entirely only to the enantiomers they produce (Borden *et al.*, 1976; Wood *et al.*, 1976; Payne *et al.*, unpublished; Birch and Lanier, personal communication).

In the southern pine beetle, frontalin is produced as 85% (1S, 5R)-(-) and 15% (1R,5S)-(+) (Stewart *et al.*, 1977). In behavioral tests in the laboratory and field, in the presence of the synergist α-pinene, beetles responded significantly more to the (-) enantiomer than to the (+) form (Payne *et al.*, unpublished). The racemic mixture, as well as the mixture naturally produced, elicited responses which were not significantly different from the (-) form. The (+) enantiomer, although not attractive, was not found to inhibit response to the (-) form.

Similar results were found at the antennal olfactory perception level (Payne *et al.*, unpublished). Electroantennograms recorded in response to stimulation by (+)-frontalin, (-)-frontalin, (±)-frontalin and the 85% (-):15% (+) mixture indicated the presence of significantly more acceptors for (-)-frontalin than for the (+) form (Figure 13). The absence of significant differences in the response to (-)-frontalin and the mixtures suggests that the (-) and (+) forms stimulate the same olfactory cells and may share some of the same acceptors. However, differential adaptation experiments showed the presence of some acceptors specific for each enantiomer alone. Adaptation to (±)-frontalin prevented response to (-)-frontalin and to (+)-frontalin. The reverse was not true. (±)-frontalin elicited response, although small, from either (-)-frontalin or (+)-frontalin adapted antennae indicating the presence of some acceptors specific for the (+) and the (-) forms respectively. Borden *et al.*, (1976) proposed a similar explanation for the interaction of S-(+) and R-(-) -6-methyl-5-hepten-2-ol (sulcatol; Byrnes *et al.*, 1974) for the ambrosia beetle *Gnathotrichus sulcatus* LeConte, although electrophysiological data are not available for the species. At the single cell level, chiral specificity was also found in the southern pine beetle in the frequency of spikes elicited in response to stimulation by (+)- and (-)-*exo*-brevicomin (Dickens and Payne, 1977).

Similar results have been found for the spruce bark beetle, *I typographus* (L.), in EAGs to the enantiomers of *cis*-verbenol (Dickens, personal communication). Significantly greater responses were elicited to the behaviorally active (S)-*cis*-verbenol over the less active (R)-*cis*-verbenol.

In contrast, the situation appears to be different for *I. pini*. Western (Idaho) beetles were found to produce only (-)-ipsdienol, whereas the eastern (New York) beetles produce a mixture of 65% (+)- and 35% (-)-ipsdienol (Stewart, 1975; Plummer *et al.*, 1976). Behaviorally, *I. pini* from Idaho responded significantly more to (-)-

ipsdienol than to the (+) form or the racemic mixture; whereas, those beetles from New York responded significantly more to the racemic mixture (Birch and Lanier, personal communication). At the antennal olfactory perception level, EAG results showed no significant differences in response to either enantiomer or the racemic mixture (Angst and Lanier, 1978). However, differences were found at the single cell level in response to the enantiomers (Mustaparta *et al.*, 1978).

The specificity that does exist in the olfactory systems of bark beetles does not appear to be extensive with regard to a one compound - one cell relationship. Even for *I. pini* it is possible that the cells found specific for a few of the compounds tested, may in fact respond to additional compounds not yet identified from the beetle since the pheromone system for the beetle has not been fully elucidated (Lanier, personal communication).

Odor Discrimination and Information Coding. The ability of bark beetles to decipher the complex olfactory messages in their environment depends on the number and specificity of acceptors for the various behavioral chemicals. Since it is probable that a single molecule can initiate a single spike (Kaissling and Priesner, 1970); whether or not a chemical message is conveyed to the central nervous system in sufficient intensity (spike frequency) to elicit a behavioral response will depend on the number of ac-

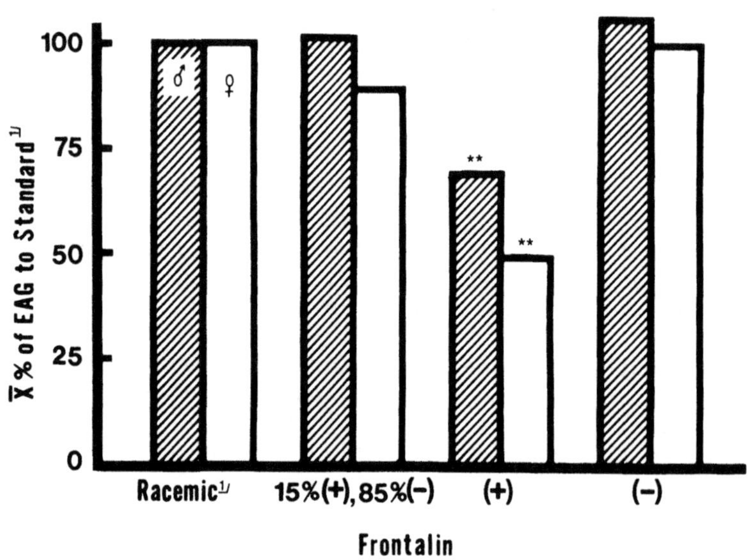

Fig. 13. Mean percent EAGs of southern pine beetles to frontalin enantiomers. ** = significantly different from all treatments/sex; 1-way ANOV.

ceptors stimulated. The apparent differences that exist in the numbers of acceptors for pheromones and host tree terpenes in general, and for individual compounds within those groups, provide the beetles with the flexibility and versatility to perceive the chemical messages and translate them into behavioral responses. In addition, the rate at which cells recover and can be re-stimulated adds additional versatility in the ability of the beetles to perceive messages.

How then, does the efficient olfactory system of the bark beetle unscramble the complex mixture of olfactory signals which continually confront the beetle? Molecule-bound information is decoded through the acceptor and transferred to the cell body as the intensity of the generator potential. The cell body transforms the information to the deutocerebrum of the beetle's brain as a frequency of spikes. The quality of the stimulus is determined by the specificity of the acceptor. The quantity is coded in the intensity of the slow potential and the frequency of spikes. Since the beetle will always be exposed to a multiplicity of olfactory stimuli in its natural environment, even if labelled lines exist, central integration of the many incoming signals must take place.

Simplistically speaking, where odor is involved, host selection behavior might occur when the olfactory signals characterizing the tree reach the central nervous system of the 'pioneer' beetle in sufficient quantity to overcome the threshold for the behavior. As the concentration of secondary attractants increases, the pattern of signals arriving in the central nervous system changes and the presence of attractant pheromones and host odors elicits aggregation behavior in emerging beetles and initiates the concentration phase. The phenomenon is continuous and as the qualitative and quantitative characteristics of the odor stimuli change, different behaviors result. As the concentration of attractant subsides and the concentration of inhibitor increases, so then changes the pattern of incoming signals to the central nervous system and the resulting behavior. Acceptors once occupied by attractant or even host odor now become increasingly stimulated by inhibitor and the responding beetles become deterred from the attacked tree and switching behavior results. As the concentrations of secondary attractants and host odors begin to increase around newly attacked trees, aggregation behavior results in the flying beetles. Stimuli which impinge on the central nervous system from other senses become integrated with those arriving from the olfactory cells and subsequently influence response to the compounds.

CONCLUSION

Bark beetles are well equipped with morphologically and physiologically efficient mechanisms for collecting molecule-bound information and transforming it into behavioral responses. Antennal olfactory sensilla with different degrees of specificity, and which occur in different numbers, provide the beetles with flexibility and versatility in their abilities to perceive pheromones and host odors. The central nervous system is adapted to decipher olfactory information through a dynamic pattern of signals which arrive from the peripheral olfactory cells.

An understanding of the events taking place at the antennal olfactory sensilla level provides insights important in the development of behavioral chemicals for use in pest management. In the southern pine beetle, for example, the fact that the acceptors for frontalin accomodate an array of other compounds, suggests that a mixture of compounds could be derived which would block response to frontalin and thus interrupt the aggregation phase of behavior in host tree attack. Presently, *endo*-brevicomin and verbenone are being evaluated for use in suppressing active infestations. The compounds do no prevent attack on trees in an active infestation. However, their presence significantly reduces the number of attacking beetles and subsequently, the number of progeny produced (Richerson and Payne, unpublished). These results are in direct agreement with what would be expected from data at the antennal olfactory sensilla level where both *endo*-brevicomin and verbenone interact with some, but not all, of the acceptors for frontalin. Thus, some frontalin acceptors are available for response to the attractant even in the presence of high concentrations of the inhibitors. It is theorized, however, that by continuous exposure to the inhibitors, fewer and fewer beetles will attack successive host trees until sufficient numbers are not available to overcome the host and the infestation will die out.

In general, however, data from the antennal olfactory perception level are far from complete for any bark beetle species. Considerably more information is needed at both the physiological and behavioral levels in order to enhance the development of pheromones and host odors in pest management. By combining information from these levels, we will begin to understand the sensory and motor aspects of the beetles and how they can be controlled or modified to our advantage.

ACKNOWLEDGMENTS

I thank J. C. Dickens, Forstzoologisches Institut der Universität Freiburg, i. Br., BRD; W. D. Seabrook, Department of Biology, University of New Brunswick, Fredericton and A. N. Starratt, Research Institute, Agriculture Canada, London. Original research funded through McIntire-Stennis project 1525 and the USDA program entitled "The Expanded Southern Pine Beetle Research and Applications Program" through TAES-CSRS grant #804-15-4. The findings, opinions and recommendations reported herein are those of the author and not necessarily those of the USDA. Texas Agricultural Experiment Station paper #14098.

REFERENCES

Adler, V. E. 1971. Physical conditions important to the reproducibility of electroantennograms. Ann. Entomol. Soc. Am. 64: 300-302.

Angst, M. E. and Lanier, G. N. 1978. Electroantennogram responses of two populations of *Ips pini* (Coleoptera: Scolytidae) to insect-produced and host tree compounds. J. Chem. Ecol. In press.

Bakke, A. 1976. Spruce bark beetle, *Ips typographus*: pheromone production and field response to synthetic pheromones. Naturwissenschaften 63: 92.

Bakke, A., Frφyen, P. and Skattebφl. 1977. Field response to a new pheromonal compound isolated from *Ips typographus*. Naturwissenschaften 64: 98.

Bedard, W. D., Tilden, P. E., Wood, D. L., Silverstein, R. M., Brownlee, R. G. and Rodin, J. O. 1969. Western pine beetle: field response to its sex pheromone and a synergistic host terpene myrcene. Science 164: 1284-1285.

Birch, M. C. 1971. Intrinsic limitations in the use of electroantennograms to bioassay male pheromones in Lepidoptera. Nature, London 233: 57-58.

Birch, M. C. and Wood, D. L. 1975. Mutual inhibition of the attractant pheromone response by two species of *Ips* (Coleoptera: Scolytidae). J. Chem. Ecol. 1: 101-113.

Birch, M. C. and Light, D. M. 1977. Inhibition of the attractant pheromone response in *Ips pini* and *I. paraconfusus* (Coleoptera: Scolytidae): field evaluation of ipsenol and linalool. J. Chem. Ecol. 3: 257-267.

Boeckh, J. 1962. Elektrophysiologische Untersuchungen an einzelnen Geruchs-Rezeptoren auf den Antennen des Totengräbers (Necrophorus, Coleoptera). Z. vergl. Physiol. 46: 212-248.

Boeckh, J., Kaissling, K. E. and Schneider, D. 1965. Insect olfactory receptors. Cold Spring Harbor Symp. Quant. Biol. 30: 263-280.

Borden, J. H. 1968. Antennal morphology of *Ips confusus* (Coleoptera: Scolytidae). Ann. Entomol. Soc. Am. 61: 10-13.

Borden, J. H. 1974. Aggregation pheromones in the Scolytidae. In Pheromones (Ed. Birch, M. C.). American Elsevier Pub. Co., N.Y., pp. 135-160.

Borden, J. H. 1977. Behavioral responses of Coleoptera to pheromones, allomones, and kairomones. In Chemical Control of Insect Behavior (Ed. Shorey, H. H. and McKelvey, J.). Wiley, N.Y., pp. 169-200.

Borden, J. H. and Wood, D. L. 1966. The antennal receptors and olfactory response of *Ips confusus* (Coleoptera: Scolytidae) to male sex attractant in the laboratory. Ann. Entomol. Soc. Am. 59: 253-261.

Borden, J. H., Chong, L., McLean, J. A., Slessor, K. N. and Mori, K. 1976. *Gnathotrichus sulcatus*: synergistic response to enantiomers of the aggregation pheromone sulcatol. Science 192: 894-896.

Borg, T. K. and Norris, D. M. 1971. Ultrastructure of sensory receptors on the antennae of *Scolytus multistriatus* (Marsh.). Z. Zellforsch. 113: 13-28.

Brand, J. M., Bracke, J. W., Markovetz, A. J., Wood, D. L. and Browne, L. E. 1975. Production of verbenol pheromone by a bacterium isolated from bark beetles. Nature 254: 136-137.

Brand, J. M., Bracke, J. W., Britton, L. N., Markovetz, A. J. and Barras, S. J. 1976. Bark beetle pheromones: production of verbenone by a mycangial fungus of *Dendroctonus frontalis*. J. Chem. Ecol. 2: 195-199.

Brand, J. M., Schultz, J., Barras, S. J., Edson, L. J., Payne, T. L. and Hedden, R. L. 1977. Bark beetle pheromones: enhancement of *Dendroctonus frontalis* (Coleoptera: Scolytidae) aggregation pheromone by yeast metabolites in laboratory bioassays. J. Chem. Ecol. 3: 657-666.

Brown, C. L. and Franklin, R. T. 1977. Southern pine beetle attractant: a criticism. J. Forest. 75: 627.

Byrne, K. J., Swigar, A. A., Silverstein, R. M., Borden, J. H. and Stokkink, E. 1974. Sulcatol: population aggregation pheromone in the scolytid beetle, *Gnathotrichus sulcatus*. J. Insect Physiol. 20: 1895-1900.

Cade, S. C., Hrutfiord, B. F. and Gara, R. I. 1970. Identification of a primary attractant for *Gnathotrichus sulcatus* isolated from western hemlock logs. J. Econ. Entomol. 63: 1014-1015.

Coster, J. E., Payne, T. L., Hart, E. R. and Edson, L. J. 1977. Aggregation of the southern pine beetle in response to attractive host trees. Environ. Entomol. 6: 725-731.

Davies, J. T. 1971. Olfactory theories. In Handbook of Sensory Physiology Vol. IV Chemical Senses, I Olfaction (Ed. Beidler, L.). Springer-Verlag, N.Y., pp. 332-350.

Dickens, J. C. and Payne, T. L. 1977. Bark beetle olfaction: pheromone receptor system in *Dendroctonus frontalis* Zimm. (Coleoptera: Scolytidae). J. Insect Physiol. 23: 481-489.

Dickens, J. C. and Payne, T. L. 1978. Structure and function of the sensilla on the antennal club of the southern pine beetle, *Dendroctonus frontalis* Zimmerman (Coleoptera: Scolytidae). Int. J. Insect Morphol. and Embryol. In press.

Dravnieks, A. 1966. Current status of odor theories. Advan. Chem. 56: 29-57.

Ernst, K. D. 1969. Die Feinstruktur von Riechsensillen auf der Antenne des Aaskäfers *Necrophorus* (Coleoptera). Z. Zellforsch. 94: 72-104.
Ferkovich, S. M., Mayer, M. S. and Rutter, R. R. 1973a. Conversion of the sex pheromone of the cabbage looper. Nature, London 242: 53-55.
Ferkovich, S. M., Mayer, M. S. and Rutter, R. R. 1973b. Sex pheromone of the cabbage looper: reactions with antennal proteins *in vitro*. J. Insect Physiol. 19: 2231-2243.
Gara, R. I. and Coster, J. E. 1968. Studies on the attack behavior of the southern pine beetle. III. Sequence of tree infestation within stands. Contrib. Boyce Thompson Inst. 24: 77-85.
Graham, K. 1968. Anaerobic induction of primary chemical attractancy for ambrosia beetles. Kansas J. Zool. 46: 905-908.
Harring, C. M. and Mori, K. 1977. *Pityokteines curvidens* Germ. (Coleoptera: Scolytidae) aggregation in response to optically pure ipsenol. Z. ang. Ent. 82: 327-329.
Hedden, R. L., Vité, J. P. and Mori, K. 1976. Synergistic effect of a pheromone and a kairomone on host selection and colonization by *Ips avulsus*. Nature 261: 696-697.
Kafka, W. A. 1970. Molekulare Wechselwirkungen bei der Erregung einzelner Riechzellen. Z. Vergl. Physiol. 70: 105-143.
Kaissling, K. E. 1971. Insect olfaction. In Handbook of Sensory Physiology, Vol. IV Chemical Senses, 1 Olfaction (Ed. Beidler, L.). Springer-Verlag, N.Y., pp. 351-431.
Kaissling, K. E. and Priesner, E. 1970. Die Riechschwelle des Seidenspinners. Naturwissenschaften 57: 23-28.
Kangas, E., Perttunen, V. and Oksanen, H. 1971. Physical and chemical stimuli affecting the behavior of *Blastophagus piniperda* L. and *B. minor* Hart. (Col., Scolytidae). Acta Ent. Fenn. 28: 120-126.
Kinzer, G. W., Fentiman, A. F., Page, T. F., Foltz, R. L., Vité, J. P. and Pitman, G. B. 1969. Bark beetle attractants: identification, synthesis and field bioassay of a new compound isolated from *Dendroctonus*. Nature, Lond. 221: 477.
Moeck, H. A. 1968. Electron microscopic studies of antennal sensilla in the ambrosia beetle *Trypodendron lineatum* (Oliver) (Scolytidae). Can. J. Zool. 46: 521-556.
Moeck, H. A. 1970. Ethanol as the primary attractant for the ambrosia beetle, *Trypodendron lineatum* (Coleoptera: Scolytidae). Can. Entomol. 102: 985-995.
Moeck, H. A. 1971. Field tests of ethanol as a scolytid attractant. Bimonthly Research Notes, Dept. of Fisheries & Forestry, Canada. 27: 11-12.
Mori, K. 1974. Synthesis of *exo*-brevicomin, the pheromone of western pine beetle, to obtain optically active forms of known absolute configuration. Tetrahedron 30: 4223.
Mori, K. 1975. Synthesis of optically active forms of frontalin. Tetrahedron 31: 1381-1384.

Mori, K., Mizumachi, N. and Matsui, M. 1976. Pheromone synthesis. Part XII - Synthesis of optically pure [1S, 4S, 5S]-2-Pinen-4-ol [*cis*-verbenol] and its antipode, the pheromone of *Ips* bark beetles. Agric. Biol. Chem. 40: 1611-1615.

Mustaparta, H. 1973. Olfactory sensilla on the antennae of the pine weevil, *Hylobius abietis*. Z. Zellforsch. 144: 559-571.

Mustaparta, H. 1974. Response of the pine weevil, *Hylobius abietis* L. (Col.: Curculionidae), to bark beetle pheromones. J. comp. Physiol. 88: 395-398.

Mustaparta, H. 1975. Responses of single olfactory cells in the pine weevil *Hylobius abietis* L. (Col.: Curculionidae). J. comp. Physiol. 97: 271-290.

Mustaparta, H., Angst, M. E. and Lanier, G. N. 1977. Responses of single olfactory cells in the pine engraver beetle *Ips pini* (Say) (Col.: Scolytidae) to its aggregation pheromone ipsdienol and the aggregation inhibitor ipsenol. J. comp. Physiol. 121: 343-347.

Mustaparta, H., Angst, M. E. and Lanier, G. N. 1978. Specialization of olfactory receptor cells to insect- and host-produced volatiles in the bark beetle species *Ips pini* (Say). J. Chem. Ecol. In press.

Moulton, D. G. and Beidler, L. M. 1967. Structure and function in the peripheral olfactory system. Physiol. Rev. 47: 1-52.

Norris, D. M., Ferkovich, S. M., Baker, J. E., Rozental, J. M. and Borg, T. K. 1971. Energy transduction in quinone inhibition of insect feeding. J. Insect Physiol. 17: 85-97.

Oksanen, H., Perttunen, V. and Kangas, E. 1970. Studies on the chemical factors involved in the olfactory orientation of *Blastophagus piniperda* (Coleoptera: Scolytidae). Contrib. Boyce Thompson Inst. 24: 299-304.

Payne, T. L. 1971. Bark beetle olfaction. I. Electroantennogram responses of the southern pine beetle (Coleoptera: Scolytidae) to its aggregation pheromone frontalin. Ann. Entomol. Soc. Am. 64: 266-288.

Payne, T. L. 1975. Bark beetle olfaction. III. Antennal olfactory responsiveness of *Dendroctonus frontalis* Zimmermann and *D. brevicomis* LeConte (Coleoptera: Scolytidae) to aggregation pheromones and host tree terpene hydrocarbons. J. Chem. Ecol. 1: 233-242.

Payne, T. L. and Dickens, J. C. 1976. Adaptation to determine receptor system specificity in insect olfactory communication. J. Insect Physiol. 22: 1569-1572.

Payne, T. L., Moeck, H. A., Willson, C. D., Coulson, R. N. and Humphreys, W. J. 1973. Bark beetle olfaction. II. Antennal morphology of sixteen species of Scolytidae (Coleoptera). Internat. J. Insect Morphol. and Embryol. 2: 177-192.

Person, H. L. 1931. Theory in explanation of the selection of certain trees by the western pine beetle. J. Forest. 31: 696-699.

Perttunen, V., Oksanen, H. and Kangas, E. 1970. Aspects of the external and internal factors affecting orientation of *Blastophagus piniperda* (Coleoptera: Scolytidae). Contrib. Boyce Thompson Inst. 24: 293-297.

Pitman, G. B. and Vité, J. P. 1969. Aggregation behavior of *Dendroctonus ponderosae* (Coleoptera: Scolytidae) in response to chemical messengers. Can. Entomol. 101: 143-149.

Pitman, G. B., Vité, J. P., Kinzer, G. W. and Fentiman, A. F., Jr. 1969. Specificity of population-aggregating pheromones in *Dendroctonus*. J. Insect Physiol. 15: 363-366.

Plummer, E. L., Stewart, T. E., Byrne, K., Pearce, G. T. and Silverstein, R. M. 1976. Determination of the enantiomeric composition of several insect pheromone alcohols. J. Chem. Ecol. 2: 307-332.

Renwick, J. A. A. 1967. Identification of two oxygenated terpenes from the bark beetles *Dendroctonus frontalis* and *D. brevicomis*. Contrib. Boyce Thompson Inst. 23: 355-360.

Renwick, J. A. A. and Vité, J. P. 1969. Bark beetle attractants: Mechanism of colonization by *Dendroctonus frontalis*. Nature, London 224: 1222-1223.

Rudinsky, J. A. 1966. Scolytid beetles associated with Douglas-fir: response to terpenes. Science 152: 218-219.

Rudinsky, J. A. and Michael, R. R. 1973. Sound production in Scolytidae: stridulation by female *Dendroctonus* beetles. J. Insect Physiol. 19: 689-705.

Schneider, D. 1955. Mikro-Elektroden registrieren die elektrischen Impulse einzelner Sinnesnervenzellen der Schmetterlingsantenne. Industrie-Elektronik, Hamburg 5: 3-7.

Schneider, D. 1957a. Elektrophysiologische Untersuchungen von Chemo- und Mechanorezeptoren der Antenne des Seidenspinners *Bombyx mori* L. Z. Vergl. Physiol. 40: 8-41.

Schneider, D. 1957b. Electrophysiological investigation on the antennal receptors of the silk moth during chemical and mechanical stimulation. Experientia 13: 89-91.

Schneider, D. 1970. Olfactory receptors for the sexual attractant (Bombykol) of the silk moth. In The Neurosciences: Second Study Program (Ed. Schmitt, F.). Rockefeller University Press, N.Y., pp. 511-518.

Schneider, D. and Hecker, E. 1956. Zur Elektrophysiologie der Antenne des Seidenspinners *Bombyx mori* bei Reizung mit angereicherten Extrakten des Sexuallockstoffes. Z. Naturforsch. 11b: 121-124.

Schneider, D. and Steinbrecht, R. A. 1968. Checklist of insect olfactory sensilla. Symp. Zool. Soc. London 23: 279-297.

Schneider, D., Steinbrecht, R. A. and Ernst, K. 1966. Antennales Riechhaar des Aaskäfers. Naturwiss. Rundsch. 3: 1.

Seabrook, W. D. 1977. Insect chemosensory responses to other insects. In Chemical Control of Insect Behavior (Ed. Shorey, H. H. and McKelvey, J.). Wiley, N.Y., pp. 15-43.

Silverstein, R. M., Rodin, J. O. and Wood, D. L. 1966. Sex attractants in frass produced by male *Ips confusus* in ponderosa pine. Science 154: 509-510.

Silverstein, R. M., Brownlee, R. G., Bellas, T. E., Wood, D. L. and Browne, L. E. 1968. Brevicomin: principal sex attractant in the frass of the female western pine beetle. Science, Washington 159: 889-894.

Slifer, E. H. 1967. Thin-walled olfactory sense organs on insect antennae. In Insects and Physiology (Eds. Beament, J. and Treherne, J.). Oliver and Boyd, Edinburgh, London, pp. 232-245.

Slifer, E. H., Prestage, J. J. and Beams, H. W. 1959. The chemoreceptors and other sense organs on the antennal flagellum of the grasshopper (Orthoptera: Acrididae). J. Morphol. 195: 145-191.

Steinbrecht, R. A. 1969. Comparative morphology of olfactory receptors. In Proceedings of the Third International Symposium on Olfaction and Taste, New York, 1968 (Ed. Pfaffman, C.). The Rockefeller University Press, N.Y., pp. 3-21.

Steinbrecht, R. A. and Kasang, G. 1972. Capture and conveyance of odour molecules in an insect olfactory receptor. In Proceedings of the Fourth International Symposium on Olfaction and Taste, Seewiesen, Germany, 1971 (Ed. Schneider, D.). Wissenschaftliche Verlagsgesellschaft MBH Stuttgart, pp. 193-199.

Stewart, T. E. 1975. Identification biological activity, and the enantiomeric composition of insect pheromones. MSc Thesis, State Univ. of N.Y. Coll. Environ. Sci. and Forestry, Syracuse, N.Y.

Stewart, T. E., Plummer, E. L., McCandless, L. L., West, J. R. and Silverstein, R. M. 1977. Determination of enantiomer composition of several bicyclic ketal insect pheromone components. J. Chem. Ecol. 3: 27-44.

Stürckow, B. 1965. The electroantennogram (EAG) as an assay for the reception of odours by the gypsy moth. J. Insect Physiol. 11: 1573-1584.

Stürckow, B. 1970. Responses of olfactory and gustatory receptor cells in insects. In Advances in Chemoreception, Vol. I, Communication by Chemical Signals (Eds. Johnston, J., Moulton, D. and Turk, A.). Appleton-Century-Crofts, N.Y., pp. 107-158.

Vité, J. P. and Pitman, G. B. 1969. Aggregation behavior of *Dendroctonus brevicomis* in response to synthetic pheromones. J. Insect Physiol. 15: 1617-1622.

Vité, J. P. and Renwick, J. A. A. 1971. Inhibition of *Dendroctonus frontalis* response to frontalin by isomers of brevicomin. Naturwissenschaften 8: 418.

Vité, J. P. and Francke, W. 1976. The aggregation pheromones of bark beetles: progress and problems. Naturwissenschaften 63: 550-555.

Vité, J. P., Hedden, R. and Mori, K. 1976a. *Ips grandicollis*: field response to the optically pure pheromone. Naturwissenschaten 63: 43-44.

Vité, J. P., Klimetyek, D., Loskant, G., Hedden, R. and Mori, K. 1976b. Chirality of insect pheromones: response interruption by inactive antipodes. Naturwissenschaften 63: 582-583.

Wood, D. L. 1972. Selection and colonisation of ponderosa pine by bark beetles. In Insect/Plant Relationships (Ed. van Emden, H. F.). Symp. Roy. Entomol. Soc., London 6: 101-117.

Wood, D. L., Browne, L. E., Ewing, B., Lindahl, K., Bedard, W. D., Tilden, P. E., Mori, K., Pitman, G. B. and Hughes, P. R. 1976. Western pine beetle: specificity amoung enantiomers of male and female components of an attractant pheromone. Science 192: 896-898.

Young, I. C., Brownlee, R. G., Rodin, J. O., Hildebrand, D. N., Silverstein, R. M., Wood, D. L., Birch, M. C. and Browne, L. E. 1973. Identification of linalool produced by two *Ips* species bark beetles. J. Insect Physiol. 19: 1615-1622.

CHEMORECEPTOR PROTEINS

Dale M. Norris

642 Russell Laboratories, University of Wisconsin,

Madison WI 53706

Even though hundreds of allomonic or kairomonic molecules have been shown to affect insect behavior through neurophysiological actions, little research attention has been given to the involved energy-transferring macromolecules in chemosensitive neurons in sensilla. Much of the consideration of receptor aspects of insect chemical communication that has occurred has involved making inferences based on the properties of the messenger molecules and the characteristics of resultant insect behavior (e.g., Kafka and Neuwirth, 1975). However, direct study of candidate receptors has involved the "sugar-splitting" enzyme activity in the receptor-containing leg of the blowfly, *Phormia regina*, as initiated by Dethier (1955). Hansen (1969) subsequently showed a correlation between the distribution of glucosidase activity and the number of chemosensilla on the tarsus of the blowfly. It was proposed that a sugar-glucosidase complex formed as the first reaction between the sugar and the receptor. Norris and coworkers (Gilbert et al., 1967) started research on the 1, 4-naphthoquinones (e.g., juglone in *Carya* spp.) as allomones for *Scolytus multistriatus* and *Periplaneta americana*, and have investigated the neural receptors for these ligands. Findings on the sugar receptor of the blowfly and the naphthoquinone (e.g., bitter) receptor of the latter two mentioned insects will be the focus of this chapter.

THE NAPHTHOQUINONE RECEPTOR

Though most scientific interest in chemical ecology in the last 15 years has involved pheromones, our research group chose some electro-chemically reversible quinone-quinol redox couples and potential one-electron donors or acceptors (King et al., 1973) as our experimental ligands for investigations of receptor-messenger interactions (Gilbert et al., 1967; Norris, 1969, 1970, 1976, 1977; Norris et al., 1970, 1977). Early studies (Norris,1969; Norris et al., 1970) strongly implicated sulfhydryl-containing lipoproteins as the energy-transducing (transferring) receptors for the quinones in the dendritic membranes of chemosensory neurons in sensilla on the antenna.

In seeking a functional understanding of the involved physicochemical "black box" mechanism of energy transfer from ligand to receptor, or visa versa, a multidisciplinary investigative approach has been taken. This included experiments using behavioral assays (Norris, 1969, 1977; Norris et al. 1970, 1977); scanning electron microscopic (SEM) characterization of the external morphology of the sensilla on the antenna (Norris and Chu, 1974); ultrastructural analyses of the various chemosensilla on the antenna (Borg and Norris, 1971; Norris and Chu, 1974; Norris, 1976);electrophysiological anayses of the neuronal effects of ligand-receptor interactions using electroantennograms (EAG's) (Norris and Chu, 1974) and extracellular recordings (Baker and Norris, 1971); binding studies between ligands and extracted receptor lipoprotein (Ferkovich and Norris, 1972; Singer and Norris, 1973a; Singer et al., 1975; Rozental et al., 1975; Norris et al., 1977a), and between ligands and hignly purified receptor (Norris et al., 1977b, 1977c); and electrochemical quantitation of ligand-receptor interactions using dropping-mercury electrode polarography(Norris et al., 1970; Rozental and Norris, 1973, 1975; Norris et al., 1977a). Data from all of these experimental approaches correlate.

Ligand Effects on Insect Behavior

Several naphthoquinones inhibited the feeding of both adult Scolytus multistriatus (Norris, 1969) and Periplaneta americana (Norris et al., 1971). The order of this relative allomonic activity among several 1,4-naphthoquinones was the same for both insect species, i.e., 5-hydroxy-1,4-naphthoquinone > 1,4-naphthoquinone > 2-hydroxy-1,4-naphthoquinone > 2-methyl-1,4-naphthoquinone. The order of relative repellency of these ligands is the same as for feeding inhibition (Rozental and Norris, 1975). These correlations between molecular structure and allomonic actions in two extremely different species of insects can be explained largely by (1) the redox potential of the naphthoquinone and (2) presence, or absence, of hydroxyl substitution. Presence of hydroxyl contributes important hydrogen-bonding capabilities to the ligand (Norris, 1976 b).

Ultrastructures of Chemosensilla

Meaningful interpretation of transduction of energy between ligand and nerve-membrane-associated chemoreceptor lipoprotein requires that the physicochemical machinery is accommodated by the ultrastructural constraints of the in vivo system (Norris, 1976a). In both experimental species, the antenna is required for the evaluated chemosensory responses (Borg and Norris, 1971; Norris and Chu, 1974a). Both uniporous and multiporous chemosensilla are present on the antenna of each species. Ultrastructure of a multiporous chemosensillum on the adult male P. americana is seen in Figure 1. This high voltage electron micrograph of a cross section of such a chemosensillum clearly demonstrates that pore tubules (Pt) extend inward from the inner extent of a cuticular pore (P) to interface tightly with a dendritic membrane (D). Microtubules (Mt) are visible within the dendrite. In each insect species, hundreds of such multiporous chemosensilla are found on each antenna. The insect responds significantly to these allmones if it has only several segments of one antenna left intact (Norris et al., 1971).

The ultrastructure of such multiporous chemosensilla indicate that a ligand molecule would seem to be limited to the confines of the particular cuticular pore-pore tubule channel it entered. It would directly, or perhaps indirectly through contents of the given channel, exchange energy with lipoprotein energy transducers in the dendritic membrane exposed to the ligand's effects through direct interface with the pore tubule. Antennal dipping in 0.9% NaCl potassium phosphate buffer, pH 7.0, yields a wash which contains lipoproteinaceous components (Singer and Norris, 1973b). These eluted lipoproteins strongly bind the allomonic naphthoquinones, classical sulfhydryl reagents, and reducing agents (Norris, 1976a, c). The saline-eluted proteins are at least partially found in the cuticular pore-pore tubule system (Norris, 1976a), and have the capabilities of determining the number of messenger molecules (ligands) which are required in the tubule system to elicit an electrochemical change in the energy-transducing lipoprotein receptor in the dendritic membrane. Thus, the contents of the pore tubule system may function to buffer the energy-transducing receptor from messenger ligands, or possibly synergize the effects of ligand on the nerve-membrane receptor (Norris, 1976c). Regardless of the state of the pore-tubule contents, some ingress of messenger ligand through the tubules to the dendrites results when the chemosensillum is pulsed in vivo with radiolabelled ligand (Figure 2).

Electrophysiology Associated with Chemoreceptors

A prior pulse with one of the feeding-inhibitory and repellent 1,4-naphthoquinones reduces the magnitude (i.e., mV) of electro-

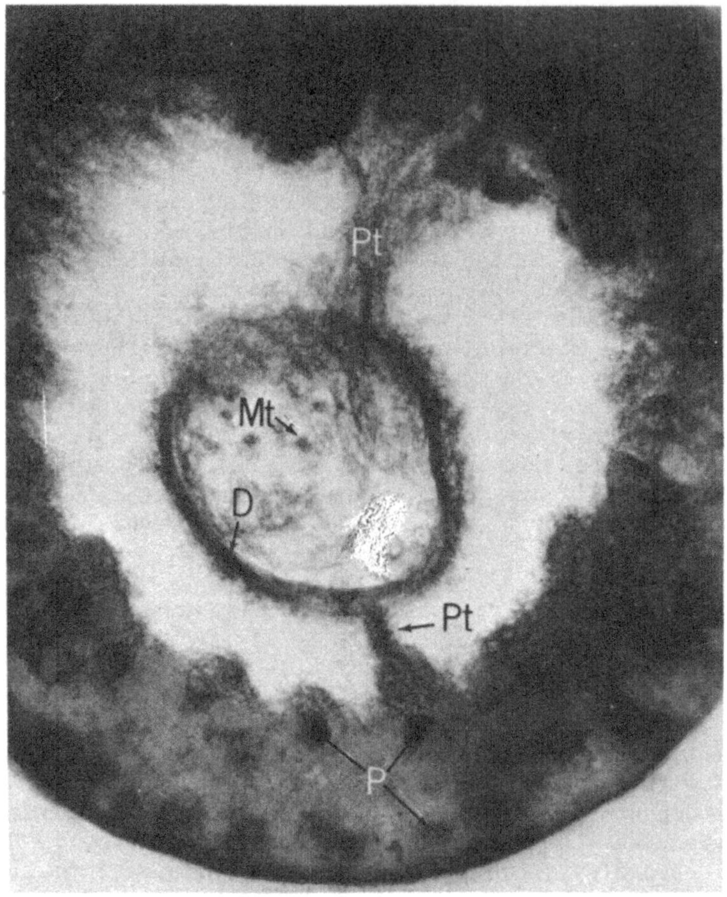

Figure 1. High voltage electron-transmission micrographic resolution of the "tight junctions" between the inner extent of the cuticular pore (P) and the pore tubule (Pt), and the Pt and the dendritic membrane (D) of this multiporous chemosensillum. Microtubules (Mt) are visible in the dendrite.

antennogram depolarization in P. americana caused by a standard pulse of the excitant, amyl acetate (Figure 3). Pulses on antennae with the naphthoquinones in the 10^{-7} to 10^{-4} M range thus tend to hyperpolarize or repolarize dendritic membranes. More extensive exposures of the chemsensory neurons to such ligands may lead to a blockage of electrophysiological action in the receptor and membrane. As with behavioral assays, the order of relative inhibition of the chemosensory neuronal electrophysiology by the studied naphthoquinones was 5-hydroxy -1,4-naphthoquinone > 1,4-naphthoquinone > 2-hydroxy-1, 4-naphthoquinone > 2-methyl-1,4-naphthoquinone (Norris and Chu, 1974b).

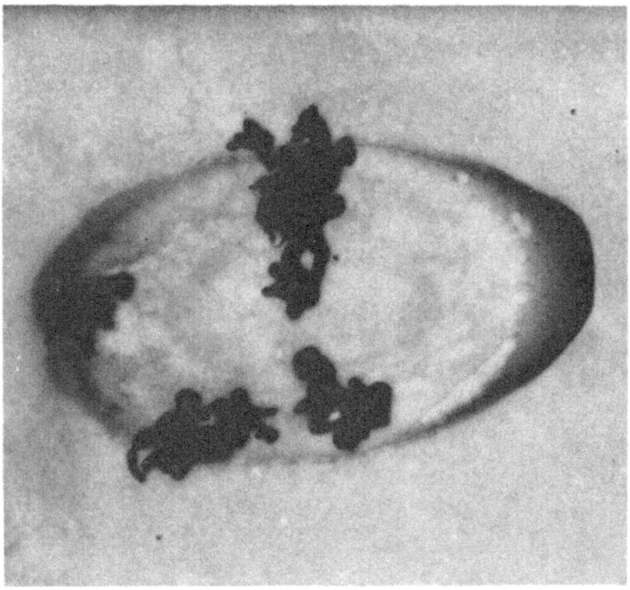

Figure 2. Transmission electron micrograph with autoradiography showing that radiolabelled messenger ligand applied in vivo to a multiporous chemosensillum penetrated the cuticular pore-pore tubule system to contact dendrites of the neurons.

Receptor Lipoprotein Binding of Ligands

When given naphthoquinones were added at 1.7×10^{-7} moles/liter to optically standardized samples of particulate fractions (F_1 -F_4) of a homogenate of adult male P. americana antennae which were seperated by density gradient centrifugation, the interaction of each naphthoquinone with each fraction caused a change in ultraviolet absorbance at 280nm which was considered as evidence of complex formation. The greater changes in absorbance occurred with fractions F_1 and F_2 which were shown by TEM monitoring to be especially rich in nerve membrane fragements. The order of relative binding among three tested naphthoquinones agreed with the orders of their activities previously discussed under behavioral assays and electrophysiology (Norris et al., 1970).

A Triton X-100 solubilized lipoprotein from the P. americana antenna, which selectively bound 2-methyl (^{14}C)-1,4-naphthoquinone (menadione) in vivo or in vitro, was subsequently resolved by disc gel electrophoresis (Ferkovich and Norris, 1972). Singer and Norris (1973a) achieved an improved disc gel electrophoretic resolution of

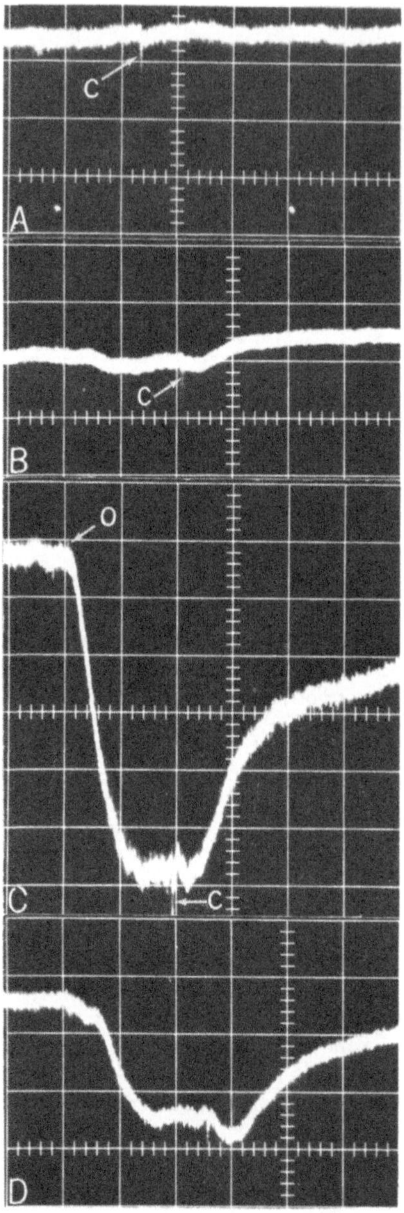

Figure 3. Electroantennograms from a male P. americana antenna elicited by the excitant amyl acetate (C), and as inhibited by a prior pulse of the antenna with the repellent 5-hydroxy-1,4-naphthoquinone (D). Baseline is shown in (A), and baseline plus standard pulse of control air in (B). (1 mV per grid unit).

the lipoprotein receptor for the naphthoquinones. The energy-transducing receptor resolved into two distinct bands with Rf's of 56 and 77 in 7 percent acrylaminde gels in pH 7.1 buffered conditions. This lipoprotein constituted about 13.7 percent of the Triton-solubilized proteins from the antenna, and it bound about $9.5 \pm 0.3 \times 10^{-5}$ µm of menadione/ug (Rozental et al., 1975). The major involvement of sulfhydryl functional groups in the receptor binding of the naphthoquinones was strongly substantiated by Singer et al. (1975). When receptor was pre-incubated for a standard time with various molar quantities of four naphthoquinones, each of the four competed with subsequently added ^{14}C-iodoacetic acid (IAA), a classical sulfhydryl reagent, for binding sites on the receptor. The order of relative competitiveness of the four ligands with the IAA again matched the order of activities observed with all previously utilized experimental approaches to analyzing the ligand-receptor interactions. Recently the receptor was further purified about 57 fold using column affinity chromatography (Norris et al., 1977b,c). Using equilibrium dialysis methods of comparison, the crude receptor protein bound $1.27 \pm 0.4 \times 10^{-5}$um of menadione per ug as compared to $7.23 \pm 0.04 \times 10^{-4}$µm per ug for the affinity-column purified receptor.

Electrochemistry of Energy Transduction by Receptor

The fact that our chemical messengers (i.e., naphthoquinones) in this study were known to react electrochemically with proteins led us to employ the dropping-mercury electrode (DME) polarograph to quantitate the involved energy interactions in millivolts. Our experiments using polarography initially employed the sulfhydryl-containing peptide, reduced glutathione (GSH) as a model receptor (Norris et al., 1970). These investigations quickly established the importance of sulfhydryls in receptor protein to the reversible complexing of naphthoquinones with receptor. Rozental and Norris (1973) expanded such electrochemical analyses of the ligand-receptor interactions to involve the Triton-extracted lipoprotein receptor and energy transducer from the insect's antennae. This work showed that a given chemical messenger - receptor interaction has a characteristic electromotive force (e.m.f.) parameter which can be determined and stated in millivolts as an $E_{1/2}$ value. The $E_{1/2}$ value in these cases represents the catalytic electromotive force which must be applied to a given amount of receptor protein in solution that has interacted with a given amount of ligand, naphthoquinone, so as to react one-half of the reactive sulfhydryls or reducible disulfides in the receptor protein (Brdička, 1933). Rozental and Norris (1973) demonstrated that the observed ligand - receptor interactions were reversible and possessed characteristics of redox systems. As an example, after a given amount of receptor from the antenna had been saturated with ligand naphthoquinone, its $E_{1/2}$ would have shifted a characteristic quantity in millivolts. By adding appropiate reducing agent (e.g., dithiothreitol or ascorbic acid), the electrochemical state of the receptor lipoprotein could be returned to its original naphthoquinone - receptive condition. The fascinating existence of genetic strains or biotypes within the species, P. americana, which have electrochemically distinct peripheral neural receptors and energy transducers (Figures 4 and 5) for the ligand naphthoquinones was documented by Rozental and Norris (1975). These genetically distinct strains, referred to as LAB and WARF, also proved to be very different behaviorally in their sensitivities to a given naphthoquinone. The LAB strain is about twice as sensitive behaviorally to 5-hydroxy -1,4-naphthoquinone as is the WARF strain (Rozental and Norris, 1975). Subsequent comparative analyses of the two strains of receptor for (1) naphthoquinone bindind capacity, (2) number of sulfhydryls, (3) susceptibility of sulfhydryls to cleavage, (4) effect of sulfhydryl cleavage on naphthoquinone binding capacity, and (5) effect of sulfhydryl cleavage on electrochemical characteristics significantly expanded our understanding of the roles of protein-sulfur chemistry in the physicochemistry of naphthoquinone-receptor interactions (Norris et al., 1977a). It seems clear that the lipoprotein receptors of the LAB and WARF strains bind comparable amounts of the ligand in vitro, but the electrochemistry of the two receptors changes in diametrically opposed fashions during their saturation with naphthoquinone (Figure 5). These significant differences in resultant electrochemistry seem largely attributable to critical sulfurs in the LAB receptor becoming progressively less accessible and/or reactive polarographically as the receptor complexes increasing amounts of ligand, and conversely critical sulfurs in the WARF receptor becoming progressively more accessible and/or reactive polarographically as the receptor complexes increasing amounts of ligand. It is interesting and mechanistically important that the in vivo saturation of the LAB recêptor in the live antenna requires only about one-half as many moles of repellent 2-methyl-1,4-naphthoquinone as with the WARF strain (Figure 6).

Other Characteristics of the Receptor Protein

Characterizations of this lipoprotein energy-transducing macromolecule in terms of molecular weights of basic subunits, amino acid compositions of the subunits, the involved lipids, sequencing of the protein, etc., have been secondary objectives as compared to our analyses of the essential electrochemical parameters of the ligand-receptor interactions. However, significant information is emerging on these more classical parameters of proteins.

The highly purified receptor has yielded three sharply resolved subunits based on SDS-disc gel electrophoresis. The estimated molecular weights of these subunits are 19,000; 23,500; and 26,000 Daltons. However, there is additional, yet ill-resolved, protein in the molecular weight range from 15,500-18,000.

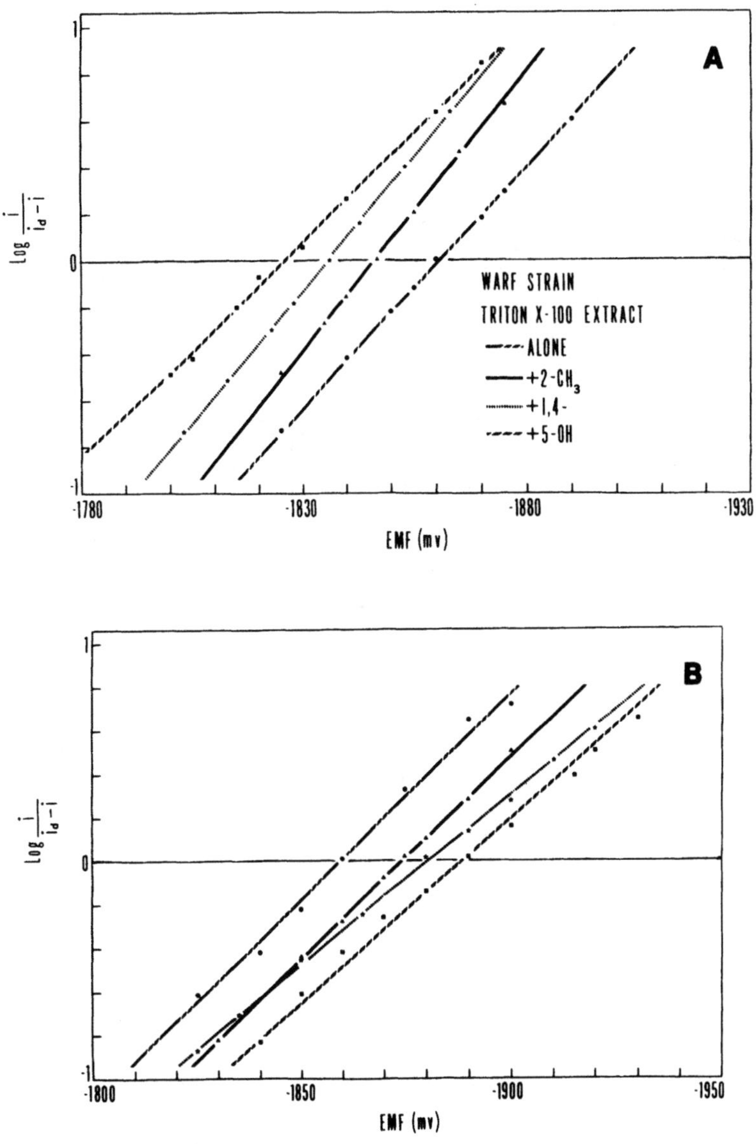

Figure 4. Logarithmic analyses of the polarographic waves obtained from the Triton-extracted lipoprotein receptor from male P. americana antennae. A. Polarograms of the WARF strain. B. Polarograms of the LAB strain. Protein alone (control) •, proteins + menadione ▲, protein + 1.4-naphthoquinone ✱, protein + 5-hydroxy-1,4-naphthoquinone ■. The potential at which the log $(i/i_d - i)$ is zero gives the half-wave potential ($E_{1/2}$), i_d is the diffussion current and i is the mean current measured from 3 replicates at the indicated potentials along the wave.

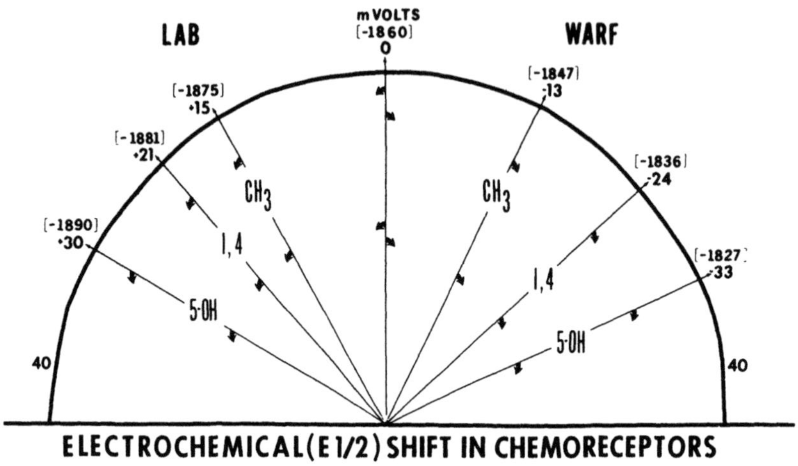

Figure 5. Diagrammatic representation of the *in vitro* quantitatively similar mV electrochemical $E\frac{1}{2}$ shifts induced in LAB and WARF receptors. However, these ligand-specific mV shifts were opposite in direction (i.e., more negative in LAB versus more positive in WARF).

Investigations of the amino acid composition of this purified receptor lipoprotein (Table 1) have, thus far, revealed 17 acids. In addition, the necessary special determination of tryptophan has not been conducted yet, and additional specially prepared samples will be analyzed for cysteine and cystine. Sequencing of the subunits is in progress using an automatic sequencer. The constituent lipids are the focus of a separate current study.

Concluding Comments

Our understanding of the physicochemistry of naphthoquinone perception by *P. americana* is still very incomplete, but it has proven adequate to enable our predictable alteration of the efficiency of energy transfer that occurs between the messenger ligand and the energy-transducing membrane-associated lipoprotein receptor (Figure 7A and 7B). Employing behavioral assays detailed by Rozental and Norris (1975) and Norris et al. (1977a), the reduced sulfhydryl-containing peptide, glutathione (GSH), was applied to the entire antennae of individual adult males of each strain by a dip in 0.9 percent NaCl containing the indicated molar concentration of the peptide. The so-treated cockroaches were then exposed in the choice test to a molar quantity of 5-hydroxy-1,4-naphthoquinone (Rozental and Norris, 1975; Norris et al., 1977a) which would normally , in the case of insects whose antennae were only dipped in the 0.9 percent NaCl (control) solutions, have repelled about 50 percent of

Table 1. Partial amino acid composition* of purified receptor protein from antennae of WARF strain of *Periplaneta americana* adult males

Amino Acid	Ratios of Amounts Present
Aspartic acid	1.75
Threonine	.750
Serine	2.178
Glutamic Acid	1.885
Proline	3.233
Glycine	3.057
Alanine	1.746
Cysteine	Off scale
Valine	.542
Methionine	.113
Isoleucine	.363
Leucine	.967
Tyrosine	.438
Phenylalanine	.435
Histidine	1.066
Lysine	.961
Ammonia	17.843
Arginine	.590

*Analyses conducted in triplicate using a Durrum Model D-500 amino acid analyzer.

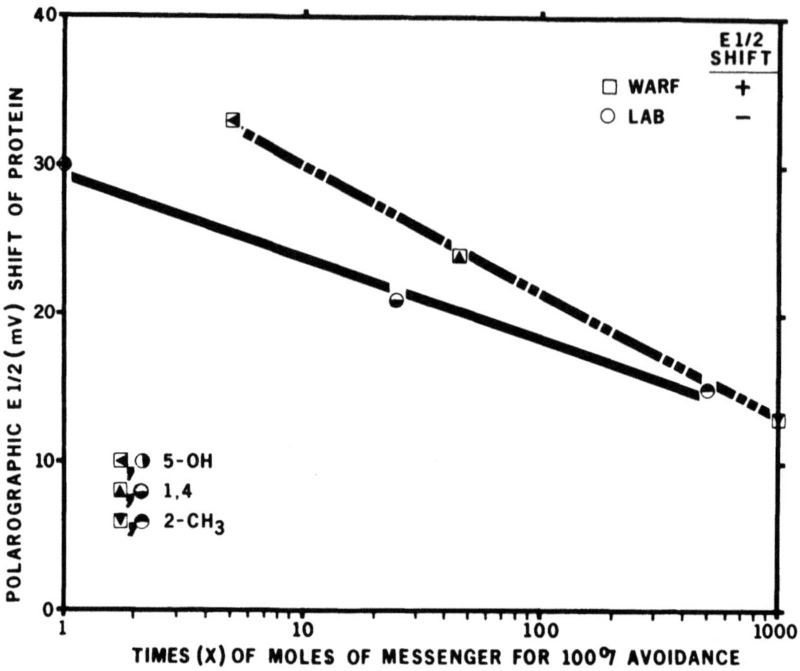

Figure 6. Relationships between the moles of each indicated 1,4-naphthoquinone required to repel behaviorally 100 percent of the assayed adult male P. americana of each strain, LAB versus WARF; and the maximum millivolt shift which saturating amounts of each repellent induces in the polarographic $E\frac{1}{2}$ value of each strain of in vitro receptor.

the insects that would enter a control tub (i.e., about 2.5 versus 5 insects would enter the treated tub) (Figure 7A). As seen in Figure 7A increasing molar quantities of weak reducing agent GSH applied as an antennal dip had opposite effects on the LAB versus WARF cockroach's subsequent behavioral sensitivities to the standard 5-hydroxy-1,4-naphthoquinone exposure. The LAB strain was progressively desensitized to the naphthoquinone, and the WARF insects were progressively hypersensitized. Antennal treatments with the mercaptid-forming reagent, PCMB, hypersensitized both strains at the lower concentrations (Figure 7B) , but then significantly desensitized both strains at higher molar concentrations. However, the more sensitive LAB strain was significantly desensitized at a lower concentration of PCMB dip than were the less sensitive WARF insects (Figure 7B). Such differing in vivo effects of sulfhydryl-affecting GSH and PCMB on the two distinct strains surely support

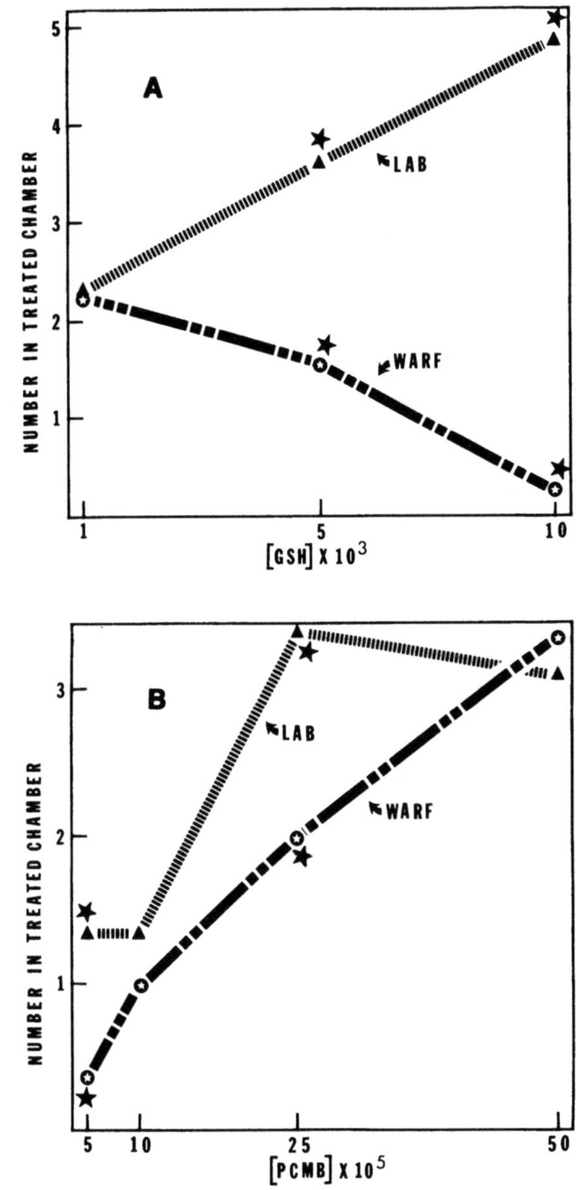

Figure 7. Alterations of the behavioral responses of LAB versus WARF cockroaches to a standard moles of the repellent 5-hydroxy-1,4-naphthoquinone per strain after a 5-minute dip of their antennae in an aqueous solution of the indicated concentration of (A) reduced glutathione, GSH; or (B) the classical sulfhydryl reagent, p-chloromercuricbenzoic acid, PCMB.

the uniqueness of the sulfur-dependent electrochemistry in each *in vivo* energy-transducing receptor. The results also seem to support strongly the roles of the peripheral receptors and involved primary sensory neurons in encoding information from the surrounding external chemical environment such as to alter predictably the behavior of the animals.

POSSIBLE SUGAR RECEPTOR OF BLOWFLY

Following up on initial interest (Dethier, 1955) in the possible roles of α-glucosidases of chemosensilla of *Phormia regina* in chemoreception by this species, Morita and coworkers (e.g., Amakawa et al., 1975) have studied insoluble forms of α-glucosidases which they propose as probably being located on the membrane system of contact chemosensory cells in the labella of the blowfly. Though these studies do not involve proteins isolated from the antennae of insects, discussion of this research is included because if represents the only other extensive investigation of possible chemoreceptor proteins in insects.

The Hansen Working Hypothesis

Much of the research conducted on α-glucosidases as possible chemoreceptors for sugars in the blowfly has been testing the interesting hypothesis of Hansen (1969). It proposed that the primary process of sugar reception involves complex formation between sugar and the receptor substance, and that the receptor is α-glucosidase. The progress of research in testing this hypothesis will be summarized rather briefly.

Behavioral Analyses

Behavioral threshold values for the receptor were established in the early work of Dethier (1955). These values are in good agreement with the Km's obtained with the crude extract of labella and with Peak III from column chromatographic separation of the extracted α-glucosidases on DEAE-Sephadex A-50 (Amakawa et al., 1972).

Ultrastructural Considerations

The vital understanding of ultrastructural constraints and characteristics which influence this chemoreception in the blowfly has been elucidated by several workers (e.g., Grabowski and Dethier, 1954; Larson, 1962). Kijimi et al. (1973) showed that α-glucosidase is located structurally at the point (i.e., tip of the sensory hair) where sugar perception supposedly occurs.

Electrophysiological Aspects

The electrophysiological constant K_b, the concentration of stimulant at which the response of the receptor is one-half of the maximum, is considered a more suitable measure for comparison with the Km of the enzyme than is the behavioral threshold value (Amakawa et al., 1972). It is intesting, and apparently biologically significant, that the Km's of Peak II, from the chromatographic separation of α-glucosidases, for sucrose and α-nitrophenyl α-D-glucopyranoside (αPNPG) are approximately equal to the K_b's, but the Km's of the crude extract, Peak I, and Peak III are much smaller than the K_b's (Amakawa et al., 1972).

Binding of Sugars to Receptor

Shimada et al. (1974) reviewed their earlier findings from analyzing the effects of receptor modification, using the sulfhydryl reagent, p-chloromercuric-benzoate (PCMB), on the response of the sugar receptor to sucrose. They concluded that PCMB reacted with sulfhydryl groups in proteins of the sugar receptor, and that the reagent reacted at a different site from the sugar-binding site. It was proposed that this PCMB binding blocked a change in the receptor membrane permeability.

These results on the marked modification of the sugar receptor in blowflies with the classical sulfhydryl reagent, PCMB, strongly support the unifying hypothesis of Norris (1971) regarding the roles of sulfhydryls and/or disulfides in the energy-transfer functions of membrane-associated proteins in neurobiology. The effects of PCMB on this sugar-receptor interaction with sucrose substantiated extensive previous experimental findings by Norris and coworkers (Norris et al., 1970, 1971) on the important roles of sulfhydryls in chemoreception in insects.

Shimada et al. (1974) demonstrated further that even though PCMB will disrupt the responses of the blowfly receptor protein to both D-glucose and D-fructose, disruption of the response to D-fructose required a longer exposure to a given concentration of the sulfhydryl reagent PCMB than for alteration of response to D-glucose. It is well known that different sulfhydryls in a protein may differ greatly in their accessibility and/or reactivity with a given sulfhydryl reagent such as PCMB. Thus, the results of Shimada et al. (1974) are very reasonable and could be expected.

GENERAL COMMENTS

Our knowledge of the physicochemistry of insect chemoreceptor-protein interactions with olfactory and taste ligands remains limited, but at least one apparently important common mechanistic principle seems to have emerged already. Sulfur chemistry seems to be fundamentally involved in the basic energy-transfer receptor processes which must occur in olfaction and taste. This seems most reasonable in view of the developing extensive body of experimental evidence showing that sulfhydryls or disulfides are functionally essential in a wide variety of membrane-associated receptors (Rozental and Norris, 1975).

ACKNOWLEDGEMENTS

Our research on chemoreception is supported by the College of Agricultural and Life Sciences, University of Wisconsin. Madison; and in part by research grants No. GB-6580, GB-8756, GB-41868, and BNS 74-00953 from the National Science Foundation. I take this opportunity to thank my many coworkers who helped along the way.

REFERENCES

Amakawa, T., K. Kawabata, H. Kijima, and H. Morita. 1972. Isozymes of α-glucosidases in the proboscis and legs of flies. J. Insect Physiol. 18:541-553.

Amakawa, T., H. Kijima, and H. Morita. 1975. Insoluble α-glucosidase: possible pyranose site of the sugar receptor of the labella of the blowfly Phormia regina. J. Insect Physiol. 21:1419-1425.

Baker, J.E., and D.M. Norris. 1971. Neurophysiological and biochemical effects of naphthoquinones on the central nervous system of Periplaneta. J. Insect Physiol. 17:2383-2394.

Benz, G., Structure - Activity Relationships in Chemoreception. Information Retrieval Limited, London, 1976.

Borg, T.K., and D.M. Norris. 1971. Ultrastructure of sensory receptors on the antennae of Scolytus multistriatus. Z. Zellforsch. 113:13-28.

Brkička, R. 1933. Coll. Czech. Chem. Comm. 5:112.

Dethier, V.G. 1955. Mode of action of sugar-baited flytraps. J. Econ. Entomol. 48:235-239.

Gilbert, B.L., J.E. Baker, and D.M.Norris. 1967. Juglone (5-hydroxy-1,4-naphthoquinone) from Carya ovata, a deterrent to feeding by Scolytus multistriatus. J. Insect Physiol. 13:1453-1459.

Grabowski, C.T., and V.G. Dethier. 1954. The structure of the tarsal chemoreceptors of the blowfly, Phormia regina Meigen. J. Morphol. 94:1-17.

Hansen, K. The mechanism of insect sugar reception, a biochemical investigation, pp.382-391,in Olfaction and Taste, III. Rockefeller University Press, New York, 1969.

Kafka, W.A., and J. Neuwirth. 1975. A model of pheromone molecule-acceptor interaction. Z. Naturforsch. 30c:278-282.

Kijima, H., T. Amakawa, M. Nakashima, and H. Morita. 1977. Properties of membrane-bound α-glucosidases: possible sugar receptor protein of the blowfly, Phormia regina. J. Insect Physiol. 23:469-479.

Kijima, I., O. Koizumi, and H. Morita. 1973. α-Glucosidase at the tip of the contact chemosensory seta of the blowfly, Phormia regina. J. Insect Physiol. 19:1351-1362.

Koizumi, Q., J. Kijima, and H. Morita. 1974. Characterization of α-glucosidase at the tips of the chemosensory setae of the fly, Phormia regina. J. Insect Physiol. 20:925-934.

Larsen, J.R. 1962. The fine structure of the labellar chemosensory hairs of the blowfly Phormia regina Meig. J. Insect Physiol. 8:683-691.

Mansour, N.A.,M.E. Eldefrawi, and A.T. Eldefrawi. 1977. Isolation of putative acetylcholine receptor proteins from housefly brain. Biochemistry 16:4126-4132.

Norris, D.M. 1977. Role of repellents and deterrents in feeding by Scolytus multistriatus, in Host Plant Resistance to Pests, ACS Symposium Ser. No. 62:215-230.

Norris, D.M. 1976a. A molecular and submolecular mechanism of insect perception of certain chemical information in their environment. Colloq. Intern. Cent. Nat. Rech. Sci. 265:81-102.

Norris, D.M. 1976b. Physico-chemical aspects of the effects of certain phytochemicals on insect gustation. Symp. Biol. Hung. 16:197-201.

Norris, D.M. 1976c. How certain insects take the bitter with the sweet. Bull. Entomol. Soc. Amer. 22:27-30.

Norris, D.M. 1970. Quinol stimulation and quinone deterrency of gustation by Scolytus multistriatus. Ann. Entomol. Soc. Amer. 63:476-478.

Norris, D.M. 1969. Transduction mechanism in olfaction and gustation. Nature 222:1263-1264.

Norris, D.M., and H.M. Chu. 1974a. Morphology and ultrastructure of the antenna of male Periplaneta americana as related to chemoreception (Blatt., Blattidae). Cell Tissue Res. 150:1-9.

Norris, D.M., and H.M. Chu. 1974b. Chemosensory mechanism in Periplaneta americana: Electroantennogram comparisons of certain quinone feeding inhibitors. J. Insect Physiol. 20-1687-1696.

Norris, D.M., S.M. Ferkovich, J.E. Baker, J.M. Rozental, and T.K. Borg. 1971. Energy transduction in quinone inhibition of insect feeding. J. Insect Physiol. 17:85-97.

Norris, D.M., and S.M. Ferkovich, J.M. Rozental, J.E. Baker, and T.K. Borg. 1970. Energy transduction: Inhibition of cockroach feeding by naphthoquinones. Science 170: 754-755.

Norris, D.M., J.M. Rozental, G. Samberg, and G. Singer. 1977a. Protein-sulfur dependent differences in the nerve receptors for repellent 1,4-naphthoquinones in two strains of Periplaneta americana. Comp. Biochem. Physiol. 57C:55-59.

Norris, D.M., M.F. Ryan, and J.J. Piotrowski. 1977b. Purification of olfactory receptor from primary sensory neurons. Trans. Amer. Soc. Neurochem. 8:191.

Norris, D.M., M.F. Ryan, J.J. Piotrowski, and J. Hageman. 1977c. Interactions between repellent quinones and purified receptor lipoprotein from Periplaneta americana sensory neurons. Proc. Intern. Symp. Olfaction Taste 6: in press.

Rozental, J.M., and D.M. Norris. 1975. Genetically variable olfactory receptor sensitivity in Periplaneta americana. Life Sciences 17:105-110.

Rozental, J.M., G. Singer, and D.M. Norris. 1975. Affinities of certain quinone repellents for detergent-solubilized proteins from Periplaneta americana antennae. Biochem. Biophys. Res. Commun. 65:1040-1046.

Shimada, I., A. Shiraishi, H. Kijima, and H. Morita. 1974. Separation of two receptor sites in a single labellar sugar receptor of the flesh-fly by treatment with p-chloromercuricbenzoate. J. Insect Physiol. 20:605-621.

Singer, G., and D.M. Norris. 1973a. A disc-electrophoresis investigation of certain characteristics of Triton X-100 solubilized proteins from male Periplaneta americana. Experientia 29: 1483-1484.

Singer, G., and D.M. Norris. 1973b. Comparative disc-electrophoretic study of different proteinaceous extracts from Periplaneta americana males. Comp. Biochem. Physiol. 46B:43-56.

Singer, G., J.M. Rozental, and D.M. Norris. 1975. Sulphydryl groups and the quinone receptor in insect olfaction and gustation. Nature 256:222-223.

MODE OF ACTION OF INSECTICIDES: INSIGHTS GAINED FROM NEURO-PHYSIOLOGICAL PREPARATIONS OF INTACT OR DISSECTED INSECTS

T. A. Miller

Department of Entomology, University of California

Riverside, California 92521

ABSTRACT

The method used for bioassay of insecticidal activity is a vital cog in the wheel of structure-activity studies. An assay preparation which consists of recording muscle potentials from the dorsolongitudinal flight muscles of house fly gives sophisticated information on neural coordination in the thoracic ganglion with a relatively simple recording procedure. Quantitative responses are measured as the time from treatment until a disruption appears in neural coordination. The type of disruption in neural coordination gives a qualitative measure of poisoning. By applying compounds directly on the exposed thoracic ganglion, or the exposed and desheathed ganglion, some measure of intrinsic activity may be obtained. Using this approach, evidence for the mechanism of resistance to insecticides has been obtained for house fly strains resistant to organophosphates or pyrethroids, and intrinsic activity has been obtained for carbamates, organophosphates, pyrethroids and other chemicals with insecticidal activity.

INTRODUCTION

When studying the actions of neurotoxic insecticides on insects, it would be a great help to know the activity in various parts of the nervous system during poisoning to follow the impairment in function. In practice such measurements usually require considerable surgical manipulation, and for this reason most of our information comes from studies on isolated parts of the nervous system. Much of our information on toxicity, insecticide metabolism, synergism, and resistance comes from work on the house fly. From the standpoint of the basic sciences, probably the most compelling reasons to investigate the actions of insecticides on the house fly lies in the enormous advantage offered by a short generation time and the ability to develop various strains with differing resistance to insecticides.

There are, by now, a multitude of strains of *Musca domestica* which have been selected on the basis of resistance to a wide variety of not only neurotoxic insecticides, but also growth regulators and other potential chemical control agents. The house fly has become to insect toxicology what *Drosophila* is to genetics.

Despite the need to study mode of action of insecticides on the house fly, really very few references are available on such preparations. Yamasaki and Narahashi (1962) measured motor nerve discharges in susceptible and DDT-resistant strains of house flies. They applied DDT directly to the exposed thoracic ganglion and found that higher doses were required to produce motor discharges in the resistant strain when compared to the susceptible strain.

A 100-fold difference in ability of the house fly thoracic ganglion to respond to directly-applied DDT was found between DDT-resistant and susceptible strains. Although the concentrations of DDT used in these studies were rather high, the nervous system in the DDT-resistant strain did appear to be less susceptible to the insecticide. A recessive gene on the second chromosome was considered responsible for low nerve sensitivity to DDT in the resistant strain (Narahahsi, 1971; Tsukamoto *et al.*, 1965).

As luck would have it, the very insect chosen by toxicologists as a worldwide standard bioassay insect, the house fly, is also one of the better insects in which to examine physiological action. Advantages offered by the house fly, besides fast generation time, include a central nervous system which is concentrated into the cephalic and thoracic ganglion masses, and a considerable background literature.

The main drawback to the use of house fly, from the neurophysiologist's standpoint, must be size; however, this is more than offset by unique advantages offered by the flight motor system. A more recently described advantage includes a rather easily accessible neuromuscular system in the maggot of cyclorrhaphous Diptera (Hardie and Irving, 1977; Irving and Miller, 1978, unpublished work); however, the maggot will not be described here.

Because of attention paid to *Drosophila melanogaster* and other Diptera, such as *Calliphora vicina* and *Sarcophaga bullata*, a considerable amount of neurophysiological and behavioral research has been reported which is applicable to *Musca domestica*. These studies include a good deal of work on the neurophysiological basis for the generation of the flight pattern (Vater, 1961; Nachtigall and Wilson, 1967; Nachtigall, 1968; Tung and Lee, 1968; Mulloney, 1969, 1970a, 1970b, 1976; Heide, 1971; Ikeda and Kaplan, 1974; Ikeda *et al.*, 1975, 1976; King, 1977; Harcombe and Wyman, 1977; Hart *et al.*, 1977; Anderson, 1978; Coggshall, 1978). Using this knowledge, the flight motor system has yielded an enormous amount of information and insight into the mode of action of insecticides.

The most recent work on poisoning symptoms using house fly shows that each category of insecticide produces characteristic poisoning behavior (Ramade, 1968). When the sounds made by whole house flies were recorded with audioactograph techniques (Ramade, 1968; Miller, 1976), it was noted that each category of insecticide produced a response which was peculiar to that class of chemical poison; in effect, the chemical was "fingerprinted" by its actograph response.

By recording the flight motor system of house flies during insecticide poisoning, the actions of different insecticides may be further categorized in terms of effects on the behavior of motor neurons. This approach has been particularly useful in providing information on the mode of action of insecticides and the method and its usefulness as a bioassay tool will be reviewed below with special attention paid to what the insecticide chemist can learn from such information.

THE FLIGHT MOTOR SYSTEM OF DIPTERA

The peculiar neuromuscular properties of insect fibrillar muscle are well known to comparative physiologists by now, and documented in texts. Much of what follows concerns fibrillar flight muscles of Diptera, and in particular the dorsal longitudinal flight muscles of *Musca domestica*.

The thorax of the house fly has two main sets of muscles which are attached between the dorsal and ventral cuticle (called the dorsoventrals) and to the anterior and posterior cuticle of the thorax (the dorsolongitudinals) (Fig. 1). Since these main flight muscles make no connection to the wing base and, therefore, do not move the wings directly, they are termed indirect flight muscles.

The dorsolongitudinal muscles in *Musca domestica* are composed of 6 pairs of muscle fibers which occupy a median position in the thorax. If the house fly is split longitudinally along the midline and the left half of the thorax removed, the inner aspect of the flight thorax reveals the 6 right dorsolongitudinal muscle fibers (Fig. 1). These muscles were numbered 1-6 starting with the most ventral by Nachtigall and Wilson (1967) and by Ikeda (personal communication) and we numbered them this way as well.

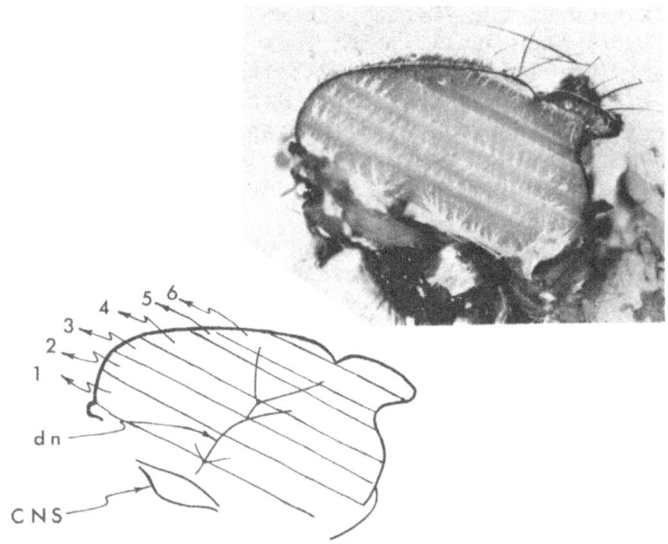

Figure 1. A photograph of the inner aspect of the right half of the thorax from *Musca domestica* split longitudinally. The right dorsolongitudinal muscle fibers are shown numbered 1-6 and the thoracic ganglion is shown in the ventrum of the thorax. Thoracic ganglion is shown in the ventrum of the thorax (CNS). Dorsal nerve (dn)

If one examines the dorsal aspect of the thorax of the intact house fly, several prominent rows of large setae are evident (Fig. 2). These bristles are described by Greenberg (1971). The bristles of the median rows are important landmarks in that they outline fairly accurately the position of the anterior cuticular attachments of the dorsal longitudinal flight muscles (Fig. 3). The position of the setae and the transverse median suture, also prominent on the dorsal thorax, serves to separate the positions of muscles 1 through 4 anterior to the suture and 5 and 6 posterior to the suture.

Muscle potentials may be recorded from the dorsolongitudinal flight muscles by puncturing the cuticle in the area of cuticular insertion of the fiber being examined. A small diameter wire of either tungsten, silver, gold, or copper (typically 25.4 μm) may be inserted just through the cuticle and waxed in place with Tacki-wax® (Central Scientific Co., CENCO) and a warm needle. Because of the large voltages recorded from the wire electrodes (up to 50 mv potentials), no preamplification is necessary, and the wires can be connected directly to the input of a pen recorder. We typically use Brush recorders (Gould, Inc., Cleveland, Ohio) because of their superior reliability and good frequency response. Alternatively, the muscle potentials may be tape recorded, and played back to a pen recorder later.

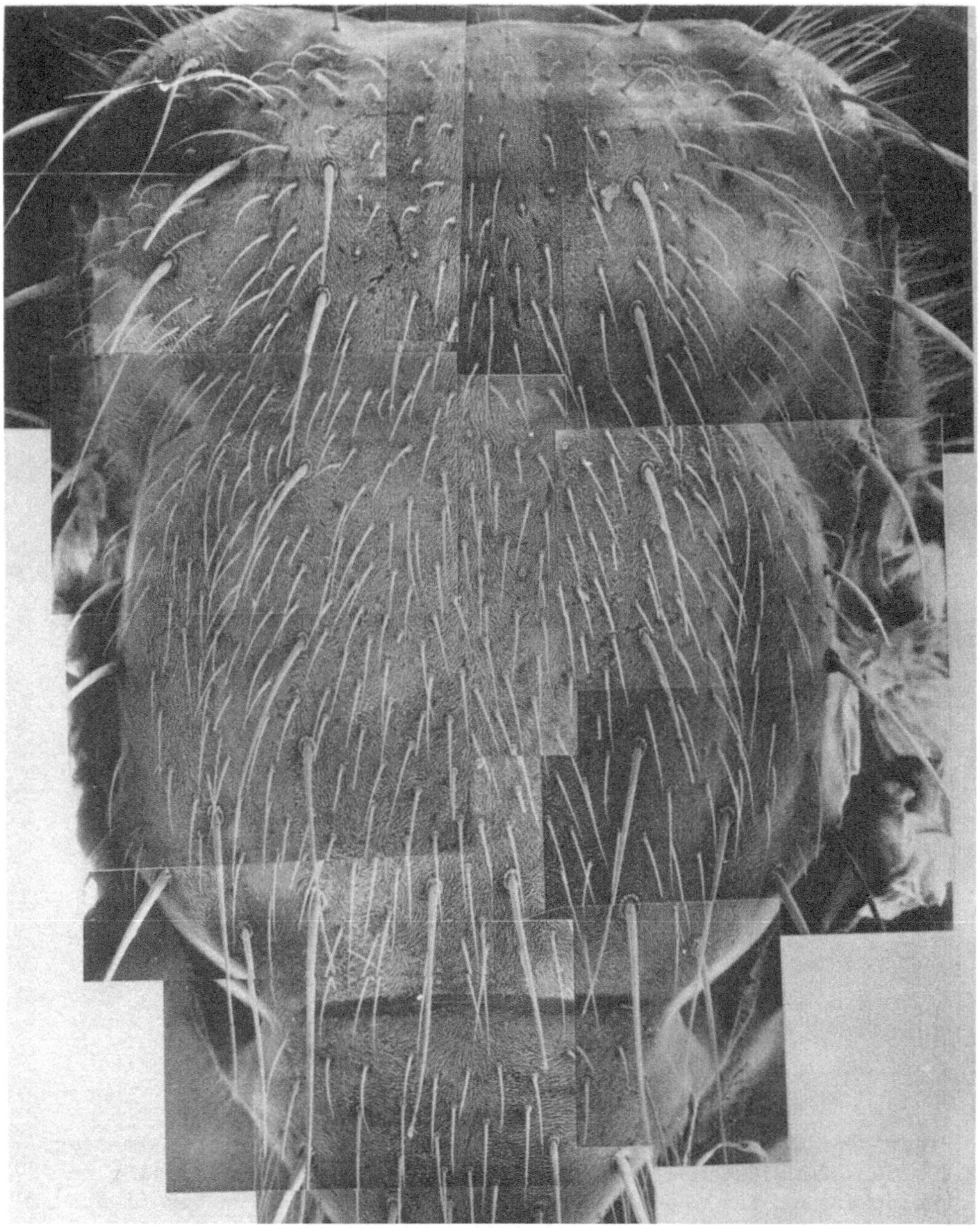

Figure 2. Scanning electron micrograph composite of the dorsal cuticle of the thorax of *Musca domestica*. The major bristles shown are used as landmarks to locate the areas of anterior insertion of the dorsolongitudinal flight muscles.

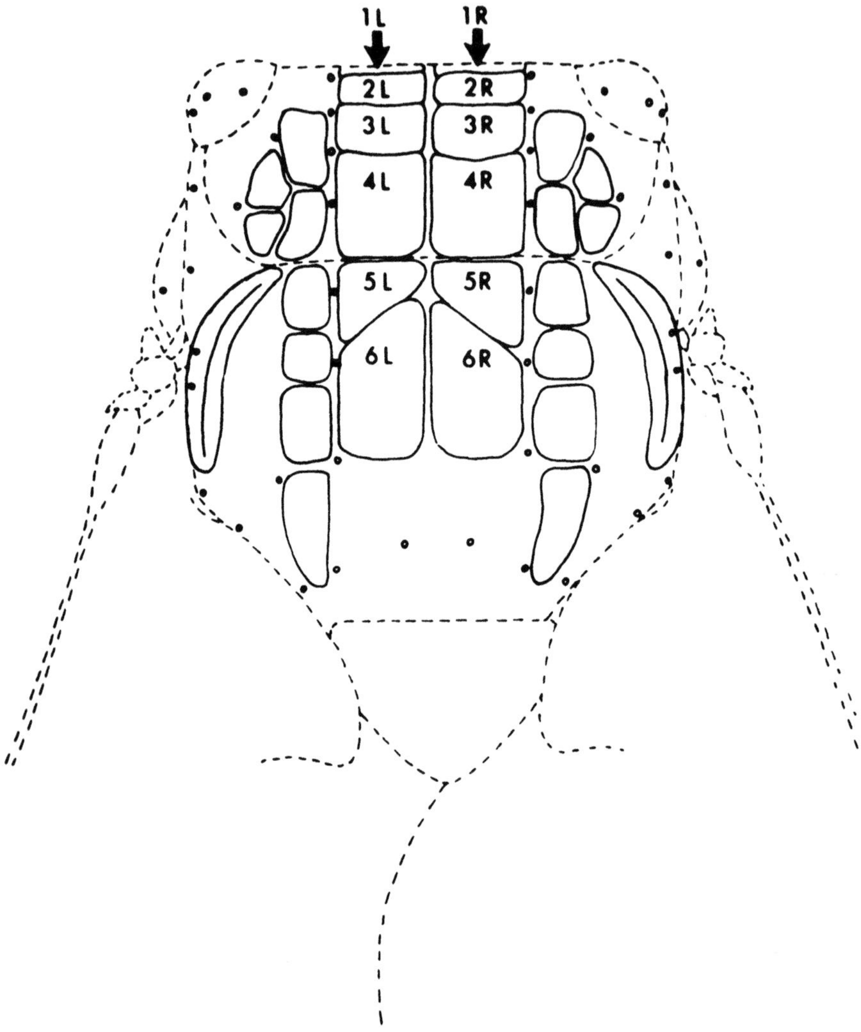

Figure 3. Dorsal view of thoracic musculature of *Musca domestica*. Two dorsolongitudinal flight muscles (DLM) run longitudinally in the dorsal thorax, each comprised of six fibers. The six fibers of each DLM are stacked unilaterally and are designated 1-6 from ventral to dorsal. Recordings were made by inserting fine wires just below the cuticle into the appropriate motor unit. The desired unit was located in relation to the major bristles depicted in the map with circles. (Redrawn from Miller and Adams, 1977.)

The wire recording method may be used with whole intact house flies by first lightly anesthetizing, then attaching a tether to the scutellum (Fig. 4). Once the tether is attached, the fly may be allowed to recover, while any number of electrodes are placed in appropriate muscle fibers.

From all evidence gathered to date, the dorsolongitudinal muscle fibers of house fly appear to be singly innervated; that is, one motor neuron appears to innervate each one of the muscle fibers. There is one major exception to this rule in that muscle fibers 5 and 6 appear to be innervated by the same neuron. Thus, for muscle fibers 1 to 4 there are 4 pairs of motor neurons in the thoracic ganglion and for muscle fibers right and left 5 and 6, there is a single pair of motor neurons in the thoracic ganglion.

It should be emphasized that the pattern of innervation described above is somewhat unique in violating the polyneuronal innervation pattern of most ordinary skeletal muscle fibers of insects.

Figure 4. Photograph of a house fly tethered at the scutellum and standing on a wheel. Four electrodes are mounted in 4 dorsolongitudinal flight muscles and a reference electrode is inserted in the abdomen (photography by Steve Manweiler and Dennis MacFarland).

However unique the flight muscle is, the muscle potential itself is not of primary concern here. The information of importance is the fact that one nervous impulse propagates from a burst initiation zone of its motor neuron and descends the axon to the muscle. Thus, the pattern of nervous impulses can be read directly from the pattern of muscle potentials recorded from the dorsal longitudinal muscles in house fly.

Flight Motor Uncoupling

Control recordings from intact house flies revealed that at 25°C or less, the flight motor of the indirect flight muscles started spontaneously without flight or movement (Fig. 5). Nachtigall and Wilson (1967) noticed this and briefly mentioned it from their work with *Calliphora*. Alternatively, the flight motor can be started by quickly removing the substrate from a tethered fly (Fig. 5, B). Reflex flight may last for extended periods; indeed, scientists at the Wellcome Research Laboratories, Berkhamsted, England, mount the greenbottle fly, *Lucilia sericata*, by using stiff tungsten needles as both electrodes and tether. The *Lucilia* then fly throughout the course of experiments after recovery from anesthesia (R. J. Hart, 1974, personal communication).

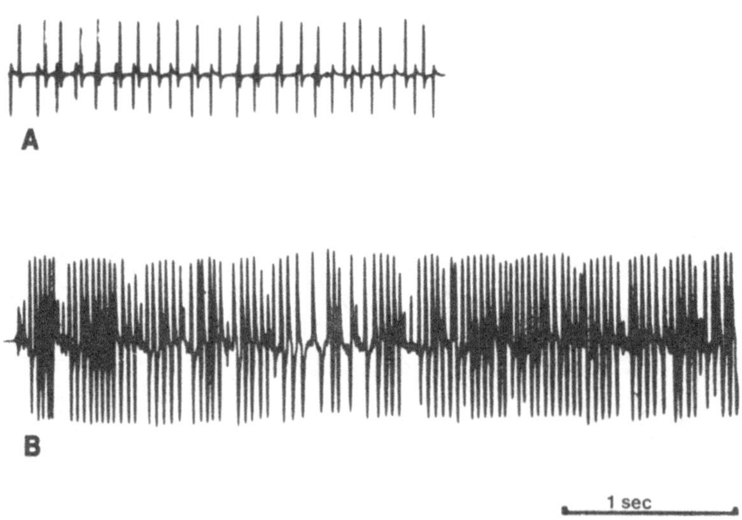

Figure 5. Flight muscle potentials recorded from house fly during no movement (A) or during flight (B). When the flight motor initiates spontaneously, there is no movement unless the thorax is held under tension by the fly.

The most noticeable feature about control recordings of Dipteran flight motor activity in the dorsolongitudinals is that all of the 5 pairs of motor units are activated at about the same rate and that the potentials rarely coincide in time. This remarkable consistency in the pattern of activation of the flight motor has lead to the hypothesis that some form of interaction exists between the motor neurons in the thoracic ganglion. Wyman has termed this interaction "lateral inhibition" (cf. Harcombe and Wyman, 1977).

Mulloney (1970a) shocked single dorsolongitudinal muscle fibers of *Calliphora* during flight to initiate antidromic nerve impulses which propagated from the muscle back into the thoracic ganglion. Under certain circumstances, these antidromic impulses caused a resetting or delay in the appearance of motor nerve impulses in other fibers of the dorsolongitudinal muscle. And this provided some evidence that some form of interconnection between the motor neurons does exist. Harcombe and Wyman (1977) found antidromic potentials in *Drosophila* also caused resetting of firing order in the same unit shocked.

The term "coupled" was used to describe the normal condition in which all ten motor units of the dorsolongitudinal muscle of *Musca domestica* fired at about the same rate, usually avoiding simultaneous potentials in any 2 units. Records of the flight motor during poisoning showed a disruption in coupling between the flight motor units for carbamate, organophosphorus, cyclodiene, and Lindane insecticides examined (Fig. 6). The disruption was termed "uncoupling" of the flight motor (Miller and Kennedy, 1972).

Some early studies with poisoning by DDT and some of its analogs and of the pyrethroid, Barthrin, showed that despite greatly increased nervous activity in convulsions, the flight motor units examined were still coupled, even several hours into poisoning (Fig. 7) (Miller *et al.*, 1971; Miller and Kennedy, 1972, 1973).

It was reasoned that if the flight motor could remain coupled through hyperactivity and convulsions during DDT poisoning, that merely drastically increasing the overall nervous activity did not insure uncoupling. It was further assumed that uncoupling had to be produced by direct action of a convulsant or insecticide on the thoracic ganglion.

Evidence for this assumption includes the following: 1) slight physical pressure on the thoracic ganglion caused reversible uncoupling; 2) excessive overdoses of DDT applied directly to the thoracic ganglion caused uncoupling of the flight motor units; 3) treatment of the thoracic ganglion by any convulsant (e.g. picrotoxin) (Miller *et al.*, 1978), if in high enough dose, will cause uncoupling; 4) no amount of damage to the peripheral nervous system

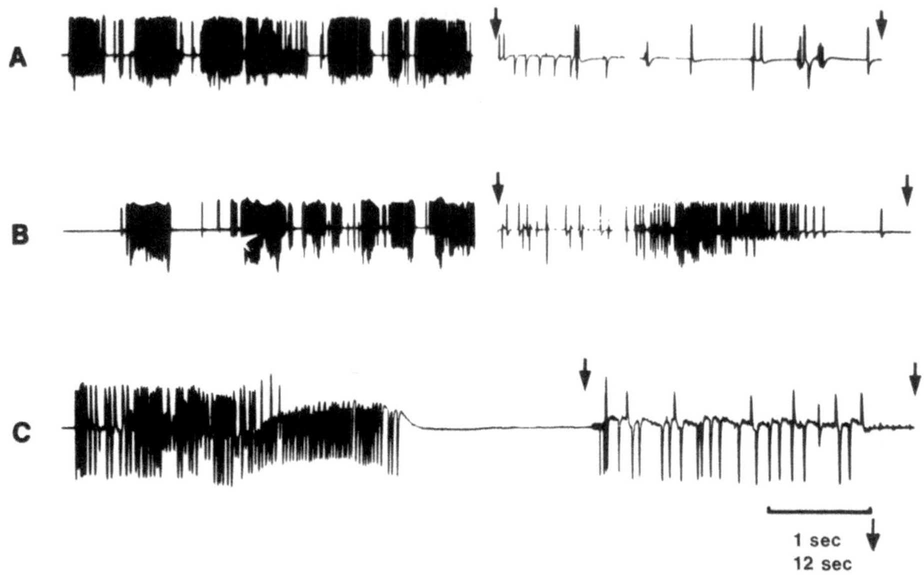

Figure 6. Flight motor uncoupling following treatment by insecticides which are thought to act on the central nervous system. A. 3.25 hours following treatment by 1 γ of Dieldrin. B. 6 hours after application of 0.2 γ ronnel. C. 10-12 minutes following treatment by 0.5 γ of *N*-thio-*p*-*t*-butylphenyl carbofuran.

will cause uncoupling. Thus, the coupling of the flight motor was considered diagnostic of peripheral or central poisoning.

The patterns of flight motor potentials obtained during pyrethroid poisoning were not consistent with the patterns recorded from either peripheral or central nervous poisons (Miller and Adams, 1977). Figure 8 shows flight muscle potentials recorded from muscle fibers 4L, 4R, 6R, 6L 15 min after an LD_{50} of bioneopynamin.

Comparison between Figs 6, 7 and 8 emphasizes that the potentials recorded in Fig. 8 are rather unique. The appearance of bursts of high frequency muscle potentials in 4L, 6R and 6L followed by a delay, then the onset of single potentials, was not recorded during poisoning by other insecticides. It should be especially

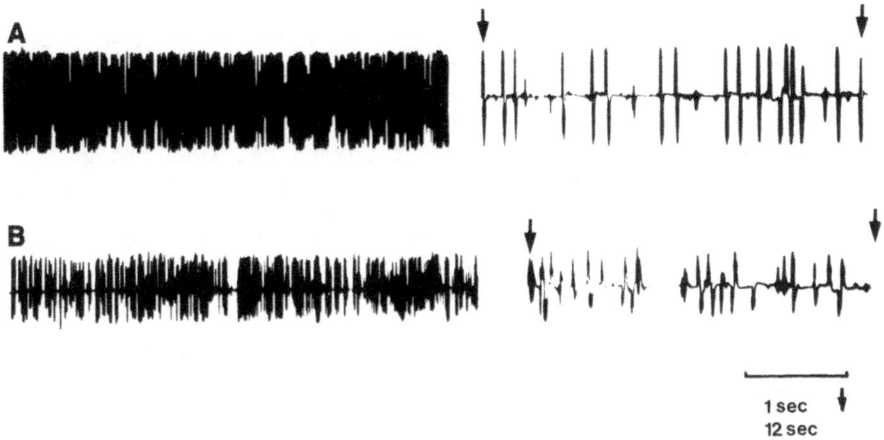

Figure 7. Flight muscle potentials recording during poisoning by the "peripherally-acting" insecticides: DDT (1 μg, A) and Barthrin (5 μg, B) (half of B is from Miller and Kennedy, 1972).

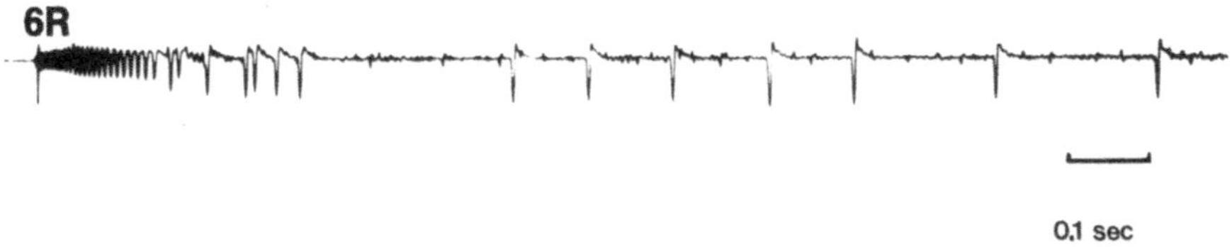

Figure 8. Flight motor recorded from house fly after poisoning by Bioneopynamin (*d-trans* isomer of tetramethrin). Amidst single muscle potentials are burst discharges in single units which occur without obvious relationship to adjacent motor units.

noted that a high frequency discharge in one unit might be ignored by adjacent units as in the second discharge in unit 6R. It should also be noted that the burst discharge itself is highly irregular and unpredictable in its appearance in recordings.

Studies in progress (Adams and Miller, 1978; M. E. Adams, personal communication) suggest that the burst discharges during bioneopynamin poisoning originate along the motor axon and propagate both ortho and antidromically. Antidromic nervous impulses may not occur in the action of low doses of DDT poisoning, even though similar multiple potential discharges are recorded. Thus, the type of discharge in DDT poisoning and the type of discharge in bioneopynamin point to fundamentally different modes of action of these 2 categories of insecticide even though the axonal site of action is similar.

With this background, it is now possible to go further and ask more specific questions about the mode of action of pyrethroids at this site of action. For example, the action of bioneopynamin may be duplicated by perfusing house fly nerve-muscle preparations with low calcium concentrations.

The first insight obtained from recordings of whole house flies had to do with the total amount and type of nervous activity present after poisoning with lethal doses. After hyperactivity and convulsions, the amount of movement gradually declined over a period of hours until overt behavior told us that all obvious movement had stopped. This *implied* that nervous activity had greatly declined.

A search of the literature was made to see if an actual decline in overall nervous activity had been reported during the terminal stages of insecticide poisoning. Brown (1951, 1963) described flaccid paralysis as distinguished from tetanic paralysis in describing the terminal stages of poisoning, indicating 2 levels of nervous activity. Camougis (1973) wrote "It is this nerve block that ultimately leads to the symptoms of paralysis observed in intact insects." Brown was referring to insecticide poisoning in general, and Camougis was referring to pyrethrum actions on arthropod nerves.

Narahashi (1971) referred to terminal stages of poisoning being dominated by an "autotoxin" in DDT poisoning: "...DDT stimulates the nerve to cause hyperactivity, and the resultant toxic substance eventually paralyzes the nerve." On pyrethrin action a similar conclusion was reached: "The toxin is presumably responsible for the paralysis of the poisoned insect together with pyrethrins themselves."

Our early conclusions from studying the late stages of poisoning included a doubt cast upon the possibility that an autotoxin was being produced coinciding with the prostrate stage of poisoning

(Miller *et al.*, 1971). The reasoning behind this conclusion was that 1) the terminal stages of poisoning included exaggerated levels on nervous activity continuing long after overt movement had declined, and 2) each insecticide suspected of producing an autotoxin in the prostrate stage of poisoning was actually producing an overtly similar condition, while the pattern of nervous activity of the flight motor system was characteristic of each compound. If an autotoxin is actually released in insecticide poisoning, then it does not have a uniform effect upon the central nervous system, but may cause subtle changes. It certainly does not cause a decline or block of nervous activity in many of the motor units examined so far. A similar conclusion was reached by Flattum (personal communication) and Gammon (this volume) who found unimpaired function of much of the ventral nerve cord of cockroaches prostrate from either organophosphorus or allethrin poisoning.

Another insight gained from studying whole insects includes the discovery of a "still period," noted earlier by Ramade (1968). Before the onset of hyperactivity, house flies entered a state characterized by a subtle decrease in overall locomotory activity (Miller *et al.*, 1971). This "still period" was common to all insecticides tested.

The "still period" for organophosphorus compounds lasted far longer on the average than for carbamates, leading to delays of up to 6 hours before the onset of hyperactivity poisoning symptoms for Phosvel®. To determine if penetration of compounds from topical treatment to appearance in the thoracic ganglion was greater for organophosphorus compounds compared to carbamates, the penetration of Monitor® and Phosvel® were compared to that of Furadan® (Miller *et al.*, 1973).

Penetration curves were similar enough, for Furadan and Phosvel, but the rather surprising fact was that Phosvel or Monitor accumulated considerably higher doses than Furadan before hyperactivity was recorded in the flight motor units.

Bioassays in which carbamate and organophosphorus compounds were perfused directly onto the thoracic ganglion also reflected a difference. Organophosphorus compounds required longer times and higher concentration to disrupt the coordination of the flight motor neurons compared to carbamates (Miller, 1976).

The lack of significant difference in penetration and the consistent differences in temporal development of symptoms between carbamates and organophosphorus compounds led to a hypothesis for mode of action. Carbamates were hypothesized to produce excitation and organophosphorus compounds were considered to produce inhibition as their initial action. The action of organophosphorus compounds included an extensive still period which was characterized by Janice

Kennedy as representing a period of reduced responsiveness to external stimuli (cf. Miller, 1976).

The condition of still period or inhibition caused by initial organophosphorus action is difficult to measure quantitatively. Possibly the best measure of such a condition involves monitoring a reliable stimulus-response behavior. We used the loss of tarsal contact stimulus to fly as a bioassay to characterize organophosphorus action (Miller, 1976); however, any stimulus-response behavior would presumably be suitable for measuring such an action.

The heartbeat of the house fly was recorded during poisoning to gain insight into the response of the circulatory system (Miller and Kennedy, 1973). Normal heartbeat rates were found to vary from 0 to over 300 beats per min before and during poisoning. However, if the heartbeat became extremely regular late in poisoning, this was a fairly reliable diagnostic indication that no recovery would occur. In effect, a steady heartbeat represented an abnormal condition which was of predictive value.

Since the interactions of insecticides on the thoracic ganglion were of primary importance as a possible site of action, it was important to establish whether treatment of compounds topically on whole house flies bore any resemblance to actions of insecticides when perfused onto the exposed and *in situ* thoracic ganglion. As a preliminary observation, it was first noted, then documented (Miller *et al.*, 1973), that tethered house flies were less susceptible to topically-applied poisons than untethered house flies. Thus, restriction of mobility in some way conveyed a protection of the house fly against poisoning. Although documented, the reasons for this are not well understood. In a similar way, it was assumed the restrictions necessary in dissecting insects for *in vitro* or *in situ* nerve preparation would give erroneous information.

Burt and Goodchild (1971) found that constantly perfused nervous systems were more susceptible to insecticide action compared to compounds applied in a static pool or bath over the cockroach ventral nerve cord. For the most part, dissected preparations can only approximate conditions found *in vivo*.

Another problem hardly ever mentioned is the saline solutions used to maintain perfused nerve tissues. Despite considerable work, there is still no consensus on a common saline for the American cockroach. The same is true of house fly. One wonders how valid data are which apply bioassay compounds made up in saline solutions which omit the huge amounts of proteins and amino acids in the blood of insects. Furthermore, the circulation of hemolymph in intact insects is never duplicated in perfused preparations, and compounds are rarely applied in a manner expected of the arrival of compound at the CNS from topical treatment.

Despite these reservations, one may obtain useful information from applications of compounds directly onto the thoracic ganglion of house flies. The flight motor system is extremely sensitive to convulsive poisons. If a certain concentration does not produce convulsions, one may be fairly confident that it represents a sublethal dose.

While the flight motor system of Diptera is probably not a primary site of action, it does give a clue as to the possible multitude of central sites of action of insecticides. By observing the activity of the flight motor neurons during poisoning, one may obtain great insight into qualitative and quantitative aspects of insecticide poisoning. Hopefully, the flight motor system will lead to central nervous studies on a level of even greater detail which will come ever closer to the primary sites and modes of action of insecticides and ultimately provide an improved bioassay system for analysis of structure-activity at the membrane or receptor levels.

REFERENCES

Adams, M. E., and T. A. Miller. 1978. Site of action of pyrethroids: Burst discharge in flight motor axons of house fly. *Pestic. Biochem. Physiol.* (in press).

Anderson, M. 1978. A microscopical study of the innervation of flight muscles in the tsetse fly. *J. Morphol.* 155: 19-34.

Brown, A. W. A. 1951. "Insect Control by Chemicals," New York: Wiley.

Brown, A. W. A. 1963. Chemical Inquiries. *In:* "Insect Pathology" Vol. I, ed. by E. A. Steinhaus, New York: Academic Press.

Burt, P. E., and R. E. Goodchild. 1971. The site of action of pyrethrin I in the nervous system of the cockroach *Periplaneta americana*. *Entomol. Exp. Appl.* 14: 179-189.

Camougis, G. 1973. Mode of action of pyrethrum on arthropod nerves. *In:* "Pyrethrum: The Natural Insecticide," ed. by J. E. Treherne, New York: Academic Press.

Coggshall, J. C. 1978. Neurons associated with the dorsal longitudinal flight muscles of *Drosophila melanogaster*. *J. Comp. Neurol.* 177: 707-720.

Greenberg, B. 1971. "Flies and Disease" Vol. I, Princeton: Princeton University Press.

Harcombe, E. S., and R. J. Wyman. 1977. Output pattern generation by *Drosophila* flight motoneurons. *J. Neurophysiol.* 40: 1066-1077.

Hardie, J., and S. N. Irving. 1977. A preparation enriched with insect somatic, excitatory, neuromuscular terminals. *Brain Res.* 120: 138-140.

Hart, R. J., C. Potter, and R. G. Wilson. 1977. Factors governing the toxicity of putative synaptic transmitters and their analogues when injected into the haemocoel of adult male *Lucilia sericata.*

Heide, G. 1971. Die Funktion der nicht-fibrillären Flugmuskel von *Calliphora.* Teil I: Muskuläre Mechanismen der Flugsteuering und ihre nervöse Kontrolle. *Zool. Jb. Abt. Allg. Zool. Physiol.* 76: 9 ⊦137.

Ikeda, K., and W. D. Kaplan. 1974. Neurophysiological genetics in *Drosophila melanogaster.* *Am. Zool.* 14: 1055-1066.

Ikeda, K., N. Hori, and T. Tsuruhara. 1975. Inhibitory command neuron releasing patterned motor activity in *Drosophila melanogaster.* *Fed. Proc.* 34: 359.

Ikeda, K., N. Hori, and T. Tsuruhara. 1976. Motor innervation of the dorsal longitudinal flight muscle of *Sarcophaga bullata.* *Am. Zool.* 16: 179.

King, D. 1977. An interneuron in *Drosophila* synapses with a peripheral nerve onto the dorsal longitudinal muscle motoneurons. *Neurosci. Abstr.* 3: 180.

Miller, T. A. 1976. Distinguishing between carbamate and organophosphate insecticide poisoning in house flies by symtomology. *Pestic. Biochem. Physiol.* 6: 307-319.

Miller, T., and J. M. Kennedy. 1972. Flight motor activity of house flies as affected by temperature and insecticides. *Pestic. Biochem. Physiol.* 2: 206-222.

Miller, T., and J. M. Kennedy. 1973. *In vivo* measurement of house fly temperature, flight muscle potentials, heartbeat and locomotion during insecticide poisoning. *Pestic. Biochem. Physiol.* 3: 370-383.

Miller, T., and M. E. Adams. 1977. Central vs. peripheral action of pyrethroids on the housefly nervous system. *In:* "Synthetic Pyrethroids," ed. by M. Elliott, ACS Symposium Series, No. 42, pp. 98-115.

Miller, T., L. J. Bruner, and T. R. Fukuto. 1971. The effect of light, temperature, and DDT poisoning on housefly locomotion and flight muscle activity. *Pestic. Biochem. Physiol.* 1: 483-491.

Miller, T., J. M. Kennedy, C. Collins, and T. R. Fukuto. 1973. An examination of temporal differences in the action of carbamate and organophosphorus insecticides on house flies. *Pestic. Biochem. Physiol.* 3: 447-455.

Miller, T., M. Maynard, and J. M. Kennedy. 1978. Structure-activity studies of picrotoxinin. *Pestic. Biochem. Physiol.* (submitted).

Mulloney, B. 1969. Interneurons in the central nervous system of flies and the start of flight. *Z. Vergl. Physiologie* 64: 243-253.

Mulloney, B. 1970a. Organization of flight motoneurons of Diptera. *J. Neurophysiol.* 33: 86-95.

Mulloney, B. 1970b. Impulse patterns in the flight motor neurons of *Bombus californicus* and *Oncopeltus fasciatus*. *J. Exp. Biol.* 52: 59-77.

Mulloney, B. 1976. Control of flight and related behavior by the central nervous systems of Insects. *In:* "Insect Flight," ed. by R. C. Rainey, New York: Wiley, pp. 16-30.

Nachtigall, W. 1968. Elektrophysiologische und kinematische Untersuchungen über Start und Stop des Flugmotors von Fliegen. *Z. Vergl. Physiol.* 61: 1-20.

Nachtigall, W., and D. M. Wilson. 1967. Neuromuscular control of dipteran flight. *J. Exp. Biol.* 47: 77-97.

Narahashi, T. 1971. Effects of insecticides on excitable tissues. *Adv. Insect Physiol.* 8: 1-93.

Ramade, F. 1968. Contribution à l'étude du mode d'action de certains insecticides de synthèse, plus particulièrement du lindane, et des phenomènes de resistance à ces composés chez *Musca domestica* L. *Ann. Inst. Nat. Agr. Paris, (N.S.)* 5: 1-268.

Tsukamoto, M., T. Narahashi, and I. Yamasaki. 1965. Genetic control of low nerve sensitivity to DDT in insecticide-resistant houseflies. *Botyu-Kagaku* 30: 128-132.

Tung, A. S.-C., and W.-Y. Lee. 1968. The comparative study of the central nervous system between larvae and adults of house flies, *Musca domestica* L. (Muscidae, Diptera, Hexapoda). *Bull. Inst. Zool. Academia Sinica* 7: 7-25.

Vater, G. 1961. Vergleichende Untersuchungen über die Morphologie des Nervensystems der Dipteran. *Z. Wiss. Zool.* 167: 137-196.

Yamasaki, T., and T. Narahashi. 1962. Nerve sensitivity and resistance to DDT in houseflies. *Jap. J. Appl. Entomol. Zool.* 6: 293-297.

AN ANALYSIS OF THE TEMPERATURE-DEPENDENCE OF THE TOXICITY OF ALLETHRIN TO THE COCKROACH

D. W. Gammon

University of Cambridge,
Dept. of Zoology
Cambridge CB2 3EJ, England

Pyrethroid insecticides have long been known to act on insect nerve (Wigglesworth, 1941; Lowenstein, 1942) and the mode of action of allethrin, the first synthetic pyrethroid, has been studied extensively using a variety of invertebrate nerve preparations (Narahashi, 1971a). One of the distinctive features of pyrethroids is their greater toxicity to insects as temperature is lowered but, since poisoning symptoms can be reversed by changing temperature, this cannot be explained solely by differences in metabolism or cuticular penetration (see Blum and Kearns, 1956). Therefore neurophysiological techniques have been employed over a range of temperatures to gain an understanding of this phenomenon since clearly, any effect of allethrin which becomes enhanced as temperature decreases, could explain the greater toxicity of pyrethroids at low temperatures.

Allethrin acts on isolated nerve in two discrete ways, depending not so much on dose, as on temperature: at high temperatures (above about 26.5°C), allethrin causes giant axons in the cockroach abdominal nerve cord to fire repetitively following single electrical stimulation (Narahashi, 1962a,b) but as temperature is reduced, these fibers become blocked. The precise mechanism by which allethrin causes this nerve blockage, with only a small fall in nerve membrane potential (Narahashi, 1962a), has been elucidated by voltage-clamp analysis (Narahashi and Anderson, 1967; Narahashi, 1971a,b; Wang _et al._, 1972; Murayama _et al._, 1972). However, the suggestion that nerve blockage is a crucial factor responsible for the negative temperature coefficient of toxicity of allethrin (Narahashi and Anderson, 1967; Narahashi, 1971b; Wang _et al._, 1972; Narahashi, 1976) fails to explain the symptomology, for insects poisoned at low temperature become hyperactive and there is no evidence that allethrin selectively blocks inhibitory neurones, to account for this anomaly.

The claim that nervous elements within ganglia are more sensitive to excitation by pyrethrin I than giant axons in the cockroach abdominal nerve cord (Burt et al., 1971; Burt and Goodchild, 1971) is difficult to incorporate within this framework of knowledge of pyrethroid action, since Burt and co-workers failed to consider, for example, the possibility of secondary effects of poisoning (see Blum and Kearns, 1956) and furthermore, conducted their experiments at only one temperature (20°C).

I have therefore developed a technique which enables nervous responses to be recorded from a free-walking cockroach. It was felt that such a preparation would remove some of the pitfalls associated with work using dissected or restrained insects and allow correlations to be made between neural abnormalities and symptoms of poisoning. Whilst studying the reasons for the negative temperature coefficient of toxicity, it was also hoped to establish some of the target sites in a whole insect.

Electrode Implantation Technique

The adult, male cockroach, *Periplaneta americana* L., is a useful insect for studies of drug and insecticide action (e.g. Twarog and Roeder, 1957). It possesses a pair of anal sense organs, the cerci, which are receptive to weak air movements (Pumphrey and Rawdon-Smith, 1937). Afferent fibres from each cercus synapse in the sixth abdominal ganglion with interneurones which ascend the abdominal nerve cord without making further synaptic contacts.

The preparation I have developed (Gammon, 1977a) involves implantation of wire electrodes into a cockroach, in the positions indicated in Fig. 1. The electrode recording from a cercus consisted of a tungsten needle, soldered onto a length of insulated 75μ copper wire. The remaining electrodes were insulated 50μ silver wires, the tips being coated with AgCl to reduce polarisation. Recordings were made from a cercal nerve (R_C), and two positions on the abdominal nerve cord (R1 & R2). The electrode in the cercus was also used for electrically stimulating a cercal nerve, a response being recorded at R1 and R2. Likewise, electrode R1, positioned between the fifth and sixth abdominal ganglia, could be switched into a circuit for electrical stimulation of the nerve cord, the (non-synaptic) response being recorded by R2. Responses obtained from the nervous system of an unpoisoned cockroach were stable for a week or more (Fig. 2) and as individuals survived for several months with electrodes implanted, it was safe to ascribe neural abnormalities principally to the effects of an applied insecticide, rather than to stressful effects due to the recording apparatus.

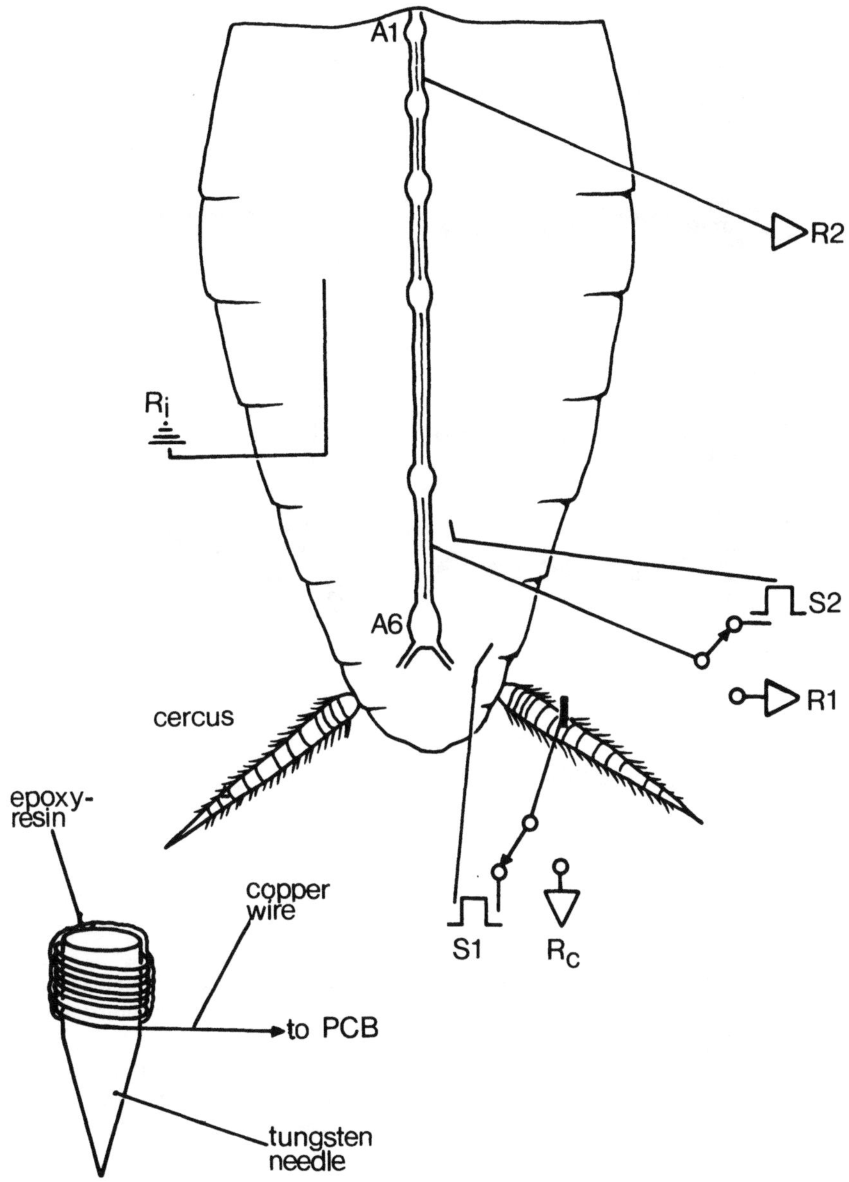

FIGURE 1. The ventral surface of a cockroach abdomen. The first (A1) and sixth (A6) abdominal ganglia are marked and the cercal nerves are shown leaving the posterior of A6. Recording electrodes are: Rc (cercus), R1 and R2 (abdominal nerve cord) and Ri (common indifferent). Nerves were stimulated electrically by applying rectangular pulses at S1, to stimulate a cercal nerve and S2, the nerve cord. Responses to stimulation at S1 were recorded by R1 and R2, and stimulation at S2 by R2 (see Fig. 2). Inset: the electrode used for stimulating and recording from a cercus. P.C.B. indicates printed-circuit-board used for electrode attachments.

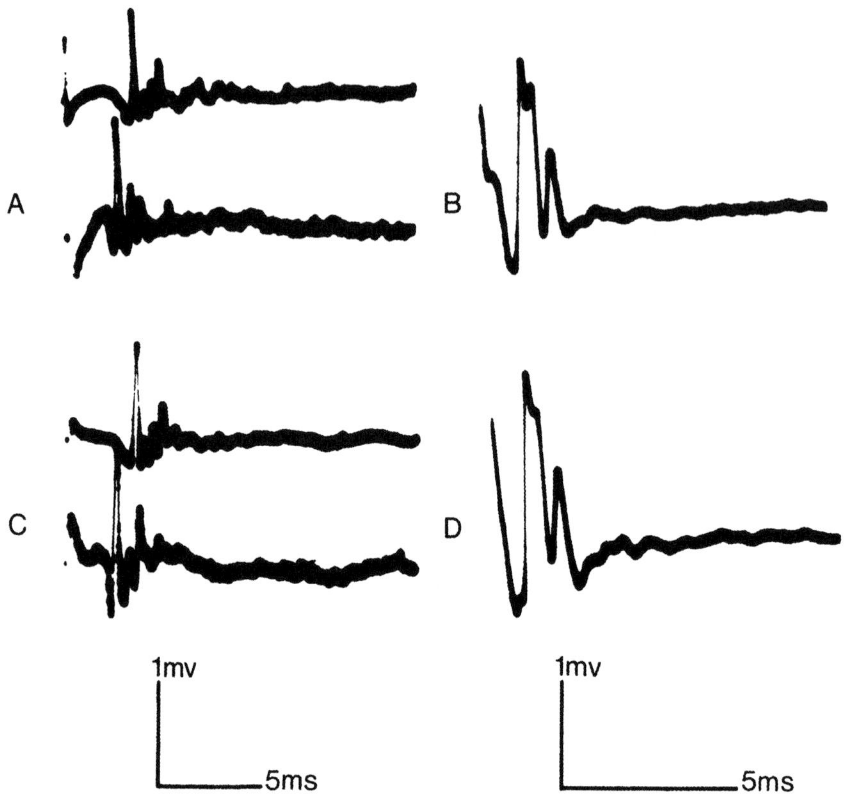

FIGURE 2.
A. Response in the nerve cord to electrical stimulation of a cercal nerve, at S1; upper trace, R2, lower trace, R1. Recorded from a free-walking cockroach on the day following electrode implantation.
B. Response in the nerve cord, at R2, to electrical stimulation of the nerve cord, at S2. Recorded from a free-walking cockroach on the day following electrode implantation.
C. as A, but recorded 5 days after electrode implantation.
D. as B, but recorded 5 days after electrode implantation.
All the recordings were taken from the same unpoisoned individual.

A more natural method of activating abdominal interneurones consists of a weak air movement near the cerci. Two types of such mechanical stimulation were used: a sound pulse, of constant intensity (200 Hz, 102dB) or the slightly stronger stimulus of an air-puff from an empty, plastic wash-bottle. Neural responses to sound pulses of different durations, applied to an unpoisoned cockroach, are given in Fig. 3. It is clear that, although there

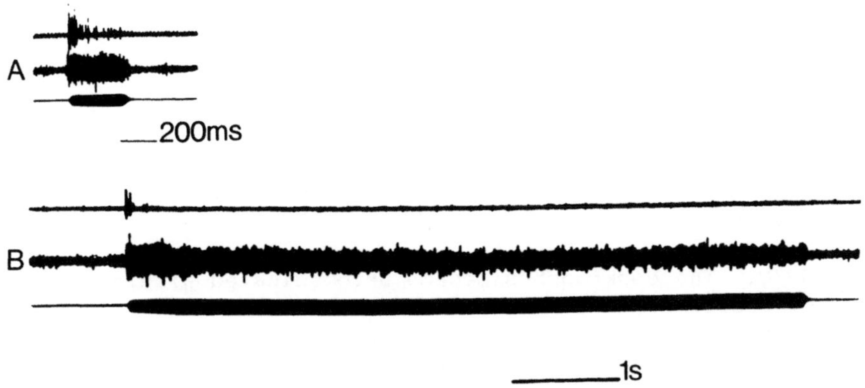

FIGURE 3. Nervous responses of an unpoisoned free-walking cockroach to sound pulses of different durations: A, a 300ms pulse and B, a 6.5s sound pulse. The top trace was recorded by R2 (nerve cord); middle trace by Rc (cercus) and the bottom trace by a sound pulse (200 Hz, 102dB) marker.

is continuous firing of tonic cercal sensory axons during a 6.5 second pulse (Fig. 3B), the discharge of interneurones within the abdominal nerve cord is not appreciably different from that evoked by a 300ms sound pulse (Fig. 3A). This is analogous to the habituation (i.e. waning of the behavioral response) which is observed if repeated mechanical stimuli are applied to a cockroach over a short time period. Neither form of mechanical stimulus (sound or air-puff), applied to an untreated cockroach, caused a nervous discharge beyond the duration of the stimulus.

Insecticide Treatment of Electrode-Implanted Cockroaches

Before treatment with an insecticide, individuals were generally left overnight to allow recovery from the effects of implantation. To ensure that the technique was viable and that the results obtained were interpretable, three insecticides which have well understood modes of action were studied first, i.e., parathion, nicotine and dinitro-o-cresol, and the results were consistent with previous work (see Gammon, 1977b).

The analysis of allethrin action using this preparation first required two experimental temperatures to be selected, 15°C and 32°C, which were below and above the critical temperature (26.5°C) at which giant axons in the isolated cockroach nerve cord fire repetitively following single electrical stimulation (Narahashi, 1962a,b). Empirical probit plots (Finney, 1971) at these two temperatures gave an approximately tenfold difference in the dose required to cause 95% mortality (LD_{95}).

Effects of Allethrin on the Free-Walking, Electrode-Implanted Cockroach

At both 15°C and 32°C, good correlations existed between neural abnormalities resulting from allethrin treatment and the observed poisoning symptoms. However, there were marked differences in both the mechanisms of action and the symptomology, between the two temperatures.

AT 32°C. The cockroaches became uncoordinated within a few minutes of receiving an LD_{95} dose of allethrin (Gammon, 1978a). This was correlated with dramatic excitant actions on peripheral and central neurones (Fig. 4). Bursts of cercal afferent spikes were evident and single electrical stimulation of a cercal nerve resulted in a prolonged discharge of abdominal interneurones (Fig. 5). However, electrical stimulation of the nerve cord above the sixth ganglion at S2 (see Fig. 1), did not result in a (similar) discharge of abdominal interneurones until later in the poisoning sequence (Fig. 5). It is therefore possible that synapses between cercal afferents and abdominal interneurones are more sensitive to excitation by allethrin than interneurones within the nerve cord. However, it is also possible that postsynaptic nerve, within the sixth ganglion but away from the synaptic zones, was being affected, and further clarification of this problem would require intracellular microelectrode investigation.

A cockroach became so uncoordinated that it was unable to right itself when it fell over, this prostrate stage being reached 1.5 to 3 hours after treatment. Over the subsequent 3 to 4 hours the insect's movements became progressively weaker (paralysis) and the abdominal response to cercal nerve stimulation often became blocked. However, electrical stimulation of the nerve cord above the ganglion and mechanical stimulation of the cerci continued to yield responses for many hours (Fig. 5 and see Gammon, 1978a). Therefore it is possible, once again, that synapses in the sixth abdominal ganglion are more sensitive to blockage following allethrin poisoning than non-synaptic neurones in the cockroach CNS.

Despite the dramatic excitant effects on the CNS (Fig. 5), it is perhaps the hyperexcitation of the peripheral nervous system by allethrin at 32°C which was even more remarkable. Fig. 6 shows responses to sound and air-puff stimuli recorded from a paralysed cockroach, 17 hours after prostration. Although the responses in the nerve cord had returned to normal, an air-puff induced a burst of cercal afferent spikes lasting for several seconds. There were also bursts of large amplitude potentials recorded from a cercus, superimposed on this afferent activity, and these have been shown to be efferent in origin (Gammon, 1978a).

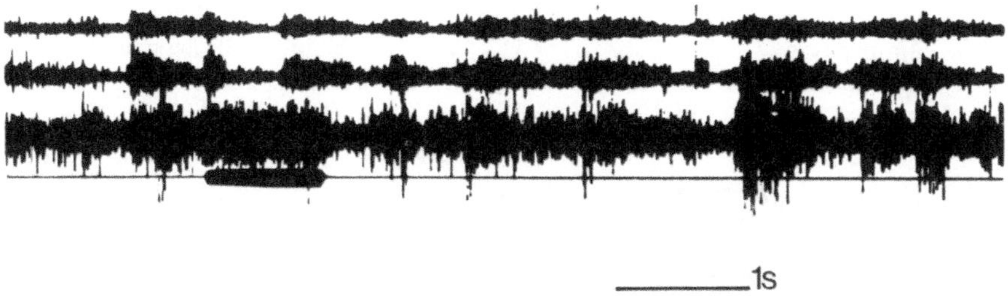

FIGURE 4. Recordings from an uncoordinated and hyperactive cockroach 60 min after receiving an LD_{95} dose of allethrin at 32°C (17.0 μg). Trains of cercal afferent, efferent and abdominal spikes are apparent. First trace, R2 (nerve cord); 2nd trace, R1 (nerve cord); 3rd trace, Rc (cercus); 4th trace, sound pulse marker.

At 15°C. The cockroaches became restless, although still well coordinated in their movements, within a few minutes of poisoning with an LD_{95} dose of allethrin. This was correlated with an abnormally high level of afferent spikes in a cercus (Fig. 7A) and sometimes a prolonged discharge of these spikes followed mechanical stimulation of the cerci (Fig. 7B). However, no abnormal activity was detected in the CNS until a cockroach's movements became uncoordinated, 40 to 60 minutes after dosing, when prolonged discharges of abdominal interneurones sometimes followed cercal nerve stimulation (Fig. 8). These were unlike the discharges occurring at 32°C in several respects. Firstly, they were never observed following stimulation of the nerve cord (S2). Secondly, they were less intense than those at 32°C, particularly in terms of spike frequency (Fig. 9). Moreover, unlike those at 32°C, the 15°C discharges were inconsistent, decaying to leave a normal response after a few stimuli at 1/second.

Fig. 10 shows chronologically responses in the nerve cord to electrical stimulation, taken from a single experiment. The abdominal discharges at 15°C (Fig. 8) were inconsistent and are therefore not shown in Fig. 10. The abdominal response to cercal nerve stimulation blocked at 27 hours in this experiment, whilst the cockroach was paralysed. As at 32°C, responses in the nerve cord to electrical stimulation of the nerve cord and in a cercus to mechanical stimulation of the cerci, persisted for several more hours, indicating that synaptic block may precede axonal block.

Although nerves sometimes became blocked soon after paralysis, there was often dramatic activity in the nervous system during this stage of poisoning. Mechanical stimulation of the cerci continued to induce discharges of abdominal interneurones and cercal afferent and efferent neurones, often for many hours after paralysis (Fig. 11).

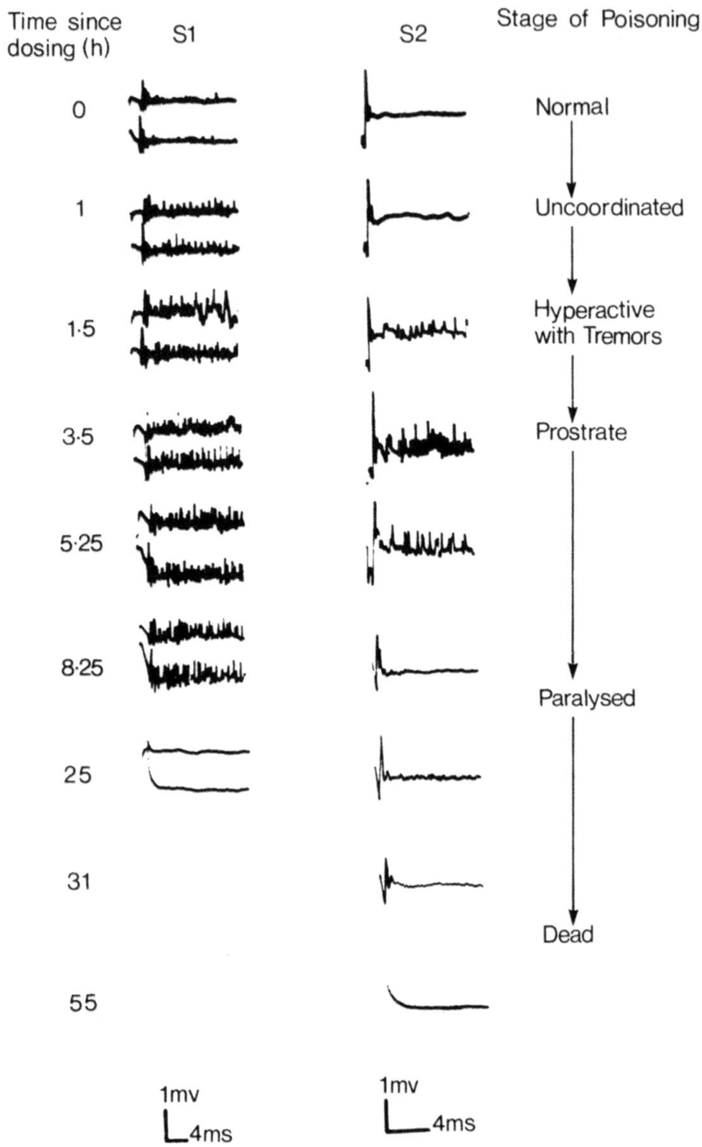

FIGURE 5. Responses in the nerve cord to single electrical stimulation taken from a cockroach dosed with allethrin (LD_{95}) at 32°C. The responses to stimulation of a cercal nerve (S1) were recorded by R2 (upper) and R1 (lower) traces and stimulation of the nerve cord (S2) recorded by R2. All recordings were taken from the same individual and the intensity of electrical stimulation was approximately constant.

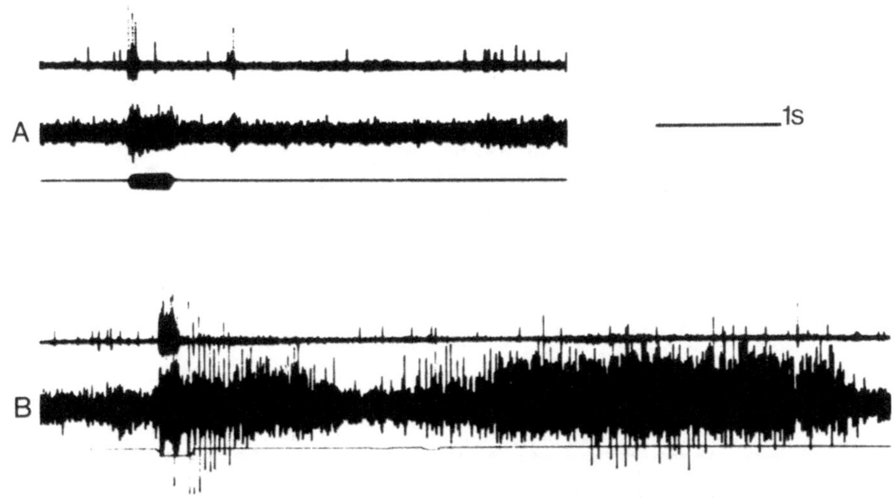

FIGURE 6. Responses from a paralysed cockroach, 17 h after prostration following allethrin dosing (LD_{95}) at 32°C. A, a sound pulse sometimes gave a normal response, although an abnormally high level of spontaneous activity is evident in the nerve cord. B, recorded 60 s after A, an air-puff caused prolonged firing of cercal afferent and efferent axons, a brief period of cercal twitching being the only discernible movement following this air-puff. The top trace was recorded by R2 (nerve cord); middle trace by Rc (cercus) and the bottom trace, sound pulse (A) and air-puff (B) markers.

Thus although the action of allethrin on the whole insect at 32°C was not unexpected from previous in vitro studies, the effects caused by an equivalent dose at 15°C were not anticipated. Even when a very large dose of allethrin was applied (a 32°C LD_{95}), the symptoms were not explicable by nerve blockage but by nerve hyperexcitation. Indeed, although cercal afferent nerve fibers were the first detectable site of action, at both 15°C and 32°C they did not become blocked until after the CNS. This suggests that nerve blockage may in fact have been a secondary rather than a primary result of allethrin treatment. In an attempt to corroborate this suggestion, I used tetrodotoxin.

Effects of Tetrodotoxin (TTX) on the Cockroach

Although a potent neuropoison (see Kao, 1966), a dose of TTX was found which, when injected into a group of cockroaches at 15°C, caused 95% prostration within 3 to 6 hours, followed by paralysis over the ensuing 24 hours but, during the next week, about 80% of the cockroaches recovered. An analysis using electrode-implanted

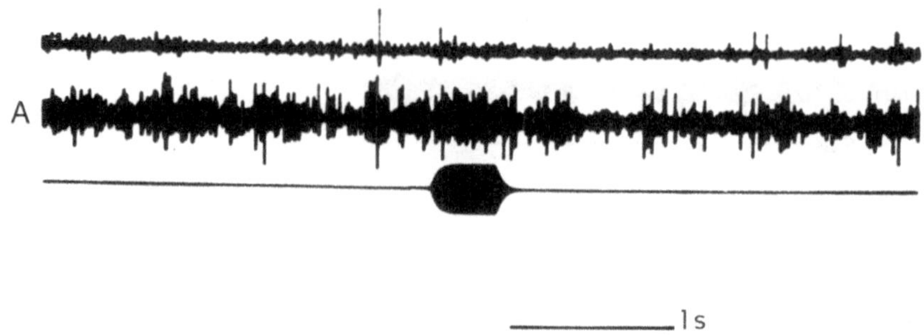

FIGURE7. Responses from a cockroach during the restless stage of poisoning after receiving an LD_{95} dose of allethrin at 15°C (1.78 μg). A, 20 min after dosing, B, 40 min after dosing. The insect was motionless whilst the recordings were taken. The top trace was recorded by R2 (nerve cord); middle trace by Rc (cercus) and the bottom trace, sound pulse marker.

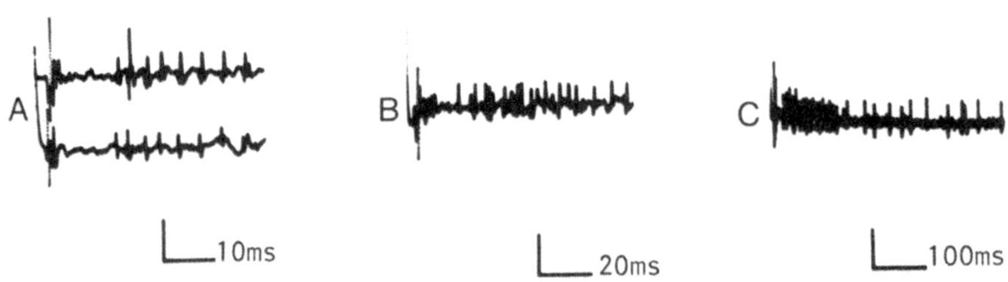

FIGURE 8. Inconsistent, ascending after-discharges in the nerve cord following single electrical stimulation of a cercal nerve (S1) after dosing with allethrin (LD_{95}) at 15°C. A, the cockroach was prostrate 200 min after dosing; top trace recorded by R2 (nerve cord), lower trace, R1 (nerve cord). B, the cockroach was uncoordinated 55 min after dosing, recorded by R2. C, the cockroach was prostrate 165 min after dosing, recorded by R2. Voltage calibration 500 μV.

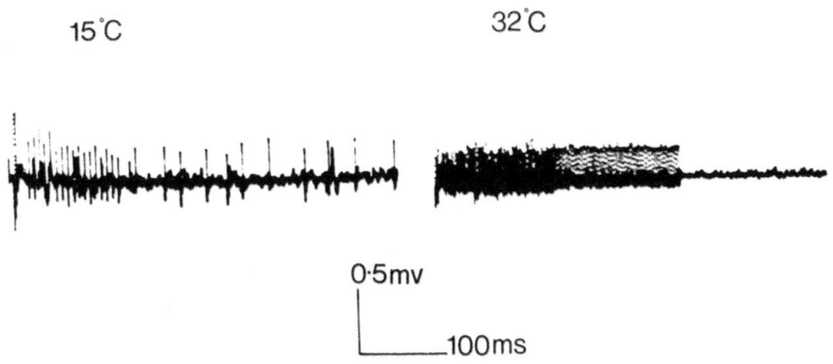

FIGURE 9. Recordings from the nerve cord (R2) of cockroaches which were prostrate following allethrin poisoning (LD_{95}) at 15 and 32°C. At 15°C, an example is shown of the inconsistent discharge which sometimes followed a single electrical stimulation of a cercal nerve (S1). At 32°C, an example is shown of the discharge following stimulation of the nerve cord (S2). Discharges following stimulation above the sixth abdominal ganglion (S2) were never observed at 15°C.

cockroaches revealed that this behavior was synchronous with reversible blockage of peripheral nerves, whilst the CNS was apparently unaffected (Gammon, 1978b). Two types of response were obtained from recovering individuals. In the majority of cases, the response in a cercus to a sound pulse was blocked at prostration or by the time of paralysis but the response in the abdominal nerve cord to electrical stimulation of a cercal nerve was unaffected (Fig. 12). In some preparations, however, there was a reduction of both these responses following TTX administration (Fig. 13). The reduction of the abdominal nerve cord response to electrical stimulation of a cercal nerve is thought to result from blockage of axons in the cercal nerve anterior to the cercus rather than in the CNS, since insects in which this response was reduced by TTX nevertheless yielded a normal response to electrical stimulation of the nerve cord, at S2 (Gammon, 1978b). Although a large proportion of cockroaches recovered from TTX injection, even those which did not regain normal behavior recovered some nervous responses to cercal nerve stimulation.

Given the apparent inaccessibility of the cockroach CNS to this dose of TTX, it was used to pharmacologically distinguish between the actions of allethrin on cockroach central and peripheral nerves, at 15°C. Firstly a toxicity test was conducted to determine whether the insecticidal potency of allethrin was diminished as a result of prior injection of TTX. The results (Table 1) show that prostration by TTX affords complete protection against allethrin

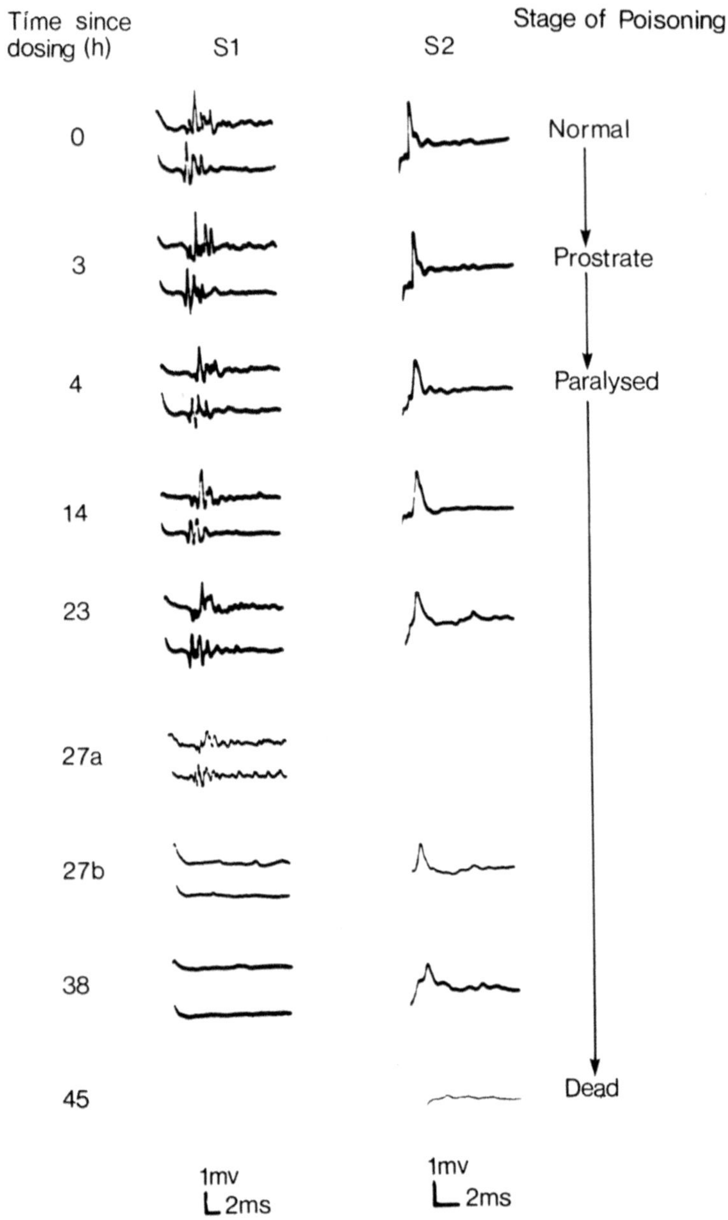

FIGURE 10. Responses in the nerve cord to single electrical stimulation taken from a cockroach dosed with allethrin (LD_{95}) at 15°C. The responses to stimulation of a cercal nerve (S1) were recorded by R2 (upper) and R1 (lower) traces and stimulation of the nerve cord (S2) recorded by R2. All recordings were taken from the same individual, 27b was recorded 30 s after 27a.

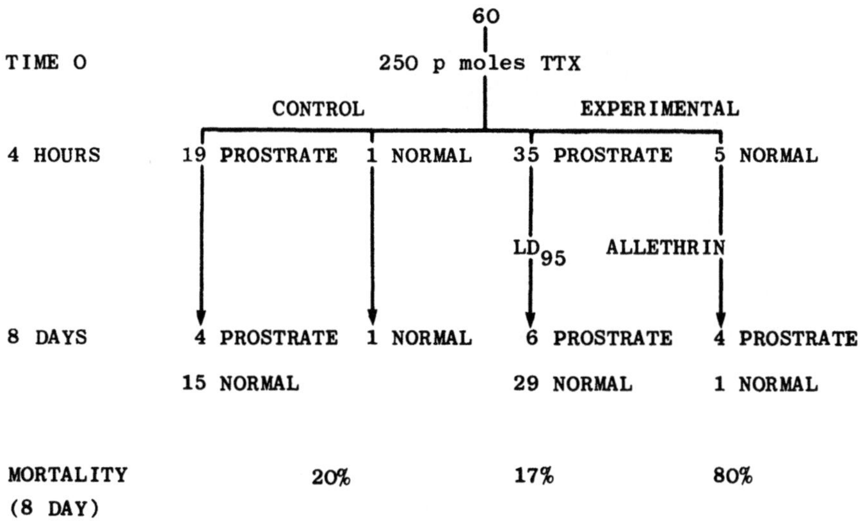

TABLE 1. 60 cockroaches were each injected with 250 pmoles of TTX (Batch 1), a dose known to cause approximately 95% prostration within 3 to 6 hours but only 20% mortality at 8 days. After 4 hours they were separated into prostrate and unaffected, and 40 individuals (as indicated) were treated topically with 1.78 µg of allethrin (24 hr. LD_{95} at 15°C). Subsequently, they were examined daily, and recovery was noted. After 8 days there was no significant difference between the TTX (control) mortality and that of cockroaches which were made prostrate by TTX and then subjected to an LD_{95} dose of allethrin. Temperature 15°C. 250 pmol of Batch 1 TTX was equivalent to 175 pmol of Batch 2 TTX (see Figs. 12 and 13 and Gammon, 1978b).

poisoning, since cockroaches which were prostrate at the time of allethrin administration showed a final mortality no higher than that of "control" insects treated with TTX alone. This finding, <u>per se</u>, suggests that hyperexcitation of peripheral nerves by allethrin may be of crucial importance in causing death of the allethrin-poisoned cockroach, at 15°C, particularly since a similar toxicity test showed no protection by TTX against DDT poisoning (Gammon, 1977b) and DDT causes hyperexcitation of peripheral and central nerves in the cockroach, across a large range of temperatures (Gammon, 1978c). This suggestion was confirmed by studying cockroaches with chronically implanted electrodes, subjected to TTX plus allethrin treatment (Gammon, D.W. - in preparation).

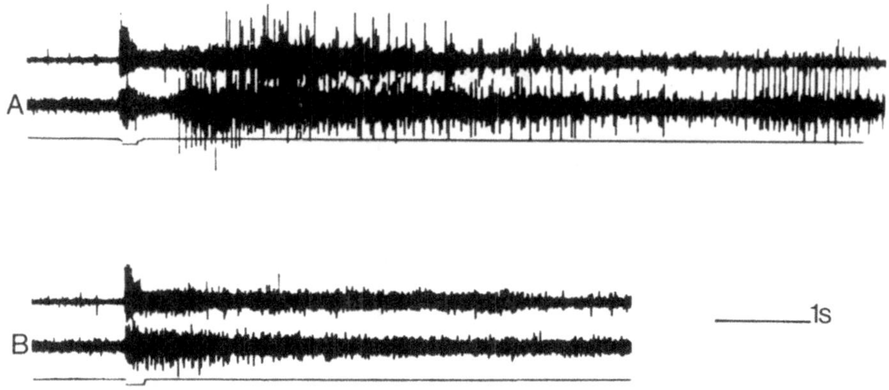

FIGURE 11. Responses from a cockroach which was paralysed, 20.5 h after dosing (LD_{95}) with allethrin at 15°C.
A, an air-puff induced prolonged firing of cercal afferent, efferent and abdominal neurones, a weak tremor of one leg being the only movement of the cockroach.
B, a second air-puff, 30 s after that in A resulted in prolonged discharges of cercal afferent and abdominal neurones, but no detectable movements of the cockroach. The top trace was recorded by R2 (nerve cord); middle trace Rc (cercus) and the bottom trace, air-puff marker.

Effects of Allethrin on the TTX-Treated, Electrode-Implanted Cockroach, at 15°C

Two individuals recovered fully and they showed none of the allethrin-induced disruptions to the nervous system already described. In both cases, the cercal nerve response to a sound pulse was blocked by TTX at the moment of allethrin application (Gammon, 1977b). A further two cockroaches were unaffected by TTX when allethrin was applied and these underwent the usual allethrin poisoning symptoms, dying within 48 hours of treatment. However, the other 18 experimental animals, although prostrate at the time of allethrin dosing, failed to recover. Fig. 14 shows nerve recordings from one such individual. Prolonged cercal afferent discharges followed a sound pulse at 9.5 and 22 hours and discharges of abdominal interneurones followed electrical stimulation of a cercal nerve at these times. However, central effects were observed only in those individuals having functioning fine cercal afferents, i.e. giving a response to a sound pulse, suggesting that such discharges in the CNS may be a secondary effect of allethrin intoxication. This contention is supported by the close similarity between discharges occurring in a TTX plus allethrin poisoned cockroach and those in a physically stressed but unpoisoned individual (Fig. 15).

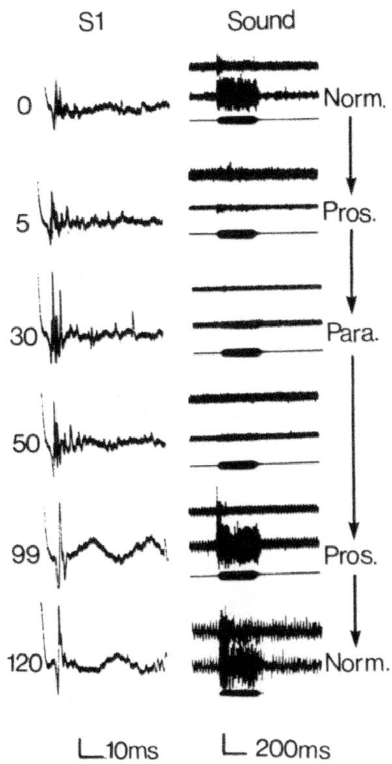

FIGURE 12. Nervous activity from a cockroach injected with 175 pmol of TTX (Batch 2) at 15°C. Responses to electrical stimulation of a cercal nerve (S1) recorded by R2 (nerve cord), are shown on the left. Responses to sound pulses were recorded on the top trace by R2 (nerve cord); middle trace Rc (cercus) and the bottom trace, sound pulse marker. The stage of intoxication is shown on the right and the numbers on the left show time (in hours) since injection of TTX. Slight variations in amplitude of the evoked responses presumably result from variations in the recording conditions around the electrodes. All recordings were taken from the same individual and were typical of responses observed in 5 other cockroaches. Voltage calibration 200 μV.

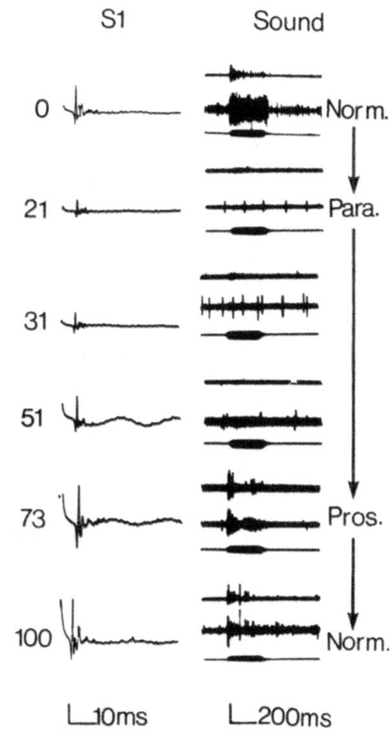

FIGURE 13. Nervous activity from a cockroach injected with 175 pmol of TTX (Batch 2) at 15°C. Responses to electrical stimulation of a cercal nerve (S1) recorded by R2 (nerve cord), are shown on the left. Responses to sound pulses were recorded on the top trace by R2 (nerve cord); middle trace Rc (cercus) and the bottom trace, sound pulse marker. The stage of intoxication is shown on the right and the numbers on the left show time (in hours) since injection of TTX. Both evoked responses were normal at prostration (5 h). All recordings were taken from the same individual and were typical of responses observed in 2 other cockroaches. Voltage calibration 500 μV.

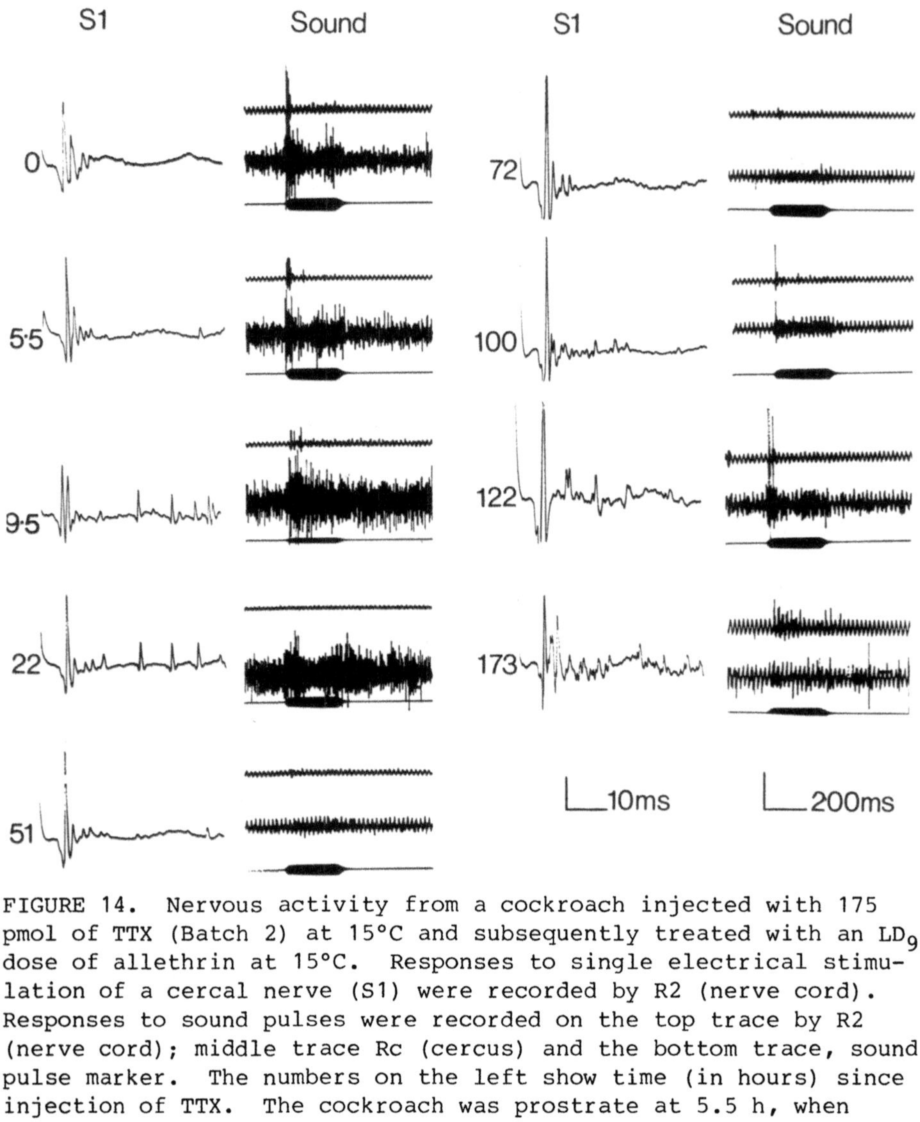

FIGURE 14. Nervous activity from a cockroach injected with 175 pmol of TTX (Batch 2) at 15°C and subsequently treated with an LD_{95} dose of allethrin at 15°C. Responses to single electrical stimulation of a cercal nerve (S1) were recorded by R2 (nerve cord). Responses to sound pulses were recorded on the top trace by R2 (nerve cord); middle trace Rc (cercus) and the bottom trace, sound pulse marker. The numbers on the left show time (in hours) since injection of TTX. The cockroach was prostrate at 5.5 h, when allethrin was applied. At 22 h, it was paralysed and this individual remained motionless until the end of the experiment. Blockage of the response to sound at 51 h was a late effect of TTX as allethrin blockage was always irreversible. At 173 h, evoked responses were still present following stimulation, although the cockroach was in *rigor mortis*. Voltage calibration 200 µV. Temperature 15°C.

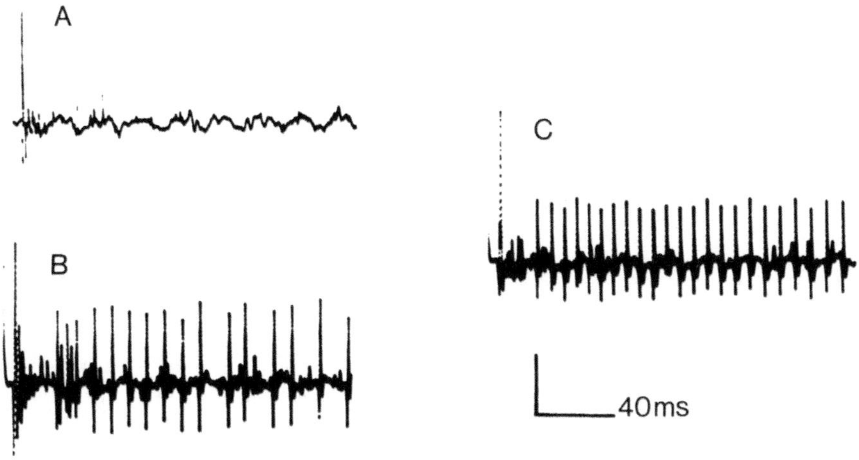

FIGURE 15.
A: Response to single electrical stimulation of a cercal nerve (S1) recorded by R2 (nerve cord) from an unpoisoned cockroach.
B: As A, but recorded 21 h after an LD_{95} dose of allethrin had been applied to a TTX-prostrate cockroach. The insect was paralysed.
C: As A, but recorded 6 days after an unpoisoned cockroach had been restrained on its back using quick drying epoxy-resin (Araldite). The insect was paralysed.
The abdominal discharges shown in B and C could not be elicited from a normal cockroach, irrespective of the stimulus parameters. Furthermore these discharges were not consistent, decaying to leave a response similar to A after a few stimuli at 1 Hz. Voltage calibration 200 μV. Temperature 15°C.

It is therefore concluded that TTX protects cockroaches by reversibly blocking those peripheral nerves normally hyperexcited by allethrin. This protection depends critically on the insect being prostrate at the moment of allethrin administration. Furthermore, augmentation of residual peripheral nerve activity by physical stress is often enough to cause death, irrespective of the presence of TTX. The poor recovery rate of electrode-implanted cockroaches, compared with unimplanted "controls" receiving identical doses of TTX and allethrin (Fig. 16), supports this conclusion. As untreated individuals have lived for months with chronically implanted electrodes and the only difference between groups 2 and 4 (Fig. 16) lay in the presence of electrodes in group 2, then the physical disturbances involved in making recordings from group 2 presumably triggered off allethrin-induced repetitive activity in peripheral nerves not blocked by TTX. This then led to secondary effects which included discharges in the CNS, nerve blockage and eventual death.

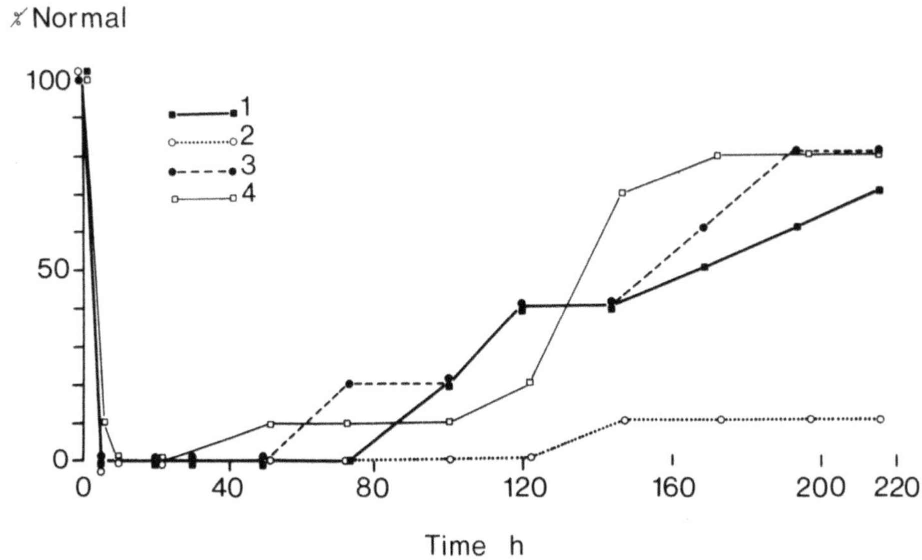

FIGURE 16. Recovery rates (% unaffected against time) for four groups of 10 cockroaches subjected to various treatments:
1. TTX injected, electrode-implanted.
2. TTX injected, allethrin treated, electrode-implanted.
3. TTX injected "control".
4. TTX injected, allethrin treated, "control".
A dose of 175 pmol of TTX (Batch 2) was injected at time 0. An LD_{95} dose of allethrin, at 15°C, was applied at 5.5 h to groups 2 and 4, when all these individuals except one were prostrate. Temperature 15°C.

Discussion

In summarising the action of allethrin on the cockroach, one important point to emerge is that good correlations exist between symptomology and effects on the nervous system, particularly during the early stages of poisoning. For example, prolonged discharges of (cercal) sensory neurones were correlated with restless behaviour without a loss of coordination whereas the onset of uncoordinated behaviour coincided with the appearance of abnormal discharges not only in (cercal) sensory neurones but also in motor neurons and the CNS.

Hyperexcitation of peripheral and central neurones were both primary effects of allethrin at 32°C (see Narahashi, 1962a,b). However, at 15°C it is concluded that hyperexcitation of the peripheral nervous system by allethrin was a direct effect but abnormalities in the CNS were secondary. At both temperatures, nerve blockage was probably a secondary consequence of the insecticide.

It has previously been suggested that particular pyrethroids applied to a whole insect act on either the central or the peripheral nervous system (see Burt et al., 1971; Burt and Goodchild, 1971; Miller and Kennedy, 1972; Burt and Goodchild, 1974; Burt et al., 1977; Miller and Adams, 1977). My work, however, demonstrates that attempts to define target sites (see Takeno et al., 1977; Narahashi et al., 1977) must consider the role of temperature. It is likely that different pyrethroids have different critical temperatures at which they will induce repetitive firing in a given nerve. Moreover, as shown by Narahashi using allethrin, giant nerve fibers in different animals can have their own discrete range of temperatures at which they fire repetitively. Thus, giant axons in the cockroach CNS fire repetitively at temperatures above 26.5°C and up to, and perhaps beyond, 33°C (Narahashi, 1962a,b), but the giant axon in the squid fires repetitively over a much narrower band of temperatures, around 20°C, the discharges diminishing on raising the temperature to 25-30°C (Narahashi, 1976).

A further consideration when studying nerve excitation by pyrethroids is provided by my own studies. I have shown that different nerves, even within the same animal, have different temperature ranges over which they fire repetitively. Allethrin, applied to the cockroach, had a direct excitant action on the CNS only at the higher temperature (32°C) whereas axons in the peripheral nervous system fired repetitively at high and low temperatures. Therefore, the greater sensitivity of the peripheral nervous system to hyperexcitation by allethrin at low temperature is the probable reason for allethrin's greater insecticidal potency as temperature is reduced.

Acknowledgements. I wish to thank Mrs. Margaret Clements for her secretarial assistance, Mr. John Rodford for help in preparing figures and Mrs. Elaine Gammon for reading the manuscript. Some of the figures are reproduced with the kind permission of Blackwell Scientific Publications Ltd.

References

Blum, M.S. and Kearns, C.W. (1956). Temperature and the action of pyrethrum in the American cockroach. J. Econ. Entomol. 49, 862-865.

Burt, P.E., Elliott, M., Farnham, A.W., Janes, N.F., Needham, P.H. and Stevenson, J.H. (1977). Evaluation of pyrethroids for insect control, in Crop Protection Agents - their Biological Evaluation (ed. McFarlane, N.R.) pp.393-411. London: Academic Press.

Burt, P.E. and Goodchild, R.E. (1971). The site of action of pyrethrin I in the nervous system of the cockroach Periplaneta americana. Entomol. Exp. Appl. 14, 179-189.

Burt, P.E. and Goodchild, R.E. (1974). Knockdown by pyrethroids: its role in the intoxication process. Pestic. Sci. 5, 625-633.

Burt, P.E., Lord, K.A., Jennifer, M., Forrest, J.M. and Goodchild, R.E. (1971). The spread of topically-applied pyrethrin I from the cuticle to the CNS of the cockroach Periplaneta americana. Entomol. Exp. Appl. 14, 255-269.

Finney, D.J. (1971). Probit Analysis. 3rd ed. Cambridge University Press.

Gammon, D.W. (1977a). Nervous effects of toxins on an intact insect: a method. Pestic. Biochem. Physiol. 7, 1-7.

Gammon, D.W. (1977b). Neural correlates of insecticide poisoning in the cockroach. Ph.D. Thesis, University of Cambridge.

Gammon, D.W. (1978a). Neural effects of allethrin on the free walking cockroach, Periplaneta americana: an investigation using defined doses at 15 and 32°C. Pestic. Sci. 9, 79-91.

Gammon, D.W. (1978b). The action of tetrodotoxin on the cockroach, Periplaneta americana: a toxicological and neurophysiological study. Physiol. Entomol. 3, 37-42.

Gammon, D.W. (1978c). Effects of DDT on the cockroach nervous system at three temperatures. Pestic. Sci. 9, 95-104.

Kao, C.Y. (1966). Tetrodotoxin, saxitoxin and their significance in the study of excitation phenomena. Pharmacol. Rev. 18, 997-1049.

Lowenstein, O. (1942). A method of physiological assay of pyrethrum extract. Nature 150, 760-762.

Miller, T.A. and Adams, M.E. (1977). Central vs. peripheral action of pyrethroids on the housefly nervous sytem, in Synthetic Pyrethroids (ed. Elliott, M.) pp.98-115. ACS Symposium 42, Washington.

Miller, T. and Kennedy, J.M. (1972). Flight motor activity of houseflies as affected by temperature and insecticides. Pestic. Biochem. Physiol. 2, 206-222.

Murayama, K., Abbott, N.J., Narahashi, T. and Shapiro, B.I. (1972). Effects of allethrin and Condylactis toxin on the kinetics of sodium conductance of crayfish axon membrane. Comp. Gen. Pharmacol. 3, 391-400.

Narahashi, T. (1962a). Effect of the insecticide allethrin on membrane potentials of cockroach giant axons. J. Cell Comp. Physiol. 59, 61-65.

Narahashi, T. (1962b). Nature of the negative after-potential increased by the insecticide allethrin in cockroach giant axons. J. Cell Comp. Physiol. 59, 67-76.

Narahashi, T. (1971a). Mode of action of pyrethroids. Bull. World Health Organ. 44, 337-345.

Narahashi, T. (1971b). Effects of insecticides on excitable tissues. Adv. Insect Physiol. 8, 1-93.

Narahashi, T. (1976). Nerve membrane as a target of pyrethroids. Pestic. Sci. 7, 267-272.
Narahashi, T. and Anderson, N.C. (1967). Mechanism of excitation block by the insecticide allethrin applied externally and internally to squid giant axons. Toxicol. Appl. Pharmacol. 10, 529-547.
Narahashi, T., Nishimura, K., Parmentier, J.L., Takeno, K. and Elliott, M. (1977). Neurophysiological study of the structure-activity relation of pyrethroids, in Synthetic Pyrethroids (ed. Elliot, M.) pp.85-97. ACS Symposium 42: Washington.
Pumphrey, R.J. and Rawdon-Smith, A.F. (1937). Synaptic transmission of nervous impulses through the last abdominal ganglion of the cockroach. Proc. Roy. Soc. B 122, 106-118.
Takeno, K., Nishimura, K., Parmentier, J. and Narahashi, T. (1977). Insecticide screening with isolated nerve preparations for structure-activity relationships. Pestic. Biochem. Physiol. 7, 486-499.
Twarog, B.M. and Roeder, K.D. (1957). Pharmacological observations on the desheathed last abdominal ganglion of the cockroach. Ann. Entomol. Soc. Amer. 50, 231-237.
Wang, C.M., Narahashi, T. and Scuka, M. (1972). Mechanism of negative temperature coefficient of nerve blocking action of allethrin. J. Pharmacol. Exp. Ther. 182, 442-453.
Wigglesworth, V.B. (1941). The effect of pyrethrum on the spiracular mechanism of insects. Proc. Roy. Entomol. Soc. Lond. 16, 11-14.

THE COMPARATIVE NEUROTOXICITY OF FORMAMIDINE PESTICIDES

Albert E. Lund, Robert M. Hollingworth, George K.W. Yim

Department of Entomology, Purdue University

West Lafayette, Indiana 47906

INTRODUCTION

The formamidine pesticides continue to interest investigators because of their novel chemistry, pharmacology, and spectrum of activity. They have been used commercially to control certain Lepidoptera and Hemiptera species and acarines including mites and ticks. Formamidines are ineffective against many other arthropods including harmless or beneficial non-target species. The representative of the formamidines receiving the most attention is chlordimeform (_N_'-(4-chloro-_o_-tolyl)-_N_,_N_-dimethylformamidine). Three reviews on the chemistry and biological actions of chlordimeform (CDM, Fig. 1) and related compounds have appeared (Hollingworth, 1976; Matsumura and Beeman, 1976; Beeman and Matsumura, 1978) revealing a broad range of biological effects for this group. This report will attempt to relate some of these pharmacological actions to toxicity of formamidines in vertebrates and invertebrates.

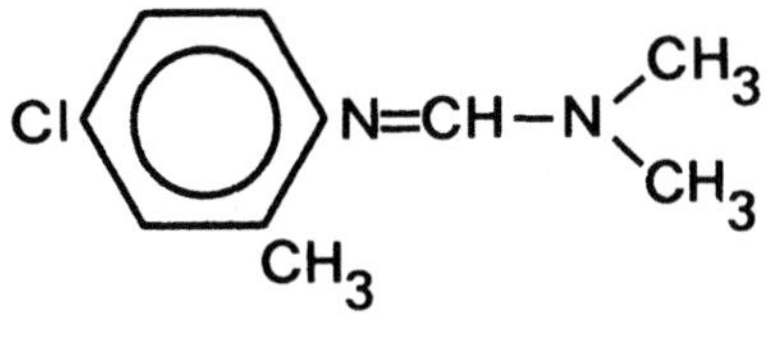

Fig. 1. Chlordimeform

CDM is moderately toxic to mammals with an oral LD_{50} to mice and rats of 220 mg/kg and 250 mg/kg respectively (Anonymous, 1975). Mice or rats intoxicated with CDM orally or intraperitoneally at lethal doses exhibit marked hyperexcitation within 5-10 minutes. A fine tremor particularly of the head and forelimbs and hyperreflexia are accompanied by occasional episodes of clonic convulsions. If a given animal dies, it generally does so during this violent convulsive stage, and usually within 1 hour. If it survives this hyperexcitation and tremor it becomes sedated, locomotion is suppressed, and it usually recovers within 24 hours. Some individuals do not recover but rather remain passive for several days and eventually die. No cholinomimetic symptoms have been observed (Matsumura and Beeman, 1976).

The insecticidal and miticidal actions of CDM are rather complex. CDM produces frank toxicity and death at low doses in only a few cases such as eggs and very young larvae (Harris and Gore, 1971) and adults (Watanabe and Fukami, 1977; Lund et al., 1978a) of some species of Lepidoptera and Acarina. Most insect species are quite resistant to the lethal action of CDM, even those which are effectively controlled in the field by the compound. This has led a number of investigators to suggest that sublethel, behavioral effects may be important in affording plant and animal protection. Phillips et al. (1971) reported that female bollworm moths, fed a dilute CDM-Honey solution (8.68 μg/ml), became hyperactive. They could not uncouple after mating and successful mating was reduced by 40%. Gladney et al. (1974) and Stone et al. (1974) showed that CDM causes detachment of cattle ticks from the host. Gemrich et al. (1976) reported that formamidines cause repellency in spider mites which manifests itself as either "walkoff" or "spinoff" depending on which compounds are used. Hirano et al., (1972) described an excitatory, antifeeding or repellent effect of CDM on the rice stem borer. Larvae crawled from the treated plant or if feeding was initiated it was not maintained. Doane and Dunbar (1973) reported that lepidopterous larvae feeding on treated leaves became hyperactive resulting in net movement away from the food source. Hirata and Sōgawa (1976) found that CDM causes a decrease in feeding of several species of Hemiptera and suggested that the mortality observed was the result of starvation and not direct toxicity of the insecticide. Thus the behavioral effects attributed to CDM may account for both plant protection and mortality.

A rather long list of biochemical or pharmacological actions have been attributed to CDM and related compounds (Table 1), but attempts to correlate such in vitro effects with symptoms observed in vivo have yielded largely negative results. Thus the biochemical basis for the several toxic responses to formamidines in vertebrates and invertebrates remains controversial.

Beeman and Matsumura (1973) and Aziz and Knowles (1973) reported that CDM and related compounds inhibited monoamine oxidase (MAO) in rat liver, and Atkinson et al. (1974) demonstrated MAO inhibition in cattle ticks. However, several observations make MAO inhibition an unlikely mode of action of CDM in mammals or arthropods. First, the development of symptoms by MAO inhibitors is in marked contrast to the rapid onset of excitation and tremors observed in CDM poisoned mammals (Matsumura and Beeman, 1976; Hollingworth et al., 1978). Secondly, Neumann and Voss (1977) found no correlation between in vitro MAO inhibitory potency and acute toxicity or symptom production in mice, nor could in vitro MAO inhibition be related to any symptoms or toxic actions in arthropods such as tick detachment from the host (Stone et al., 1974), tick toxicity (Holden and Hadfield, 1975), or excitation in Lepidoptera (Watanabe and Fukami, 1977). Thirdly, depletion of brain serotonin or norepinephrine with p-chlorophenylalanine or 1-methyl-p-tyrosine had no effect on the toxicity of CDM to rats (Robinson, et al., 1975; Robinson and Smith, 1977). Fourthly, CDM causes only moderate, reversible inhibition of MAO in vivo even at lethal doses, a degree of inhibition which other established MAO inhibitors considerably exceed without apparent poisoning symptoms (Hollingworth et al., 1978). Finally, MAO is relatively scarce in insects and no physiological function has been established for it. This makes it difficult to hypothesize how inhibition of this enzyme can lead to poisoning symptoms in insects (Matsumura and Beeman, 1976).

Beeman and Matsumura (1974) reported an interference in tryptamine metabolism by CDM in cockroaches and Nishimura et al. (1975) attributed this interference to inhibition of tryptamine N-acetyltransferase. Although this enzyme is reasonably abundant in insect nervous tissue (Dewhurst et al., 1972; Evans and Fox, 1975a, b) the high I_{50} value of 4.4×10^{-4}M suggests that inhibition of this enzyme is not of primary importance in producing symptoms at low doses.

Abo-Khatwa and Hollingworth (1972, 1973) showed that CDM uncoupled electron transport and oxidative phosphorylation in rat liver and German cockroach mitochondria with a potency similar to that of 2,4-dinitrophenol. Again, however, certain poisoning symptoms such as the rapid onset of tremors and convulsions could not be explained by the uncoupling of respiration alone.

Wang et al. (1975) reported a decrease in end-plate sensitivity to acetylcholine in frog sartorius muscle at 10^{-4}M CDM. Watanabe et al., (1975) found that 10^{-5}M CDM decreased the acetylcholine-induced contraction of the frog rectus abdominis muscle, but also reported that higher concentrations (10^{-3}M) of CDM induced a contraction by depolarizing the muscle membrane. Later Watanabe et al. (1976) reported that toxicity of formamidines to

TABLE 1

Biochemical and Pharmacological actions of chlordimeform.

Action	Preparation	Effective Concentration	Reference
MAO inhibition	Rat liver	I_{50}=1.4x10^{-5}M	Aziz and Knowles (1973)
		I_{50}=3.2x10^{-5}M	Beeman and Matsumura (1973)
	Mouse liver	I_{50}=4.7x10^{-5}M	Hollingworth et al. (1978)
	Mouse brain	I_{50}=1.2x10^{-4}M	
	Whole cattle ticks	I_{50}=2.3x10^{-5}M	Atkinson et al. (1974)
	Rat brain	I_{50}=6.0x10^{-5}M	Benezet and Knowles (1976)
N-acetyltransferase inhibition	American cockroach nerve	I_{50}=6.0x10^{-5}M 4.4x10^{-4}M	Nishimura et al.(1975); Matsumura and Beeman (1976)
Respiratory Uncoupling	Rat liver	I_{50}=9.0x10^{-5}M	Abo-Khatwa and Hollingworth (1973)
	German cockroach	I_{50}=6.0x10^{-5}M	Abo-Khatwa and Hollingworth (1972)
Decreases muscle end-plate sensitivity to transmitter	Frog sartorius muscle	Threshold=10^{-4}M	Wang et al. (1975)
	Frog rectus abdominis muscle	Threshold=10^{-5}M	Watanabe et al. (1975)
	Waxmoth larval muscle	Threshold=2x10^{-3}M	Yamamoto and Fukami (1976)
Induces muscle contraction	Frog rectus abdominis muscle	Threshold=10^{-3}M	Watanabe et al. (1976)
Decreases action potential amplitude (local anesthetic)	Frog sciatic nerve	15 min. ED_{50}= 1.5x10^{-2}M	Chinn, et al. (1976); Pfister et al. (1978a)
	Tobacco hornworm motor nerve	Threshold=10^{-3}M	Lund et al. (1978a)

TABLE 1 (CONTINUED)

Action	Preparation	Effective Concentration	Reference
Prostaglandin synthetase inhibition	Bovine seminal Vesicle microsomes	$I_{50}=3.4\times10^{-5}M$	Yim et al. (1978)
DNA, RNA, protein synthesis inhibition	Human cell culture (HeLa cells)	$I_{50}=10^{-3}M$	Murakami and Fukami (1974)
Depression of glucose stimulated insulin release	Mouse pancreatic islets	$I_{50}=5\times10^{-7}M$	Yau et al. (1977)

frogs correlated with the formamidine-induced depression of the acetylcholine-dependent contraction, but not with the contraction caused by formamidines at high doses. Thus the latter appears to be a non-specific effect that is unrelated to development of toxicity. No other cholinomimetic or cholinolytic actions (Matsumura and Beeman, 1976) or anticholinesterase activity (Dittrich, 1967; Hirano et al., 1972) have been described for CDM in vertebrates or invertebrates, and it is not obvious how a decrease in muscle end-plate sensitivity can account for the hyperexcitatory symptoms in mammals especially since this effect appears only at fairly high doses (10^{-4}M CDM).

Of the other actions outlined in Table 1, the decrease of muscle end-plate sensitivity to glutamate in waxmoth larvae and the decrease in action potential amplitude in hornworm larvae caused by CDM occur only at very high concentrations and are therefore not likely candidates as explanations of the very potent behaviorial effects in insects which in any case seem to be excitatory rather than depressant. The possible role of prostaglandin synthetase inhibition of depression of glucose-stimulated insulin release, or of DNA, RNA, and protein synthesis inhibition in the acute toxicity in mammals has not been investigated, but again the rapid onset of tremors and convulsions suggests that they are not of major significance.

It should be evident from the above discussion that although chlordimeform produces a wide range of effects at various doses in various organisms, no unified hypothesis has emerged that can plausibly relate one or more biochemical lesions to mammalian or arthropod toxicity or to the protection of plants or animals against attack by some arthropods. We have approached this problem by first identifying the physiological lesion (in the case of mammals) or the behavioral aberration (in the case of arthropods) produced by CDM and then systematically searching for a pharmacological explanation of the observed effects. The modes of action in mammals and arthropods will be considered separately.

THE MODE OF ACTION IN MAMMALS

Beeman and Matsumura (1974) suggested that the cardiovascular system might be involved in the acute toxicity of CDM when they reported that a dose of 200 mg/kg CDM, i.p., decreased the mean arterial blood pressure in the rabbit anesthetized with pentobarbital. Since cardiovascular and respiratory collapse always accompany the death of a mammal, we felt that investigation of the effects of CDM on these systems would yield useful results. A detailed account of these experiments can be found in Lund et al. (1978b, c) so only a brief summary will be given here.

CDM at doses of 1 - 30 mg/kg i.v. caused a biphasic blood

pressure response in dogs anesthetized with pentobarbital (Fig. 2). The primary depressor and subsequent secondary pressor responses were accompanied by a decrease and increase in cardiac contractility respectively and qualitatively similar, but small changes in heart rate. Another set of experiments (not pictured) using the dog perfused hind limb preparation showed that the depressor and pressor responses were also accompanied by a decrease and increase, respectively, in the vascular resistance. Thus depression of the cardiac contractility and vascular resistance account for the primary depression of blood pressure while increases in cardiac contractility and vascular resistance account for the secondary increase in blood pressure. Under light anesthesia, hyperventilation and tremors were associated with the transition from the depressor to the pressor responses. Lethal doses of CDM (50 mg/kg, i.v.) caused rapid and irreversible hypotension resulting from severe depression of both the heart and the peripheral vasculature. Respiratory arrest followed cardiovascular collapse within a few seconds. Artificial respiration did not protect the animals against hypotension and death suggesting that cardiovascular

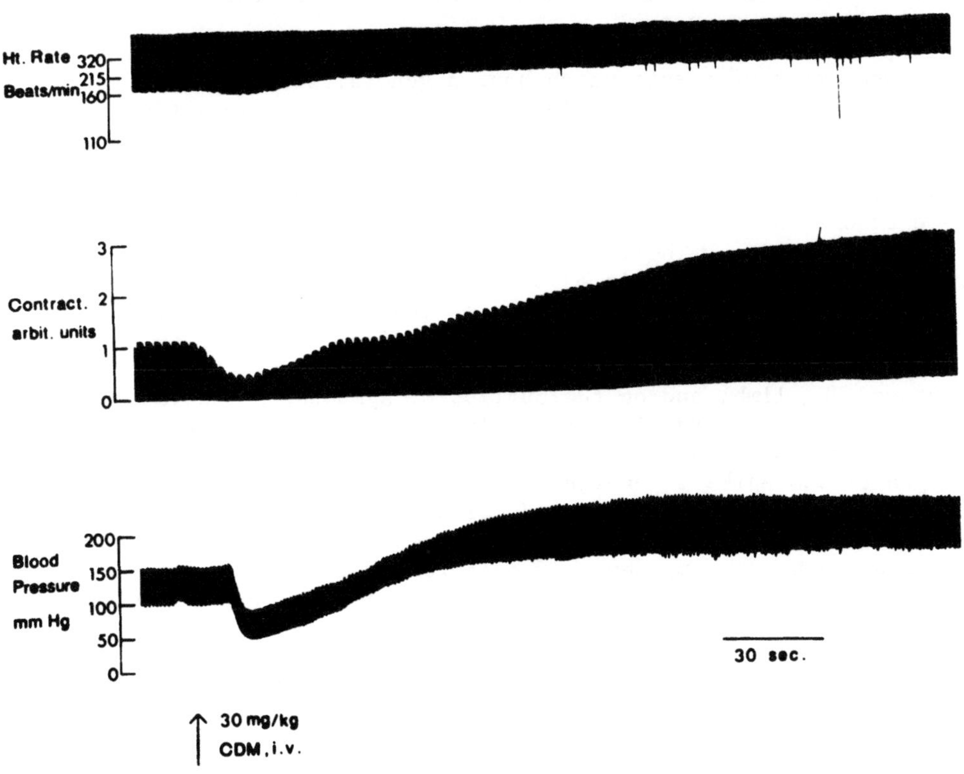

Fig. 2. Response of heart rate, cardiac contractility and blood pressure to 30 mg/kg CDM i.v. Arbit. units = Arbitrary units. Reproduced from Lund et al. (1978)

collapse is a primary cause of death and not a secondary consequence of respiratory arrest. Thus, although CDM depresses both the cardiovascular and respiratory systems, cardiovascular collapse is probably the primary cause of death in dogs.

Since the cardiovascular system is controlled by the autonomic nervous system, the autonomic involvement in the expression of the blood pressure response to CDM intoxication was investigated through a series of pharmacological manipulations. Tables 2 and 3 summarize the results of those experiments. Interference with the function of the sympathetic nervous system at the level of the brain, sympathetic ganglia or neuroeffector junction blocks the secondary pressor response indicating that this part of the CDM response is an autonomic consequence of an action of CDM on the brain. Blockade of the parasympathetic nervous system or histaminergic receptors had no effect on the primary depressor response to CDM nor did CDM itself have any blocking effect on the sympathetic nervous system. Thus the primary depression of blood pressure is the result of a non-specific direct action on the cardiac and vascular smooth muscle and is independent of the autonomic nervous system.

One group of agents which have a generalized depressant effect on nerve and muscle are the local anesthetics. An investigation conducted simultaneously in this laboratory revealed that CDM does, in fact, have local anesthetic activity on frog nerves (Chinn, et al., 1976; Pfister, et al., 1978a) and that the structure of CDM is similar to that of many therapeutically useful local anesthetics. Therefore, a study was undertaken to compare the cardiovascular toxicities of CDM and a representative local anesthetic lidocaine. The results of that investigation indicated that the effects of lidocaine on the cardiac contractility in the dog and isolated rabbit heart, on the vascular resistance in the perfused dog limb, and on the outflow of sympathetic nervous activity in cats were all qualitatively similar to the effects produced by CDM. Chinn, et al. (1977) have since found similar amygdalar EEG spike discharges, overt seizures, and blood pressure responses when either CDM or lidocaine were injected into the lateral ventricle of rats, further suggesting that both the autonomic and somatic components of seizures produced by these two compounds originated centrally and were probably caused by similar mechanisms.

The comparison between the toxic actions of CDM and local anesthetics has subsequently been extended by Pfister, et al. (1978b). They reported that the toxicity of CDM in mice was antagonized by anticonvulsants which also antagonize local anesthetic toxicity, but not by those which are ineffective in reducing local anesthetic toxicity.

TABLE 2

The effect of various pharmacological treatments on the blood pressure response to CDM in the dog.

Treatment	Effect on CDM Response: Primary Depressor	Effect on CDM Response: Secondary Pressor
α-adrenergic blocker (phentolamine)	none	blocked
β-adrenergic blocker (propranolol)	none	partially blocked
ganglionic blocker (hexamethonium)	none	blocked
central sympathetic blocker (diazepam)	none	blocked
sympathectomy[a]	none	blocked
partial parasympathectomy[b]	none	none
muscarinic blocker (atropine)	none	none
histaminergic blocker (tripelennamine)	none	none

[a]denervated heart responses measured using the isolated rabbit heart; denervation of the hind limb by sciatectomy.

[b]denervation of the heart by vagotomy

TABLE 3.

The effect of 10 or 30 mg/kg i.v. on the responses of the blood pressure to various pharmacological treatments.

Treatment	Control Response of Blood Pressure	Effect of CDM on Blood Pressure Response
ganglionic stimulant (DMPP[a])	increase	none
α-adrenergic agonist (norepinephrine)	increase	none
β-adrenergic agonist (isoproterenol)	decrease or biphasic response	none
muscarinic agonist (acetylcholine)	decrease	none

[a]1,1-dimethyl-4-phenylpiperazinium

In summary, many of the pharmacological and toxic actions of CDM on the cardiovascular and central nervous systems are due to actions that are common to local anesthetics, and these actions can plausibly account for the acute lethality of CDM through cardiovascular collapse.

THE MODE OF ACTION IN INSECTS

As mentioned above, no one has successfully related the _in vitro_ actions of formamidines to their toxicity to insects or to plant protection. A potential weakness in several investigations is that they were performed on non-target species, e.g. cockroaches, at unreasonably high doses. Therefore, we studied the behavior of formamidine-intoxicated tobacco hornworm larvae (_Manduca sexta_), a target species (Anonymous, 1975), and then looked for a pharmacological explanation of the behavior. Parts of this investigation have been submitted for publication (Lund, et al., 1978a) where more details of the methods are given.

Chlordimeform is only moderately lethal to tobacco hornworm eggs (LC_{50}=0.1% by dipping). Yet when aqueous solutions of CDM hydrochloride were sprayed onto tomato plants, larval feeding damage was reduced at rather low rates of application. By measuring leaf area before and after feeding for 24 hours, we found that feeding was reduced by 50% at a spray concentration of 0.01% CDM (AC_{50}), but that no mortality occurred in 24 hours. Thus direct mortality does not account for plant protection. Closer examination of the larvae feeding on treated foliage revealed that after a few minutes of feeding, the larvae showed fine tremors. As they continued to feed for several more minutes, the tremors became more severe, and the larvae ultimately crawled or fell from the plant. These same larvae recovered within one hour, crawled back onto the plant, initiated feeding and again precipitated the same series of events. Identical results were observed when larvae were injected with CDM and allowed to feed on untreated leaves except that after they recovered, they fed normally. Thus it appears that excitation and tremors may make it impossible for the larvae to feed normally.

The origin of the tremors was investigated by making an incision in a treated larva and pinning the cuticle back so that the twitching muscles could be observed. Cutting the motor nerves abolished the twitching indicating that the convulsions are of nervous rather than muscular origin. Next, the fourth abdominal ganglia along with the left anterior lateral motor nerve were excised from untreated 5th instar larvae, bathed in a well with the saline solution of Yamasaki and Narahashi (1959) and motor neurone activity was recorded with a suction electrode (Fig. 3a). Usually, the nerve was active initially but quieted down in 20 - 30 minutes (Fig. 3b). At this time irrigation with 10^{-7} - 10^{-6}M CDM within a few minutes caused firing in two motor cells distinguished by different spike sizes (Fig. 3c), again suggesting that the tremors are produced by some excitatory effect of CDM on the central nervous system.

The relationships between plant protection and tremors and between tremors and neuronal excitation are further supported by structure-activity studies. The concentration that reduced leaf feeding by 50% (AC_{50}), the dose that produces tremors in 50% of larvae injected (TD_{50}), and the lowest concentration at which neural excitation occurs in excised fourth abdominal ganglia (CNS excitation threshold) were measured for a series of N'-substituted-phenyl-N,N-dimethylformamidines. Table 4 shows that changes in the ring substituents markedly affect feeding, the production of tremors, and the production of neural excitation and that good correlations ($r^2 > 0.95$) are observed between the parameters.

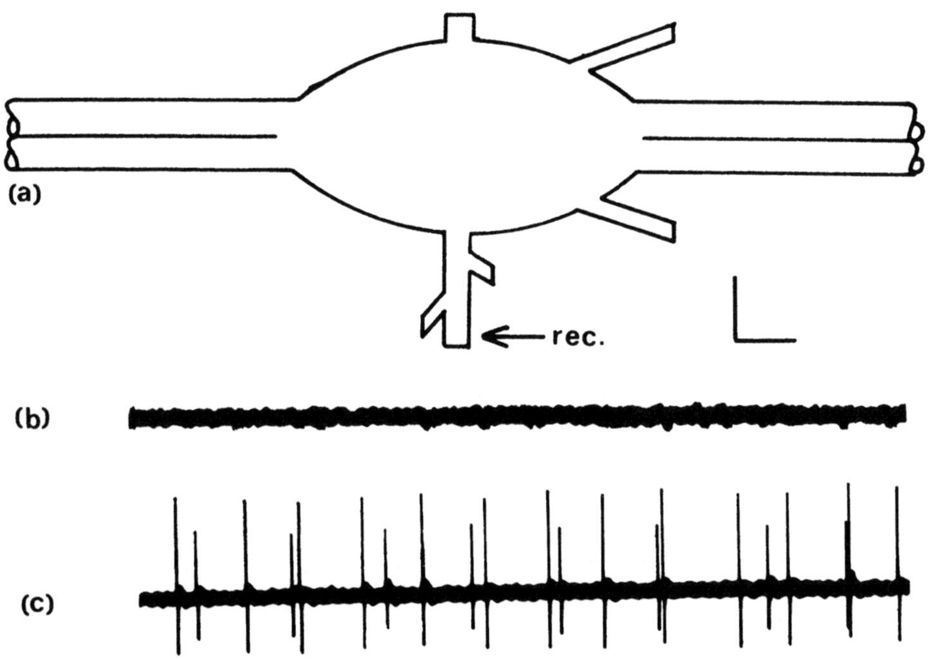

Fig. 3. (a) Diagram of a fourth abdominal ganglion from a tobacco hornworm larva rec. = recording site (b) Control record from the anterior lateral motor nerve bathed in physiological saline solution (c) Action potentials recorded from the same nerve 4 minutes after beginning irrigation with 10^{-6}M CDM. Calibration : 200 msec, 75 mV.

TABLE 4

The relative potency of selected formamidines in reducing leaf feeding (AC_{50}), producing tremors (TD_{50}) and in causing neuronal excitation (CNS excitation threshold).

X	AC_{50} (%)[a]	TD_{50} (μg/larva)[b]	CNS Excitation Threshold (M)
2,4-diCl	0.005	0.49	8×10^{-7}
4-Cl,2-CH_3 (CDM)	0.011	0.42	1×10^{-6}
2,4-diCH_3	0.3	21.0	2×10^{-5}
4-Cl	0.8	124	5×10^{-4}
2-CH_3	0.6	175	1×10^{-4}

[a]r^2= 0.96, [b]r^2 = 0.93

Recently, we noticed another very interesting relationship. Chang and Knowles (1977) studied the toxicity of a long list of formamidines to twospotted spider mites. We have measured the effects of some of these compounds on causing neuronal excitation in hornworm fourth abdominal ganglia. Figure 4 shows the relationship between the LC_{50} for mites and the neural excitation threshold in hornworm ganglia. The excellent correlation (r^2 = 0.93) is surprising considering the two sets of data come from different species and different laboratories and suggests that a common effect on the nervous system may account not only for excitation and tremors in species that are not readily killed by formamidines, but may also be related to frank lethality in species such as mites which are readily killed.

Certainly, neural excitation caused by pesticides is not a new strategy for controlling pest organisms (Narahashi, 1971, 1976; Shankland, et al., 1978). Some compounds, such as DDT and pyrethroids interfere with axonal ionic conductance mechanisms which results in repetitive firing of action potentials. Others interfere with cholinergic synaptic transmission by stimulating postsynaptic receptors (e.g. nicotine), by blocking enzymatic breakdown of the transmitter (e.g. organophosphates and carbamates), or by causing excessive release of the transmitter (e.g. lindane and cyclodienes). Therefore, experiments were performed to determine if CDM caused neural excitation by one of these mechanisms.

The possibility of an axonal involvement in excitation was studied by removing a pair of central connectives between two ganglia, stimulating one end, and recording the compound action potential at the other. High concentrations of CDM (10^{-3}M) caused a local anesthetic-like decrease in the amplitude of the action

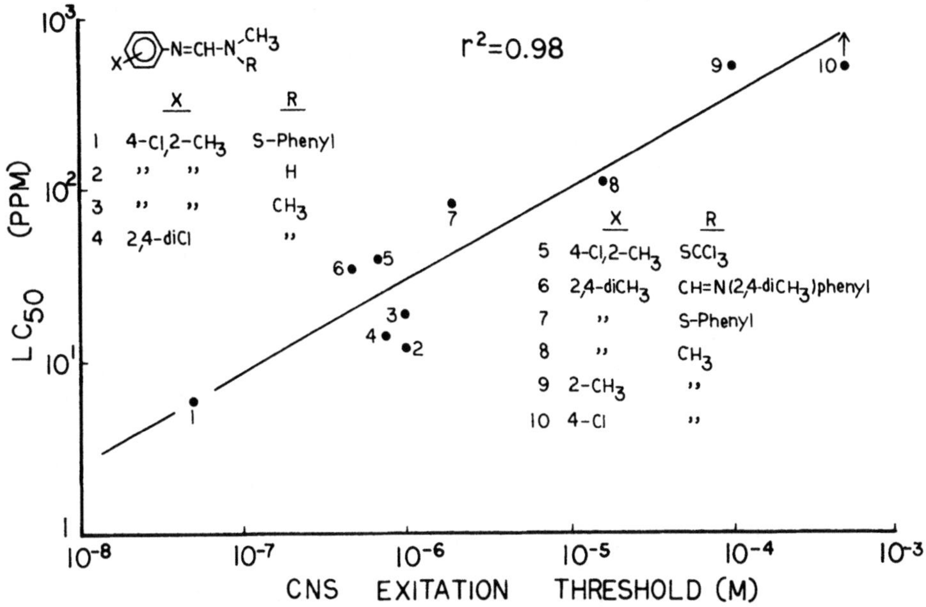

Fig. 4. Correlation between LC_{50} of CDM to twospotted spider mites (Data from Chang and Knowles, 1977) (Ordinate) and central nervous system excitation threshold in tobacco hornworm larvae (Abscissa) for the formamidines listed. The line was calculated by linear regression analysis.

TABLE 5

The effect of nereistoxin (NTX) on the neuronal excitatory actions of carbachol and CDM.

Treatment	Concentration (M)	Maximum frequency (ips, $\bar{X} \pm SE(n)$)[a] Saline pretreated	NTX pretreated
Carbachol	5×10^{-8}	0	0
	1×10^{-7}	18 ± 6.1 (4)	0
	5×10^{-7}	22 ± 6.0 (6)	0
	1×10^{-6}	29 ± 4.1 (3)	22 ± 6.1 (3)
CDM	5×10^{-7}	0	0
	1×10^{-6}	16 ± 3.5 (5)	11 ± 3.1 (5)
	1×10^{-5}	18 ± 2.1 (3)	14 ± 4.2 (3)

[a]impulses/second, mean $\pm$ standard error of the mean (sample size)

potential, but no effect was observed in 30 minutes at 10^{-4}M. No repetitive firing was observed indicating that CDM is not DDT or pyrethroid like. Identical results were obtained with isolated peripheral nerves which contain only motor and sensory axons.

To examine whether there is cholinergic involvement in the CDM response, we studied the effect of various drug treatments on the spike frequency recorded from the fourth abdominal ganglion motor nerve as in Figure 3. Treatment of the preparations with various concentrations of the cholinergic agonist carbamylcholine (carbachol) or with CDM resulted in the maximum neuronal firing rates shown in Table 5. Pretreatment of preparations with 10^{-5}M nereistoxin oxalate (NTX), a non-depolarizing cholinergic receptor blocker (Sakai, 1967) abolished the responses to the lower doses of carbachol as expected, but had little or no effect on the response to CDM. Therefore, the concentrations of CDM used in this study appear to have no effect on cholinergic transmission in this ganglion, and the excitatory actions of CDM probably are induced at non-cholinergic sites. The lack of cholinomimetic actions of formamidines in vertebrates has already been described.

Thus the mode of action of CDM in insects is novel for two reasons. First, it can confer plant protection by a sub-lethal, behavioral mechanism that reduces plant feeding, and, secondly, the neuronal excitation that drives the behavioral aberration is caused by some synaptic action other than at cholinergic synapses. Since the neuronal excitation is not related to local anesthesia, the mechanism by which CDM achieves plant or animal protection from arthropods is different from the mechanism by which CDM produces cardiovascular toxicity in mammals.

SUMMARY OF NEUROTOXIC ACTIONS OF CDM

Based on the above discussion we might divide the actions of CDM into two fairly distinct categories. (1) Depressant actions are probably the result of a rather non-specific membrane stabilizing effect of CDM. A characteristic of these depressant actions is that they require fairly high doses. This category includes the decrease in cardiac and smooth muscle contractility in mammals, the decrease in end-plate sensitivity to acetylcholine in frogs or to glutamate in Lepidoptera, and the decrease in axonal transmission in frog and hornworm nerves. In addition to these effects on excitable membranes, the importance of membranes and membrane-bound enzymes in the coupling of electron transport and oxidative phosphorylation, DNA, RNA, and protein synthesis, prostaglandin synthesis, and monoamine oxidation, it is conceivable that the mechanisms by which CDM interferes with many of these processes are related to some common physico-chemical properties of formamidines which disturb membrane dependent processes.

(2) In some cases excitatory actions are produced at lower doses. The best example of excitation observed so far is neuronal excitation in hornworms. A similar action may also account for the excitation and toxicity observed in other arthropods (ticks and mites). Further, the excitatory action of CDM may not be limited to invertebrates. Pfister, et al. (1978b) reported that low doses of CDM (10 mg/kg) caused increased feeding in rats and Pfister, et al. (1978a) suggested that CDM has a CNS excitatory effect that cannot be explained by a local anesthetic-like mechanism, which may contribute to the hyperexcitation and tremors in mice. Certainly, further study of neuronal excitatory effects observed at low doses with the formamidines is indicated in both mammals and arthropods. Thus we cannot rule out the possibility that the kind of potent neuronal action we have observed in hornworms also occurs in mammals.

LIST OF REFERENCES

Abo-Khatwa, N. and Hollingworth, R.M. (1972) Chlordimeform: The relation of mitochondrial uncoupling to toxicity in the German cockroach. Life Sci. 11: 1181-1190.

Abo-Khatwa, N. and Hollingworth, R.M. (1973) Chlordimeform: Uncoupling activity against rat liver mitochondria. Pestic. Biochem. Physiol. 3: 358.

Abo-Khatwa, N. and Hollingworth, R.M. (1974) Pesticidal chemicals affecting some energy-linked functions at rat liver mitochondria *in vivo*. Bull. Environ. Contam. Toxicol. 12: 446-454.

Anonymous (1975) Chlordimeform acaricide-insecticide. 3rd Report; Shering A.G., Berlin.

Atkinson, P.W., Binnington, K.C., Roulston, W.J. (1974) High monoamine oxidase activity in the tick *Boophilus microplus*, and inhibition by chlordimeform and related pesticides. J. Austral. Entomol. Soc. 13: 207-210.

Aziz, S.A. and Knowles, C.O. (1973) Inhibition of monoamine oxidase by the pesticide chlordimeform and related compounds. Nature 242: 417-418.

Beeman, R.W. and Matsumura, F. (1973) Chlordimeform: A pesticide acting upon amine regulatory mechanisms. Nature 242: 273-274.

Beeman, R.W. and Matsumura, F. (1974) Studies on the action of chlordimeform in cockroaches. Pestic. Biochem. Physiol. 4: 325-336.

Beeman, R.W. and Matsumura, F. (1978) Formamidine pesticides--actions in insects and acarines. In Pesticide and Venom Neurotoxicity, p. 178-188, Eds. D.L. Shankland, R.M. Hollingworth, and T. Smyth, Jr., Plenum Press, N.Y. 283 pp.

Benezet, H.J. and Knowles, C.O. (1976) Inhibition of rat brain monoamine oxidase by formamidines and related compounds. Neuropharmacol. 15: 369-373.

Chang, K. and Knowles, C.O. (1977) Formamidine acaricides. Toxicity and metabolism studies with twospotted spider mites, Tetranychus urticae Koch. J. Agr. Food Chem. 25: 493-501.

Chinn, C., Lund, A.E., and Yim, G.K.W. (1977) Central actions of lidocaine and a pesticide chlordimeform. Neuropharmacol. 16: 867-872.

Chinn, C., Pfister, W.R. and Yim, G.K.W. (1976) Local anesthetic-like actions of the pesticide chlordimeform. Fed. Proc. 35: 729.

Dewhurst, S.A., Crocker, S.G., Ikeda, K., and McCaman, R.E. (1972) Metabolism of biogenic amines in Drosophila nervous tissue. Comp. Biochem. Physiol. 43B: 975-981.

Dittrich, V. (1967) A formamidine acaricide as an ovicide for three insect species. J. Econ. Entomol. 60: 13-15.

Doane, C.C. and Dunbar, D.M. (1973) Field evaluation of insecticides against the gypsy moth and the elm spanworm and repellent action of chlordimeform. J. Econ. Entomol. 66: 1187-1189.

Evans, P.H. and Fox, P.M. (1975a) Comparison of various biogenic amines as substrates of acetyltransferase from Apis mellifera (H.) CNS. J. Insect Physiol. 21: 343-353.

Evans, P.H. and Fox, P.M. (1975b) Enzymatic N-acetylation of indolealkylamines by brain homogenates of the honeybee Apis mellifera. J. Insect Physiol. 21: 343-353.

Gemrich, E.G., Kaugars, G., and Rizzo, V.L. (1976) Insecticidal and miticidal activity of aryl thioformamidines. J. Agr. Food Chem. 24: 593-595.

Gemrich, E.G., Lee, B.L., Tripp, M.L., and VandeStreek, E. (1976) Relationship between formamidine structure and insecticidal, miticidal, and ovicidal activity. J. Econ. Entomol. 69: 301-306.

Gladney, W.J., Ernst, S.E., and Drummond, R.O. (1974) Chlordimeform: A detachment-stimulating chemical for three-host ticks. J. Med. Entomol. 11: 569-572.

Harris, C.R. and Gore, F. (1971) Toxicological studies on cutworms. VIII. Toxicity of three insecticides to the various states in the development of the darksided cutworm. J. Econ. Entomol. 64: 1049-1050.

Hirano, T., Kawasaki, H., and Shinohara, H. (1972) Studies on some biological activities of N-(2-methyl-4-chlorophenyl)-N', N'-dimethylformamide (Galecron) to the rice stem borer, Chilo suppressalis Walker. Botyu-Kagaku 37: 135-142.

Hirata, M. and Sōgawa, K. (1976) Antifeeding activity of chlordimeform for plant-sucking insects. Appl. Entomol. Zool. 11: 94-99.

Holden, J.S. and Hadfield, J.R. (1975) Chlordimeform and its effect on monoamine oxidase activity in the cattle tick, Boophilus microplus. Experientia 31: 1015-1017.

Hollingworth, R.M. (1976) Chemistry, biological activity, and uses of formamidine pesticides. Environ. Health Perspect. 14: 57-69.

Hollingworth, R.M., Leister, J. and Gahli, G. (1978) Mode of action of formamidine pesticides: An evaluation of monoamine oxidase as the target. Chem. Biol. Interact. (Submitted).

Lund, A.E., Hollingworth, R.M. and Shankland, D.L. (1978a) Chlordimeform: Plant protection by a sublethal, non-cholinergic action on the central nervous system. Pestic. Biochem. Physiol. (In press).

Lund, A.E., Shankland, D.L., Chinn, C., and Yim, G.K.W. (1978b) Similar cardiovascular actions of the pesticide chlordimeform and lidocaine. Toxicol. Applied Pharmacol. (In press).

Lund, A.E., Yim, G.K.W., and Shankland, D.L. (1978c) The cardiovascular toxicity of chlordimeform: A local anesthetic-like mechanism. Pesticide and Venom Neurotoxicity, p. 171-177, Eds. D.L. Shankland, R.M. Hollingworth, T. Smith, Jr., Plenum Press, N.Y., 283 pp.

Matsumura, F. and Beeman, R.W. (1976) Biochemical and physiological effects of chlordimeform. Environ. Health Perspect. 14: 71-82.

Murakami, J. and Fukami, J. (1974) Effects of chlorphenamidine and its metabolites on Hela Cells. Bull. Environ. Contam. Toxicol. 11: 184.

Narahashi, T. (1971) Effects of insecticides on excitable tissues. Adv. Insect Physiol. 8: 1-93.

Narahashi, T. (1976) Effects of insecticides on nervous conduction and synaptic transmission. In Insecticide Biochemistry and Physiology, p. 327-352, Ed. by C.F. Wilkinson, Plenum Press, N.Y., 768 pp.

Neumann, R. and Voss, G. (1977). MAO inhibition, an unlikely mode of action for chlordimeform. Experientia 33: 23-24.

Nishimura, K., Fujita, T. and Nakajima, M. (1975) Catabolism of tryptamine by cockroach head enzyme preparation. Pestic. Biochem. Physiol. 5: 557-565.

Pfister, W.R., Chinn, C., Noland, V. and Yim, G.K.W. (1978a) Similar pharmacological actions of chlordimeform and local anesthetics. Pestic. Biochem. Physiol. (In press).

Pfister, W.R., Hollingworth, R.M., and Yim, G.K.W. (1978b) Increased feeding in rats treated with chlordimeform, and related formamidines, a new class of appetite stimulants. Psychopharmacol. (In press).

Robinson, C.P. and Smith, P.W. (1977) Lack of involvement of monoamine oxidase inhibition in the lethality of acute poisoning by chlordimeform. J. Toxicol. Environ. Health 3: 565-568.

Sakai, M. (1967) Studies on the insecticidal action of nereistoxin, 4-*N*,*N*-dimethylamino-1,2-dithiolane. V. Blocking action on the cockroach ganglion. Botyu-Kagaku 32: 21.

Shankland, D.L., Hollingworth, R.M., and Smyth, T. Jr. (1978) Pesticide and Venom Neurotoxicity. Plenum Press, N.Y., 283 pp.

Stone, B.F., Atkinson, P.W. and Knowles, C.O. (1974) Formamidine structure and detachment of the cattle tick *Boophilus microplus*. Pestic. Biochem. Physiol. 4: 407-416.

Wang, C.M., Narahashi, T. and Fukami, J. (1975) Mechanism of neuromuscular block by chlordimeform. Pestic. Biochem. Physiol. 5: 119-125.

Watanabe, H. and Fukami, J. (1977) Stimulating action of chlordimeform and desmethyl-chlordimeform on motor discharges of armyworm, Leucania separata Walker (Lepidoptera: Noctuidae). J. Pestic. Sci. 2: 297-302.

Watanabe, H., Ishibashi, S. and Fukami, J. (1976) Chlordimeform and its metabolites; toxicity and inhibition of acetylcholine-induced contraction of frog muscle. J. Pestic. Sci. 1: 301-305.

Watanabe, H., Tsuda, S. and Fukami, J. (1975) Effects of chlordimeform on Rectus Abdominis muscle of the frog. Pestic. Biochem. Physiol. 5: 150-154.

Yamamoto, D. and Fukami, J. (1976) Effect of chlordimeform on nerve-muscle system of the larva of the waxmoth, Galleria mellonella. J. Insect Physiol. 22: 1511-1516.

Yamasaki, T. and Narahashi, T. (1959) The effects of potassium and sodium ions on the resting and action potentials of the cockroach giant axon. J. Insect Physiol. 3: 146-158.

Yau, E.T., Pfister, W.R., Yim, G.K.W., Mennear, J.H..(1977) Effect of chlordimeform (CDM) on carbohydrate metabolism in mice. Pharmacologist 19: 218.

Yim, G.K.W., Holsapple, M.P., Pfister, W.R., and Hollingworth, R.M. (1978) Prostaglandin synthesis inhibition by formamidine pesticides. (In preparation).

ACTION OF DIELDRIN AND RELATED COMPOUNDS ON SYNAPTIC TRANSMISSION

D. L. Shankland

Department of Entomology

Mississippi State, Mississippi 39762

INTRODUCTION

The symptoms of poisoning by dieldrin (HEOD) and related bicyclodiene insecticides in vertebrates (Gowdey *et al.*, 1952; 1954; Gowdey and Stavraky, 1955; Khairy, 1960; Natoff and Reiff, 1967; Revzin, 1968), and insects (Gianotti *et al.*, 1956; Yamasaki and Narahashi, 1958), and of lindane in insects (Yamasaki and Ishii, 1954; Narahashi, 1971) suggest a similar excitatory action somewhere in the central nervous system by all of these compounds.

Action reported on peripheral nerves has shown inconsistent correlation with symptoms of poisoning, and was not sufficient in any case to account for the ataxia, hyperexcitability and convulsion associated with poisoning (Lalonde and Brown, 1954; Wang *et al.*, 1971). In the American cockroach, axonal membrane in the central nervous system has been eliminated as a site of action, (Ryan and Shankland, 1971) but ganglia have been implicated (Gianotti *et al.*, 1956; Wang *et al.*, 1971). Most significantly, the effects of dieldrin on the giant fiber system in the sixth abdominal ganglion, which is known to be cholinergic (Shankland *et al.*, 1971) are sufficient to suggest that they reflect the neurological basis for symptoms of poisoning.

This report will review the evidence that at least in the cockroach, dieldrin and probably other bicyclodiene insecticides and lindane do, in fact, act on presynaptic terminals of cholinergic synapses to cause excessive and spontaneous release of acetylcholine. Furthermore, I will discuss evidence that dieldrin, and not one of its metabolites, is responsible for this action, and that the action has been shown to correlate with the lethality of a number of the compounds under consideration.

EXPERIMENTAL EVIDENCE ON DIELDRIN ACTION

Synaptic Function and Pharmacology

The critical evidence on the action of dieldrin came from experiments on cholinergic cercal nerve-giant fiber synapses of the American cockroach. These synapses are in the sixth abdominal ganglion and mediate input from sensory structures called cerci on the tip of the abdomen to the ascending giant fibers which have connections to the motor system. This pathway forms part of a mechanism for rapid escape from strong mechanical stimuli from behind. A brief description of the process and pharmacology of cholinergic transmission will make the evidence on dieldrin and related compounds more understandable.

The mechanisms of junctional transmission have been comprehensively treated by several workers (Eccles, 1964; McClennan, 1970; Pappas and Purpura, 1972), and the cercal nerve-giant fiber synapses in the cockroach appear to be consistent in this regard with other kinds of cholinergic synapses. The architecture of synaptic machinery varies from simple contacts between cells to an elaborate multibranched structure forming many contacts with large dendritic trees of one or more post synaptic cells. For the present purpose, the junction can be depicted in simple form, as in Figure 1, with emphasis on those features which have relevant toxicological implications.

The presynaptic terminal has a specialized presynaptic membrane through which acetylcholine (ACh) is released from vesicles into the synaptic cleft. The vesicles have been well established as repositories of ACh, which is released in quantal units ranging from about 1500 to 60,000 or so molecules in synapses in cerebral cortex and neuromuscular junctions respectively. Transmitter release is initiated by a depolarization of the terminal membrane by an invading nerve action potential, often called a spike. Current theory holds that the depolarization leads to inward diffusion of Ca^{2+} which facilitates the attachment of vesicles to active zones on the presynaptic membrane. Subsequently the vesicular membrane and presynaptic membrane fuse, producing a hole through which ACh diffuses into the synaptic cleft. The fusion of the vesicular and presynaptic membranes results in an expansion of the latter, with new vesicles being formed from cisternae which arise from the membrane a short way back from the terminal.

The ACh, once released into the synaptic cleft diffuses as if in free solution, and in so doing may come in contact with

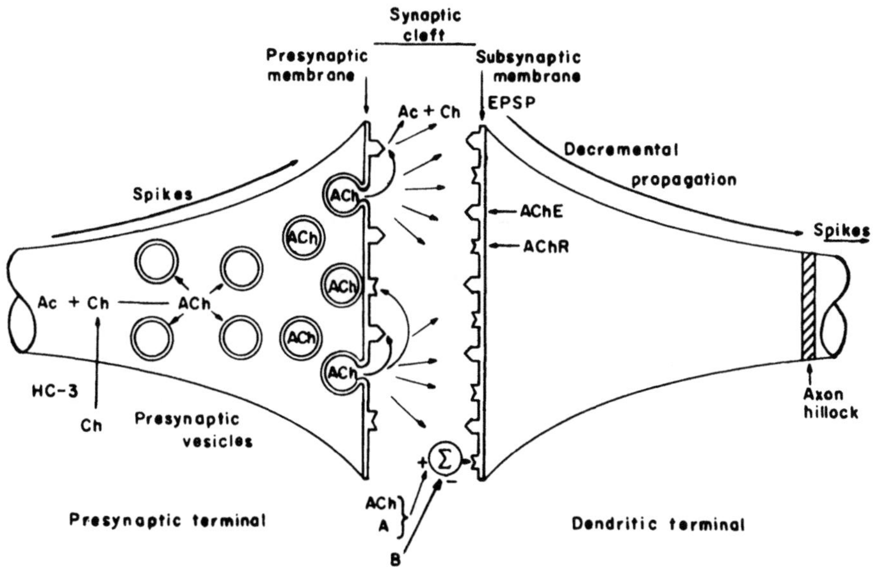

Figure 1. Schematic diagrams of cholinergic synapse illustrating toxicologically relevant elements. See text for details.

either the pre- or subsynaptic membrane. In vertebrate nerve muscle junction it has been estimated that the subsynaptic membrane contains about equal numbers of membrane bound acetylcholinesterase (AChE) and acetylcholine receptor (AChR) molecules, so diffusing molecules of ACh have approximately equal probability of colliding with a receptor or enzyme molecules. Collision with AChR results in activation of the receptor, presumably involving a conformational change in the latter, which leads to selective changes in the ionic permeability of the subsynaptic membrane. In excitatory synapses such as the one under discussion, the change generally involves Na^+ and K^+, and leads to diffusion of these two ions through the subsynaptic membrane to cause a depolarization of the electrical potential of the postsynaptic cell. The amplitude of the depolarization is proportional to the amount of transmitter released, and is called the excitatory postsynaptic potential or EPSP. The EPSP is propagated decrimentally to a specialized region of the cell called the axon hillock, at the base of the axon, or fiber, which communicates with the next cell in the network. At the axon hillock propagated all-or-none nerve action potentials (spikes) are produced at a repetition rate which is proportional to the amplitude of the EPSP. Subsequent to receptor activation, ACh dissociates from AChR. A subsequent collision may occur with an AChE molecule, in which case ACh is hydrolyzed to choline and acetate, neither of which is pharmacologically active on the AChR. The AChE on both pre- and subsynaptic membranes ensures that few molecules of ACh survive more than one or two receptor occupations, and thus prevent undue prolongation of transmitter action.

Some of the choline released by hydrolysis of ACh is taken up by the presynaptic terminal via an active choline transport system, and is used with endogenous acetate in the synthesis of new presynaptic stores of ACh. This reuse of choline is called transmitter turnover, and is important to the maintenance of synaptic transmission.

Drugs and poisons can exert excitatory or depressant action on a synapse in a variety of ways. Depressant drugs might prevent the release of transmitter, or block AChR to prevent activation by ACh. The latter drugs are called antagonists and are represented by B in Figure 1. Excitatory drugs might cause excessive release of transmitter, or inhibit AChE causing prolongation of transmitter action, or mimic the action of ACh on the receptor. Drugs having this last action are called agonists in the general case, represented by A in Figure 1, or cholinomimetics in the case of cholinergic synapses. The carbamate and organophosphate insecticides act mainly by inhibiting AChE, and nicotine and muscarine are cholinomimetics.

HEOD and related compounds are known to have excitatory effects on ganglia as described earlier. In order to determine which of the possible actions could account for this, the synapse can be manipulated pharmacologically to deactivate certain parts of the machinery and thus localize the site of action of an active compound. For example, the drug hemicholinium-3 (HC-3 in Figure 1) blocks choline transport, preventing transmitter turnover. Prolonged presynaptic stimulation in the presence of this drug leads to depletion of presynaptic stores of transmitter. Drugs which act by causing excessive release of transmitter or by inhibiting AChE have no excitatory effects on synapses so treated. Drugs which act as cholinomimetics are still active, however.

Alternatively, since Ca^{2+} is essential for transmitter release, and because Mg^{2+} antagonizes Ca^{2+} in this action, drugs which act by enhancing transmitter release might be antagonized by low Ca^{2+} or high Mg^{2+}. On the other hand, certain poisons such as black widow spider venom cause enhanced transmitter release by a Ca^{2+} independent action, and are not prevented from acting by low Ca^{2+} (Ing Kao and price, 1976). Drugs and poisons which act by AChE inhibition have also consistently shown such inhibitory action in vitro (Koelle, 1963).

Therefore, of the three main types of excitatory action, evidence of several kinds can be used to determine which is exerted by an excitatory drug or poison.

The Action of Dieldrin on Synaptic Transmission

The ventral nerve cord with the cercal nerves still attached to the sixth abdominal ganglion can be completely removed from the body of the cockroach without destroying the functional integrity of the cercal nerve-giant fiber synapses. If placed in a trough irrigated with physiological saline solution it will maintain its vitality for two or three days. In the experimental work described here an outer sheath which covers the sixth abdominal ganglion and which serves as a barrier to penetration of solutes was removed. This facilitates the penetration of dieldrin and other drugs to the synapses.

The functional state of the cercal nerve-giant fiber synapses was monitored by conventional electrophysiological techniques (Shankland et al., 1971). Stimulating electrodes were connected to both cercal nerves and a recording electrode to the ventral connectives anterior to the sixth abdominal ganglion. This provides for close observation of postsynaptic response to controlled presynaptic input.

Figure 2 illustrates the normal and pathological responses of the giant fibers to stimulation of the cercal nerves. The upper trace shows a compound action potential made up of a few small and one large deflection. The latter represents the synchronous firing of perhaps eight or ten giant fibers. The lower trace shows the response to the same kind of stimulus but after the ganglion had been exposed to 1×10^{-6} dieldrin for 2 hours.

Normal and poisoned nerves also show various levels of spontaneous activity, i.e. electrical activity which occurs without stimulation. Figure 3 illustrates some representative oscillograms. The top trace shows activity in a normal nerve cord. A few small spikes can be seen above the amplifier noise, and occasionally, larger spikes will occur. The second trace shows considerably more small fiber activity, and this will appear in response to treatment with dieldrin and other poisons including carbamate and organophosphate insecticides. However, the transmitter in these small fiber systems is not known. It is only the giant fiber system which has been shown to be cholinergic. The third trace shows patent giant fiber activity which appears after treatment with dieldrin. In quantitating the effects of dieldrin on spontaneous activity in isolated nerve cords, Shankland and Schroeder (1971) used a gated counting circuit which counted pulses which were larger than an adjustable threshold. Each nerve cord was standardized by determining the voltage of responses to supramaximal cercal nerve stimulation, and setting the threshold at 10%

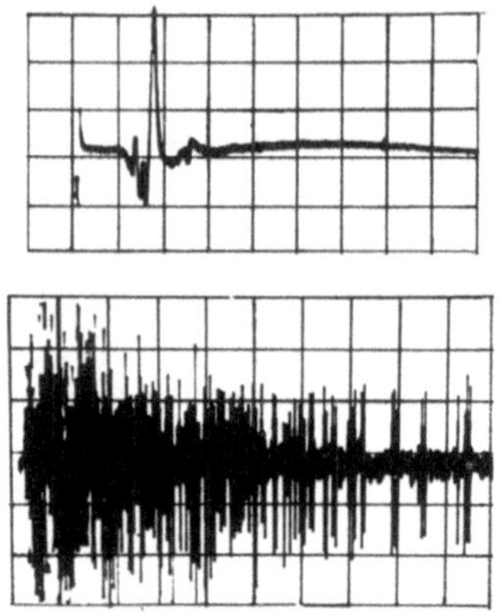

Figure 2. Responses of giant fibers to cercal nerve stimulation in normal (upper) and dieldrin poisoned (lower) nerve cords.

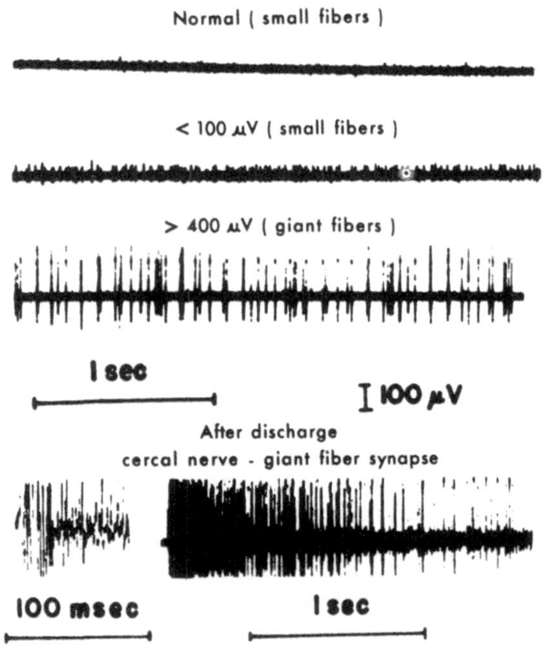

Figure 3. Representative oscillograms of spontaneous activity (upper three traces) and synaptic after discharge (bottom two traces) in cockroach nerve cords.

of that value. This was generally about 70 μV, and exceeded the usual maximal voltage of small fiber activity by about 20 μV. The bottom two traces in Figure 3 simply show the extent of after discharge occurring early (left) and late (right) in dieldrin poisoning. Cords were treated by irrigating them with physiological saline solutions containing dispersions of dieldrin. The latter were made by squirting ethanolic solutions of dieldrin through a 30 gauge hypodermic needle into the saline solution. The concentration of ethanol in these mixtures was 2% v/v.

Table 1 summarizes the effects of dieldrin alone and in combination with other drugs on spontaneous giant fiber activity. Cords treated with 1×10^{-6} M dieldrin did not differ from controls during the first hour of treatment. During the succeeding three hours, however there was a significant increase in the number of spontaneous spikes produced.

Cords treated with 5 mM Mg^{2+} produced no spikes of threshold value. Cords treated with a combination of 5 mM Mg^{2+} and 1×10^{-6} M dieldrin produced more spikes than controls but fewer than cords treated with dieldrin alone. Mg^{2+} at 10 mM completely antagonized

Table 1. Effects of HEOD alone and in combination with other drugs on spontaneous activity in giant fibers of the American cockroach.

Treatment	No. of experiments	Period of observation hours	Spontaneous activity spikes/hour (means)
Controls	13	1st-4th	156 a*
HEOD 10^{-6}M	13	1st	108 a
HEOD 10^{-6}M	13	2nd-4th	16310 b
Mg^{2+} 5 mM	9	1st-4th	0
HEOD 10^{-6}M Mg^{2+} 5 mM	13	2nd-4th	2569 c
HEOD 10^{-6}M Mg^{2+} 10 mM	3	2nd-4th	0
HEOD 10^{-6}M HC-3 10^{-4}M	4	2nd-4th	0

*Numbers followed by different letters are significantly different at P = 0.05.

the effects of 1 x 10^{-6} M dieldrin. Thus, Mg^{2+} shows a dose dependent graded antagonistic action against dieldrin. Dieldrin was completely without effect on ganglia that had been depleted of presynaptic stores of ACh by prolonged presynaptic stimulation in the presence of 1 x 10^{-4} M hemicholinium-3 (HC-3). All of these ganglia were responsive to 1 x 10^{-5} M nicotine, which indicates that the postsynaptic elements were functional.

The results of these observations are consistent with two possible modes of excitatory action: (1) enhanced release of transmitter, and (2) anticholinesterase activity. Cholinomimetic action is eliminated by the fact that dieldrin was not excitatory on ganglia which were responsive to this kind of action by nicotine. The possibility that anticholinesterase activity could account for the excitatory effects was eliminated by an in vitro assay of cockroach head AChE activity in presence of 1 x 10^{-5} M dieldrin. No measurable inhibitory action was found.

The conclusion from these observations is that dieldrin or a metabolite produces the observed effects on isolated nerve cords through calcium dependent enhancement of transmitter release.

Dieldrin and Isomeric Aldrin Diols: Neurotoxicity and Insecticidal Action

Dieldrin poisoning of isolated nerve cords and intact cockroaches has been characterized by a delay in the appearance of symptoms (Wang *et al.*, 1971; Shankland and Schroeder, 1973; and Schroeder *et al.*, 1977). This delay, ranging from 30 to 90 minutes or more, combined with the finding that 6,7-*trans* dihydroaldrindiol (t-AD), a metabolite of dieldrin acted faster than dieldrin prompted Wang *et al.* (1971) to suggest that the diol and not dieldrin is responsible for the toxic action. The picture was further complicated by the finding that 6,7-*cis* dihydroaldrindiol (c-AD) is also a major metabolite of dieldrin and may be an intermediate between dieldrin and t-AD (Nelson and Matsumura, 1973; Matthews and McKinney, 1974). Schroeder *et al.* (1977) studied the neurotoxic and insecticidal effects of dieldrin and the two isomeric aldrindiols in an attempt to resolve this issue.

The results of a simple assessment of the toxicities of the three compounds are summarized in Table 2. Doses were administered to adult male American cockroaches by injection of alcoholic solutions of the three respective compounds. Mortalities were determined 7 days after treatment. An approximate LD_{50} for dieldrin was 0.45 mg/kg. Doses of t-AD and c-AD 100 fold greater produced zero to slight mortality, while doses 255 times greater produced less than 80% mortality.

Table 2. Toxicity of HEOD and aldrin diols injected into adult American cockroaches.

Treatment	No tested	Dose (mg/kg)	Mortality after 7 days (%)
Control (EtOH)	32	0	6
HEOD	25	0.45	52
	10	2.0	100
t-AD	20	45	0
	10	115	70
c-AD	25	45	16
	15	115	53

The effects of these compounds on isolated nerve cords are summarized in Table 3. Effects on spontaneous activity were assessed according to two criteria: (1) the amplitude of spikes produced spontaneously, and (2) the frequency with which bursts of spikes occurred. Spikes greater than 100 μV represented patent giant fiber activity. Dieldrin poisoned nerves generally show bursty activity like that shown in the third trace in Figure 2. Effects on after discharge in response to cercal nerve stimulation were also assessed according to two criteria: (1) amplitude of spikes involved, and (2) duration of after discharge. These assessments were made 3 hours after treatment began.

Dieldrin at 1×10^{-6} M consistently caused spontaneous spikes of greater than 400 μV to appear in bursts of more than 5/sec., and to cause after discharge of large spikes lasting well over 300 msec, and in some cases for 2 sec. Even at concentrations 100 fold greater than that the effects of both isomeric diols were markedly less on spontaneous activity and after discharge.

Thus, the two diols were considerably less toxic than dieldrin to intact cockroaches and isolated nerve cords.

It might be argued that the diols were not toxic under the conditions used because they did not penetrate to a site of action in the nervous system. Such an argument would have to hold also, of course, that dieldrin did penetrate, and that the metabolic conversion to one or the other or both of the diols occurred at the site of action. To explore this possibility, Schroeder et al. (1977) poisoned cockroaches with injected doses of 40 μg/cockroach of dieldrin, t-AD and c-AD respectively. The dieldrin poisoned

Table 3. Spontaneous activity and after discharge in poisoned isolated nerve cords of the American cockroach.

Treatment	No. of experiments	Spontaneous activity amplitude freq. of bursts	After discharge amplitude duration
HEOD 10^{-6} M	13	> 400μ V > 5/sec	> 400 μV > 300 msec
t-AD, c-AD 10^{-6} M	10	< 100 μV < 1/sec	< 400 μV < 100 msec
t-AD 10^{-5} M	5	< 100 μV < 1/sec	< 400 μV < 100 msec
c-AD 10^{-5} M	5	100 - < 400 μV < 1/sec	< 400 μV < 100 msec
t-AD, c-AD 10^{-4} M	10	100 - < 400 μV < 1/sec	> 400 μV < 100 msec

cockroaches were sacrificed when they became prostrate, which event required from 5 to 16 hours. The nerve cords were excised, rinsed quickly in cold ether, and stored in sealed vials in ether at -18°C to await further processing. Previous studies had shown that few of the aldrindiol poisoned cockroaches would become prostrate at any time, so they were sacrificed and processed between 11 and 16 hours after poisoning.

To analyze for the presence of the subject compounds in the cords, they were homogenized in tissue grinders in the ether in which they were stored, and the homogenates extracted with hexane. The hexane extracts were reduced to dryness by mild heat in the presence of Na_2SO_4, and the dry residue silylated to render the diols detectable by an electron capture detector head on a gas chromatograph. The silylated residue was taken up in hexane and held for G.C. analysis.

Figure 4 illustrates representative chromatograms. A is a silylated standard containing 1 ng each of dieldrin (peak 2), t-AD (peak 3) and c-AD (peak 4). B is a silylated tissue blank which shows a small peak (1) which has a retention time close to that of dieldrin. C is an extract of nerve cords from dieldrin poisoned cockroaches showing the presence of all three subject compounds (peaks 2, 3, and 4). D is an extract of cords from t-AD poisoned cockroaches, showing the presence of t-AD only. Finally, E is an extract of cords from c-AD poisoned animals which contains c-AD

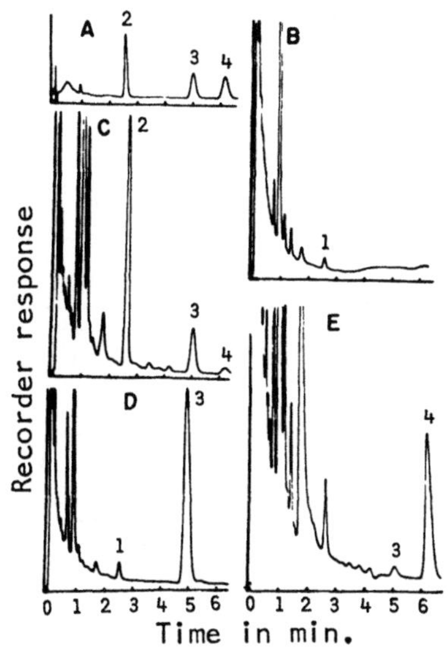

Figure 4. Gas chromatograms of: A. standard solution of HEOD (peak 2), t-AD (peak 3) and c-AD (peak 4). B. tissue blank showing peak 1 which is close to the retention time for HEOD. C, D, and E extracts from HEOD, t-AD, and c-AD respectively poisoned cockroaches.

and a trace of t-AD.

Table 4 summarizes the results of this experiment. Approximately 0.22% of the injected dose of dieldrin was recovered as dieldrin, about 0.04% as t-AD and a trace as c-AD. About 0.19% of the injected dose of t-AD was recovered as t-AD, and none as the other two compounds. Finally, 0.13% of the dose of c-AD was recovered as that compound, with less than 0.025% as t-AD and none as dieldrin.

These results indicate that all of these compounds penetrate the nervous system in approximately equal amounts, and that the low toxicity of the two diols or the relatively modest symptomatic expression of poisoning by them can not be accounted for by a failure to penetrate the nervous system. It might be argued that some barrier within the neuropil, such as that suggested by O'Brien (1967) protects synapses from the diols but not from dieldrin. If such is the case in vivo, it is not the case in vitro, as evidenced by the direct effects of the diols on isolated nerve cords described by Wang et al. (1971) and Schroeder et al. (1977). Joy (1977) working with central sensory and motor pathways in the cat also

Table 4. Recovery of HEOD, t-AD, and c-AD from nerve cords of cockroaches poisoned by these respective compounds.

Treatment (40 μg/insect injected)	Compounds Recovered ng/cord ± SD % of injected dose		
	HEOD	t-AD	c-AD
HEOD	88 ± 10.6 0.22	17 ± 14.3 0.04	< 6 < 0.002
t-AD	ND* 0	73 ± 11.3 0.19	ND 0
c-AD	ND 0	< 10 < 0.025	52 ± 28 0.13

*ND: not detectable.

concluded that dieldrin is directly active on the vertebrate CNS to produce convulsive episodes, while t-AD is without effect in this regard even at relatively high doses.

The results described here are also compatible with the notion that both t-AD and c-AD are metabolites of dieldrin, and that c-AD may be an intermediate between dieldrin and t-AD (Nelson and Matsumura, 1973; Matthews and McKinney, 1974). All three compounds were recovered from dieldrin poisoned animals, both diols from c-AD poisoned ones, and only t-AD from those poisoned with that compound which appears to be an end product.

The findings by Uchida *et al.* (1975 a, b) with lindane on the cockroach sixth abdominal ganglion parallel those with dieldrin on the same preparation by Shankland and Schroeder (1971) and Schroeder *et al.* (1977), suggesting a similar mode of action by that compound.

Ackermans et al. (1973, 1974) showed that t-AD caused an increase in the rate of transmitter release from presynaptic terminals of frog neuromuscular junction and that dieldrin was without effect. This action appears to be exactly comparable to that described for dieldrin in the cockroach, but in frog muscle t-AD is evidently the more active compound. The bulk of the evidence discussed here has been on cockroach, with some corroborative evidence from the cat. Since there is evidence for distinct pharmacological differences among cholinergic junctions in various animals and various parts of the same animal (Hubbard and Quastel, 1973), it would be reasonable to expect that there might be different orders of effectiveness of the various bicyclodienes depending on the specific junction concerned.

However, the evidence to date still favors the hypothesis that the symptoms of dieldrin poisoning are attributable to direct action on presynaptic cholinergic junctions in the CNS, and that the action is exerted through enhanced transmitter release. Although there is a good correlation in the cockroach between symptom production and lethal effects it can not be concluded that death is attributable directly to the synaptic effects of active compounds. Secondary effects of this action might well be more directly related to death, particularly in insects which are not so vitally dependent as higher animals on nervous integrity for respiratory function. The delays in appearance of symptoms in isolated nerve cords and intact cockroaches could well be due to slow penetration because of very effective penetration barriers in the CNS (O'Brien, 1967). Such delays are probably not important in considering differences among isomeric aldrindiols and dieldrin, because even at lethal doses it takes hours or days for symptoms to appear.

REFERENCES

Ackermans, L. M. A., van der Zalm, J. M., and van den Bercken, J., 1973, Is aldrin-transdiol the active form of the insecticide dieldrin, Arch. Int. Pharmacodyn. Ther. 206, 363.

Ackermans, L. M. A., van den Bercken, J., van der Zalm, J. M., and Stratten, H. W. M., 1974, Effects of dieldrin (HEOD) and some of its metabolites on synaptic transmission in the frog motor end-plate, Pestic. Biochem. Physiol. 4, 313.

Eccles, J. C., 1964, The Physiology of Synapses, Springer, New York.

Gianotti, O., Metcalf, R. L. and March, R. B., 1956, The mode of action of aldrin and dieldrin Periplaneta americana, Ann. Entomol. Soc. Amer. 49, 588.

Gowdey, C. W., Graham, A. R., Sequin, J. J., Stavraky, G. W. and Waud, R. A., 1952, A study of pharmacological properties of the insecticide aldrin, Can. J. Med. Sci. 30, 520.

Gowdey, C. W., Graham, A. R., Sequin, J. J. and Stavraky, G. W., 1954, The pharmacological properties of the insecticide dieldrin, Can. J. Biochem. Physiol. 32, 498.

Gowdey, C. W. and Stavraky, G. W., 1955, A study of the autonomic manifestations seen in acute aldrin and dieldrin poisoning, Can. J. Biochem. Physiol. 33, 272.

Hubbard, J. L., and Quastel, D. M. S., 1973, Micropharmacology of vertebrate neuromuscular transmission, Annu. Rev. Pharmacol. 13, 199.

Ing Kao, D. P. D. and Price, D. L., 1976, Botulinum toxin: mechanism of presynaptic blockade, Science 193, 1256.

Joy, R. M., 1977, Contrasting action of dieldrin and aldrin-transdiol, its metabolite, on cat CNS function. Toxicol. Appl. Pharmacol. 42, 137.

Khairy, M., 1960, Effects of chronic dieldrin ingestion on the muscular efficiency of rats, Brit. J. Ind. Med. 17, 146.

Koelle, G. B. (ed.), 1963, Cholinesterases, Handb. Exp. Pharmacol. Vol. 15, Springer Verlag.

Lalonde, D. and Brown, A. W. A., 1954, The effects of insecticides on the action potentials of insect nerve, Can. J. Zool. 32, 74.

Matthews, H. B. and McKinney, J. D., 1974, Dieldrin metabolism to cis-dihydroaldrindiol and epimerization of cis- to trans-dihydroaldrindiol by rat liver microsomes, Drug Metabol. Dispos. 2, 333.

McClennan, H., 1970, Synaptic Transmission, 2nd ed., Saunders, Philadelphia.

Narahashi, T., 1971, Effects of insecticides on excitable tissues, Adv. Insect Physiol. 8, 1.

Natoff, I. L. and Reiff, B., 1967, The effect of dieldrin (HEOD) on chronaxie and convulsion thresholds in rats and mice, Brit. J. Pharmacol. Chemother. 31, 197.

Nelson, J. O. and Matsumura, F., 1973, Dieldrin (HEOD) metabolism in cockroaches and house flies, Arch. Environ. Contam. Toxicol. 1, 224.

O'Brien, R. D., 1967, Barrier systems in insect ganglia and their implications for toxicology, Fed. Proc. 26, 1056.

Pappas, G. D. and Purpura, D. P. (eds.), 1972, Structure and Function of Synapses, Raven Press, New York.

Revzin, A. M., 1968, Effects of endrin on telencephalic function in the pigeon, Toxicol. Appl. Pharmacol. 9, 75.

Ryan, W. H. and Shankland, D. L., 1971, Synergistic action of cyclodiene insecticides with DDT on the membrane of giant axons of the American cockroach, Periplaneta americana (L.), Life Sci. 10, Part I, 193.

Schroeder, M. E., Shankland, D. L. and Hollingworth, R. M., 1977, The effects of dieldrin and isomeric aldrin diols on synaptic transmission in the American cockroach and their relevance to the dieldrin poisoning syndrome, Pestic. Biochem. Physiol. 7, 403.

Shankland, D. L., Rose, J. A. and Donninger, C., 1971, The cholinergic nature of the cercal nerve giant fiber synapse in the sixth abdominal ganglion of the American cockroach, Periplaneta americana (L.), J. Neurobiol. 2, 247.

Shankland, D. L. and Schroeder, M. E., 1973, Pharmacological evidence for a discrete neurotoxic action of dieldrin (HEOD) in the American cockroach, Periplaneta americana (L.), Pestic. Biochem. Physiol. 3, 77-86.

Uchida, M., Irie, Y., Fujita, T. and Nakajima, M., 1975a, Effects of nereistoxin on the neuroexcitatory action of insecticides, Pestic. Biochem. Physiol. 5, 253.

Uchida, M., Irie, Y., Kurihara, N., Fujita, T. and Nakajima, M., 1975b, The neuroexcitatory, convulsive and lethal effects of lindane analogs on Periplaneta americana (L.), Pestic. Biochem. Physiol. 5, 258.

Wang, C. M., Narahashi, T. and Yamada, M., 1971, The neurotoxic action of dieldrin and its derivatives in the cockroach, Pestic. Biochem. Physiol. 1, 84.

Yamasaki, T. and Ishii, T.,1954, Studies on the mechanism of action of insecticides (x). Nervous activity as a factor of development of γ-BHC symptoms in the cockroach, Botyu-Kagaku 19, 106.

Yamasaki, T. and Narahashi, T., 1958, Nervous activity as a factor of development of dieldrin symptoms in the cockroach, Botyu-Kagaku 23, 47.

EFFECTS OF PARATHION ON THE MAMMALIAN NERVOUS SYSTEM

D. E. Woolley, J. R. Chernobieff, and L. W. Reiter

Departments of Animal Physiology and Environmental Toxicology
University of California, Davis, California 95616

SYMPTOMS OF ACUTE PARATHION EXPOSURE

Parathion is a widely used, highly toxic organophosphate insecticide. It is oxidatively desulfurated *in vivo* to its active metabolite, paraoxon, by liver and extrahepatic tissue in a number of mammalian and other vertebrate species (Hitchcock and Murphy, 1971; Poore and Neal, 1972; Whitehouse and Ecobichon, 1975). Paraoxon is a powerful inhibitor of acetylcholinesterase (AChE) (for example, see reviews by O'Brien, 1967; Corbett, 1974; Main, 1976). For this reason, paraoxon is believed to produce hyperactivity at cholinergic synapses and junctions by reducing the rate of breakdown of acetylcholine (ACh), which in turn results in a buildup of ACh (DuBois *et al.*, 1949; Stewart, 1952; Wecker *et al.*, 1977). The fact that the symptomatology results from excess ACh, rather than from the inhibited AChE *per se*, was demonstrated by Schaumann (1960), who produced nearly 100 percent inhibition of AChE in mice with a number of different anticholinesterases, including parathion, and was able to save the animals with high doses of atropine to protect against the accumulating ACh.

The exact symptomatology following parathion poisoning may vary somewhat in different mammalian species, depending on the relative stimulation of muscarinic versus nicotinic and central versus peripheral ACh receptors (AChR). Autonomic ganglionic synapses and neural-skeletal muscle junctions are nicotinic, whereas parasympathetic postganglionic synapses are muscarinic. Both nicotinic and muscarinic AChR's, as well as an intermediate type, occur in brain or spinal cord (Schleifer and Eldefrawi, 1974; Snyder *et al.*, 1975). Peripheral muscarinic symptoms of parathion poisoning include salivation, lacrimation, diarrhea, constricted pupils, bronchostriction, pulmonary edema, and excessive pulmonary secretion. Nicotinic symptoms include muscular twitching, fasciculation, and generalized muscular weakness. Tremoring and convulsions are due primarily to effects on central synapses. Death is due to respiratory failure because of effects on central respiratory neurons and/or paralysis of respiratory muscles aggravated by the effects on the lungs and bronchi mentioned above and by bradycardia (Doull, 1976). Failure at the respiratory neuromuscular junctions is due to the nicotinic effects of ACh whereby low doses stimulate and high doses block junctional transmission because of persistent depolarization of the postsynaptic membrane (Dolivia and Koelle, 1970).

It has been our experience that mice and rats demonstrate more readily than do monkeys the parasympathetic effects, such as lacrimation and salivation, of parathion poisoning (Reiter et al., 1973, 1975). In fact, we have never observed lacrimation and salivation in any monkey after administration of sublethal doses of parathion; pupillary constriction was observed once in a monkey given a near lethal (20 mg/kg) dose of parathion subcutaneously (sc) (although therapy later saved the animal). In a monkey which died within an hour of receiving a lethal oral dose of parathion (20 mg/kg), rapid loss of motor control, prostration, marked muscular fasciculations, profuse salivation, followed by respiratory arrest and continued beating of the heart were noted (Serrone et al., 1969).

Another example of the relative differences caused by parathion exposure in various mammals is the observation that atropine raised the lethal dose of paraoxon 50-fold in monkeys, 10-fold in dogs, and by only 2- to 3-fold in mice, rabbits and cats. The lethal dose of paraoxon alone was similar in the same species (DeCandole and McPhail, 1957). These investigators pointed out that the relative ease or difficulty with which atropine protected against paraoxon poisoning corresponded closely with the relative importance of central versus peripheral mechanisms in respiratory failure. In the monkey failure of respiration is due almost entirely to central effects and these are readily treated by atropine because most central cholinergic synapses are of the muscarinic type. On the other hand peripheral neuromuscular block is the predominant cause of respiratory failure in the rabbit, and atropine is not effective at this synapse.

TIME COURSE OF AChE INHIBITION AND RECOVERY

The steps involved in the inhibition of AChE by paraoxon have been described in detail (for example, see O'Brien, 1967; Corbett, 1974; and Main, 1976). Briefly, paraoxon combines with AChE because of attraction between the OH^- of the seryl residue at the active esteratic site of AChE and the diethylphosphate group of paraoxon. The diethylphosphate group is transferred from paraoxon to AChE and forms a covalent bond with the seryl residue. The rest of the paraoxon molecule is hydrolyzed off as p-nitrophenol and leaves to be excreted. Thus, p-nitrophenol is referred to as the "leaving" group. As long as the active site on AChE is phosphorylated, it is inhibited. Recovery of AChE activity can be accomplished either by resynthesis of AChE or by dephosphorylation of AChE before the leaving group has left, i.e., before "aging" of the inhibited enzyme occurs. Dephosphorylation may occur either spontaneously or through the use of oximes which have a strong attraction for the diethylphosphate group.

Some controversy still exists concerning the rate of recovery of brain AChE activity after parathion administration in a number of mammalian species. In one of the first studies on this subject 5 mg/kg parathion was administered intraperitoneally (ip) to adult male rats (DuBois et al., 1949). Brain AChE activity was measured using a manometric procedure and was found to be reduced to 5.8% of control activity by 30 minutes, and to have returned to control levels by 4 hours after administration. On the other hand Davison (1953, 1955) also used a manometric technique and adult male rats, but observed that brain AChE activity took far longer to recover. Brain AChE activity was 10-20% of normal immediately after intravenous injection of paraoxon (0.25 mg/kg), was 30-40% of normal after 24 hours, and was about 75% of control levels 14 days after administration. Davison (1953, 1955) used a number of different substrates and inhibitors to distinguish between "true" AChE activity, which specifically hydrolyzes ACh, and "pseudocholinesterase" (ChE), which demonstrates far less substrate specificity and is not believed to be involved in synaptic and junctional transmission. Brain ChE activity had recovered to 40-50% of control levels 1-4 days after administration, whereas ChE in serum, jejunum, heart and diaphragm recovered in about one day. Thus, ChE activity in brain and non-neural tissue showed different recovery patterns, but in all tissues recovered faster than did brain AChE activity. The much longer time required for recovery of brain cholinesterase activity in the studies by Davison than in the study by DuBois et al., is difficult to explain. Although Davison distinguished between AChE and ChE activities and DuBois et al. did not, this would not seem to be a sufficient explanation, because the latter did use ACh as a substrate in the analysis and ACh is hydrolyzed preferentially by AChE. Also, the level of ChE activity in brain is very low and so should not have contributed substantially to the hydrolysis of ACh.

Several other studies indicate that after administration of either parathion or paraoxon in rats or mice, full recovery of inhibited brain AChE activity requires at least a week or more. In one study (Wecker et al., 1977) a single administration of 0.75

mg/kg paraoxon ip to adult male rats resulted in maximal inhibition of brain AChE activity to 20% of control levels in 15 minutes. No recovery occurred during the first 2 hours but recovery to 50% of control levels had occurred by 24 hours. Full recovery required 12 days from the time of the injection.

Two studies in mice demonstrate that after administration of either parathion or paraoxon, brain AChE activity remains inhibited for several days. In the first study 0.7 mg/kg paraoxon was injected sc and brain AChE activity was 25% of control levels one hour later and was 55% of control activity five days later (Kewitz and Nachmansohn, 1957).

In another study (Benke and Murphy, 1974) adult male mice were injected ip with 13 mg/kg parathion, and AChE activity in brain and muscle (diaphragm) was measured using the Ellman et al., (1961) colorimetric procedure. This method utilized ASCh or BuSCh (butyrylthiocholine) as substrates, and differential enzyme inhibitors to distinguish between ChE and AChE activity. It was observed that maximal inhibition in brain and muscle was reached at about two hours, although it occurred sooner, was more marked, and lasted longer in muscle than in brain. By 3 days AChE activity was back to 80% of control levels in muscle, but to only 60% of control activity in brain. By comparison the onset of AChE inhibition and the time required for recovery was much longer in fish than in mice, presumably because of the slower metabolic activation or lesser sensitivity of AChE to paraoxon in fish than in mice (Benke and Murphy, 1974).

Time for recovery of inhibited AChE activity may vary in different brain areas. After AChE inhibition following intramuscular injection of DFP (diisopropyl fluorophosphate) in the rat, the half-time for recovery was 6 days in the septum and 17 days in the hippocampus, as determined by both histochemical and biochemical procedures (Chippendale et al., 1974). This difference in recovery times for the two brain areas corresponds with the times required for the resynthesized enzymes to be carried by the slow component of axoplasmic flow from cholinergic neuronal cell bodies in the midbrain to the septum, and from the latter to the hippocampus via the septohippocampal pathway (see Shute and Lewis, 1967; Lewis and Shute, 1967 for description of cholinergic pathways in brain).

AChE may exist as several isozymes with different turnover rates (Davis and Agranoff, 1968). Furthermore, different isozymes predominate in neural versus non-neural tissues. Following inhibition of retinal AChE with DFP in adult male rats, a differential rate of recovery of the isozymes was noted (Davis and Agranoff, 1968). One isozyme demonstrated an unusually short half-life of about 3 hours, which, perhaps not coincidentally, corresponds with the rapid recovery from inhibition after parathion noted by DuBois et al. (1949). Other isozymes had a half-life of 1-2 weeks, which corresponds with the recovery noted by Davison (1953, 1955) and Wecker et al. (1977) in the rat, and by Kewitz and Nachmansohn (1957) and Benke and Murphy (1974) in the mouse.

Vijayan and Brownson (1975) used polyacrylamide gel electrophoresis to determine the rate of recovery of AChE isozymes after ip administration of 2.5 or 1.25 mg/kg of parathion in 15- and 30-day-old rats. Brain isozymes were most inhibited 2 hours after administration and had recovered by 24 hours, whereas skeletal muscle isozymes were most inhibited 4 hours after administration and were still markedly inhibited after 24 hours. Thus, either the isozymes which demonstrated a 1-2 week turnover rate in the work of Davis and Agranoff (1968) were not evident in the isozymes studied by Vijayan and Brownson (1975) or rapid reactivation, i.e., dephosphorylation, of these inhibited isozymes occurred after parathion administration in the latter study, but not after DFP administration in the former study.

Brownson et al. (1977) summarized histochemical and biochemical studies in their laboratory which demonstrated that after parathion administration in 15- and 30-day-old rats inhibited AChE activity in cell bodies and synapses of brain stem neurons and in the myoneural junction had recovered by 24 hours.

Thus, some studies have shown that inhibited brain AChE may recover to control levels 4 to 24 hours after administration of parathion (DuBois et al., 1949; Vijayan and Brownson, 1975; Brownson et al., 1977), whereas other studies indicate that 2-4 weeks are required for full recovery of brain AChE activity in rats (Davison, 1953, 1955; Wecker et al., 1977) and mice (Kewitz and Nachmansohn, 1957; Benke and Murphy, 1974). The discrepancies between these studies cannot be readily resolved.

Work in our laboratory has measured AChE activity in brain and other tissues of rats with time after both ip and sc injection of parathion in an effort to resolve some of these conflicts (Woolley, 1976, 1978; Chernobieff and Woolley, unpublished observations). AChE and ChE activities were determined using a colorimetric procedure, ASCh or BuSCh as the substrate, and specific enzyme inhibitors (Ellman et al., 1961). In the first study the effects of ip and sc administration of 2.0 mg/kg parathion dissolved in polyethylene glycol (carbowax 300) on AChE activity in three brain areas, spinal cord and blood and on ChE activity in plasma were determined (Figure 1). Tissue samples were analyzed at 1, 4, 9 and 24 hours and at 3 and 7 days after parathion administration. Although the effects of ip versus sc administration were similar, ip administration tended to cause the more rapid and more pronounced AChE and ChE inhibition, especially in blood and plasma.

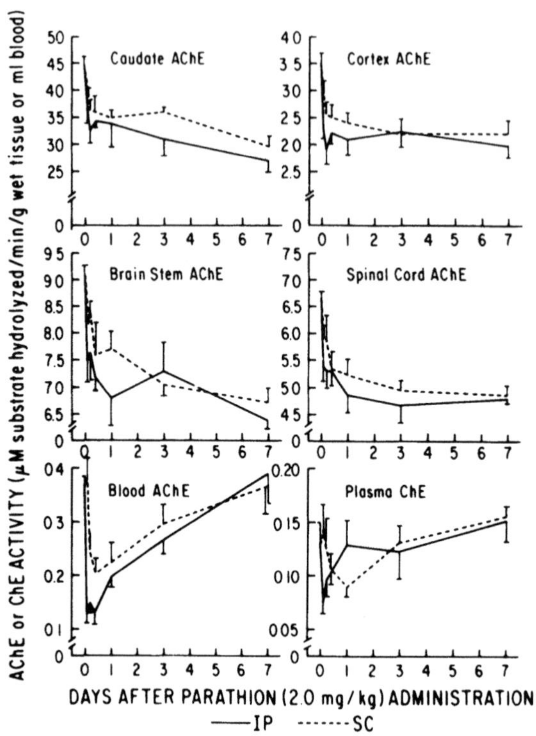

Figure 1. Effects of ip and sc administration of parathion in the rat on AChE activities in the brain (caudate nucleus, cerebral cortex, brain stem), spinal cord and blood and on ChE activity in plasma. Each point represents the mean, and the vertical bracketed lines represent the standard errors about the mean for 6 animals.

Peak inhibition of blood AChE activity occurred at 9 hours post parathion and was 36% of control levels after ip injection and 54% of control activity after sc administration. Blood AChE activity levels were 76% of controls after 3 days and had fully recovered a week later. Brain and spinal cord AChE activity levels were maximally inhibited 9-24 hours post parathion and, in contrast with activity levels in blood, demonstrated no recovery whatsoever during the remaining week of this experiment. The pattern of AChE inhibition did not differ substantially between the caudate nucleus, cerebral cortex, brain stem (midbrain, pons and medulla) and spinal cord, even though control activity levels were about 13-fold higher in the caudate than in the cortex. Control AChE activity levels, measured as μM substrate hydrolyzed/min/g wet tissue or ml whole blood, were as follows: caudate 44.5, brain stem 9.1, spinal cord 6.6, cortex 3.5, whole blood 0.36, and erythrocytes 0.87. Control ChE activity in plasma was only 0.13 and showed greater variability than did blood AChE activity.

Thus, this first experiment did not support some previous observations that inhibited brain AChE activity recovered within 24 hours after parathion administration.

A second study was conducted to determine the time course of recovery of the inhibited brain AChE activity after administration of parathion (2.25 mg/kg ip) in the rat. The pattern of inhibition and recovery of AChE activity was similar in the three brain areas (caudate, cortex and brain stem) studied (Figure 2). Brain AChE activity in all areas was slightly more inhibited at one week than at six hours after parathion administration. Brain stem AChE activity had returned to normal (94% of control activity) two weeks after parathion administration, whereas AChE activity in caudate and cortex was still significantly depressed at this time, but had recovered fully by four weeks. A more rapid recovery of AChE activity in brain stem than in cortex and caudate would be consistent with the concept that AChE is synthesized in cell bodies of cholinergic neurons in the midbrain portion of the brain stem and travels via slow axoplasmic flow to cortex and caudate (see above). AChE activity in blood was relatively more inhibited than was AChE activity in brain, but recovered sooner. Blood AChE activity was back to normal in one week, as also observed in Figure 1. ChE activity was inhibited more than AChE activity in brain stem at six hours and one week after parathion administration; both were back to normal two weeks post parathion administration. ChE activity recovered faster in plasma than in brain stem, an observation which agrees with findings by Davison (1953, 1955) that ChE activity in non-neural tissue recovers faster than ChE activity in brain.

The preceding study clearly agrees with previous observations that full recovery of inhibited brain AChE activity requires 2-4 weeks (Davison, 1953, 1955; Wecker _et al._, 1977), but not with those that recovery occurs in 24 hours (DuBois _et al._, 1949; Vijayan and Brownson, 1975; Brownson _et al._, 1977). Therefore, still a third experiment was conducted in which 14C-labeled parathion (2,4-ring labeled) was administered sc in the rat, and the levels of radioactivity and AChE activity were determined in aliquots of the same brain tissue homogenate or in samples of blood and plasma (Figure 3). Radioactivity levels in fat were also determined. The results showed that AChE activity in brain (brain stem and cortex) again was significantly inhibited at 6 hours and at 2 and 7 days after parathion administration, whereas blood AChE activity had recovered by 7 days. By contrast radioactivity levels in brain stem, cortex, blood and plasma were indetectable 2

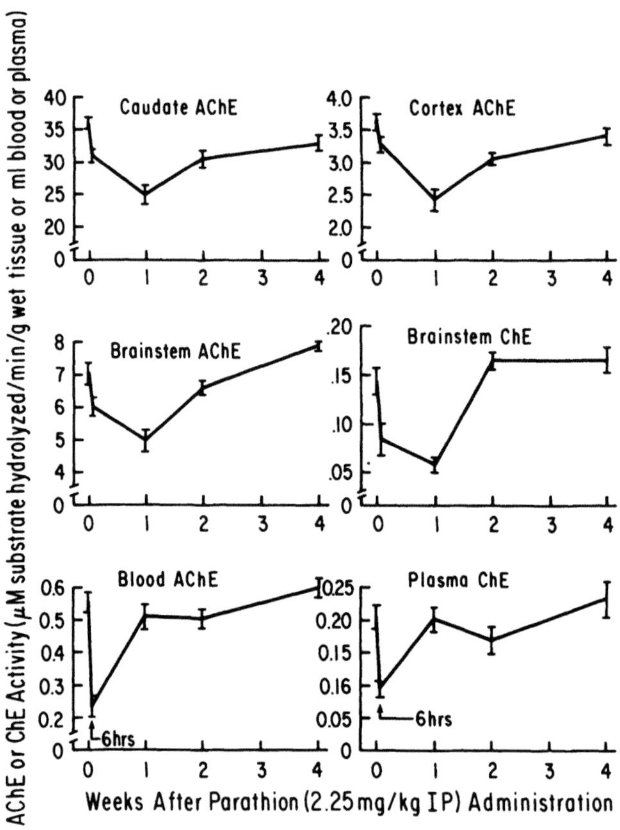

Figure 2. Effects of ip administration of parathion on inhibition and recovery of AChE and ChE activities in various brain areas, blood or plasma in the rat.

days after parathion administration even though radioactivity levels in fat were very high at this time. By 7 days radioactivity levels in fat had also fallen. Because the administered parathion was ring-labeled and because during AChE inhibition the ring portion of paraoxon leaves and is excreted as *p*-nitrophenol, absence of radioactivity in brain tissue combined with AChE inhibition rules out the possibility that AChE inhibition was the result of an homogenization artifact. During an homogenization artifact paraoxon located in a tissue compartment separated from AChE *in vivo* comes into contact with AChE during homogenization and causes AChE inhibition (Kewitz and Nachmansohn, 1957; O'Brien, 1967). Should this occur, then AChE activity would be falsely low. The

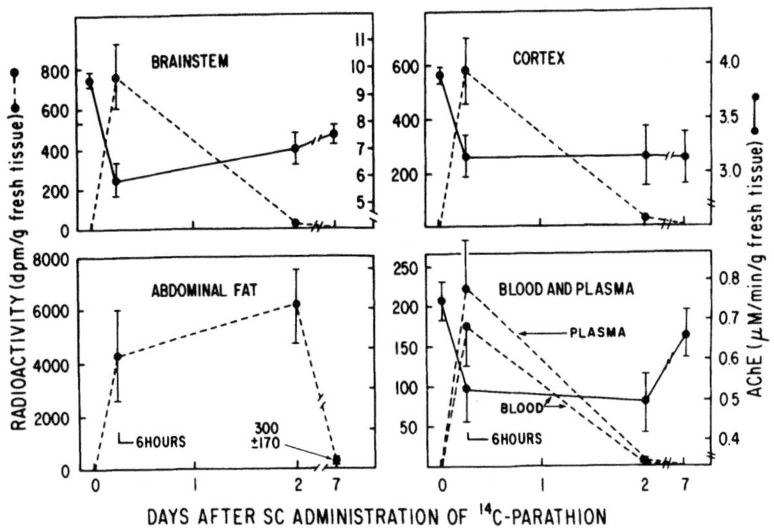

Figure 3. AChE activity and radioactivity in brain, fat, blood and plasma after ip administration of 14C-parathion (2,4-ring labeled) in the rat. Note that AChE remains inhibited in brain stem and cortex at 7 days post parathion when radioactivity levels in the same areas are no longer detectable.

present results appear to exclude this artifact as an important factor in the recovery of the inhibited brain AChE activity.

To date there are no studies on the time course of inhibition and recovery of AChE activity in the brain of subhuman primates after exposure to parathion. Such studies would have great value for helping to understand the mechanisms of action of parathion exposure in subhuman primates with the hope that the findings could be more readily extrapolated to man than could studies in lower mammals. The obvious reason that such studies have not been carried out is that the cost of sacrificing monkeys or other subhuman primates to obtain brain samples for analysis is prohibitive, plus these valuable animals are becoming increasingly too scarce for terminal experiments. Therefore, we devised a procedure whereby 50 mg samples of parietal cortex were removed serially from the same monkey before and at various times after administration of parathion (Figure 4). Three bonnet (Macacca radiata) monkeys were used. A single dose of parathion, either 3, 4, or 5 mg/kg, was administered sc. A dose-response effect on blood and cortex AChE activities is evident. As in the experiments with rats, AChE inhibition was initially greater in blood than in cortex at each of the three doses. The lowest dose produced a marked inhibition of

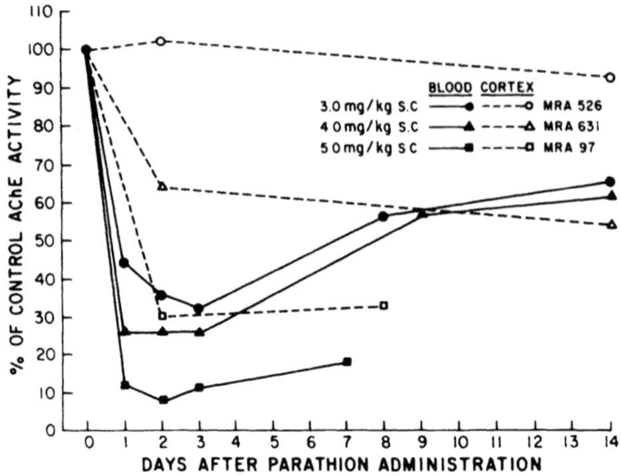

Figure 4. Effects of sc administration of parathion on AChE activities in blood and small samples of parietal cortex collected serially in each of 3 different Macacca radiata (abbreviated MRA) monkeys. A dose-response relation is indicated with blood AChE activities inhibited to a greater extent than brain AChE for each dose. Considerable recovery of blood AChE activity can be observed two weeks after parathion administration with no recovery apparent in the cortex.

blood AChE activity but had no effect on cortex AChE activity. The two higher doses produced some further inhibition of blood AChE activity but produced an even greater increase in inhibition of cortical AChE activity. The animal which received the highest dose died on the eighth day of the experiment after displaying severe anorexia. Autopsy revealed an intestinal obstruction. Whether the obstruction could be related to the parathion administration remains unknown. Within the period of the study considerable recovery of AChE activity occurred in blood, but not in cortex. Although the data are obviously limited in number, the findings suggest that in the monkey as in the rat, blood AChE activity recovers before brain AChE activity does so.

In conclusion, it appears that there are clear discrepancies in the literature regarding the rate of recovery of brain AChE inhibited by paraoxon. Most of the biochemical studies have emphasized that full recovery requires 2-4 weeks, whereas some, though not all, of the histochemical and the isozyme studies have emphasized that recovery may occur in 24 hours. In evaluating the reasons for these differing results the following should be considered: 1. Undoubtedly there are various isozymes of AChE which have differing half-lives and vary in their ability to be reactivated, i.e., dephosphorylated, following inhibition by paraoxon. 2. Different homogenization and enzyme extraction procedures may preferentially isolate one or another of these isozymes. 3. In some histochemical procedures perhaps the substrate fails to reach the AChE in its protected position at the synapse or junction.

BEHAVIORAL AND OTHER NEUROLOGICAL EFFECTS OF PARATHION ADMINISTRATION

A major goal of neurotoxicology is to relate the effects of a toxicant on the nervous system with some biochemical effect which may help to understand the mechanism of action of the poison. With this objective in mind our laboratory has spent several years comparing the effects of parathion administration on brain electrical activity and behavior with changes in AChE activity in mice, rats and monkeys (Reiter *et al.*, 1973, 1975; Woolley, 1976, 1978). Some of these findings will be summarized below.

Studies in the Rat

In one of the first studies the effects of parathion administration (3 mg/kg sc) on brain electrical activity were determined in six unanesthetized, awake and unrestrained rats with chronically implanted brain electrodes. Electrodes were positioned in the visual cortex, superior colliculus, midbrain reticular formation and cerebellum. Flash evoked potentials were elicited by a brief (10 μsec) light flash produced at the rate of 1/sec by a Grass photostimulator. Responses were averaged on-line with a Northern Scientific (model no. 544) averaging computer to produce one averaged visual evoked potential. Rats were recorded in their individual home cages. Parathion was dissolved in polyethylene glycol (Carbowax 300) in a concentration of 4 mg/ml.

The typical effects of parathion administration on visual evoked potentials recorded in the superior colliculus (Figure 5) and visual cortex (Figure 6) were to increase the latency, *i.e.*, the time from the flash until the beginning of the response, and to decrease the amplitude of the waves of the response. Latency was increased 40% or more in responses recorded from the visual cortex and superior colliculus, whereas latency of responses recorded simultaneously in the cerebellum and reticular formation were affected less. Amplitudes of the main components of the responses of the visual cortex and superior colliculus were reduced to one-third or less of control amplitudes. Amplitudes of the visual evoked responses recorded in the cerebellum and reticular formation also tended to be reduced, but variability was greater for these responses than for those recorded in the cortex and

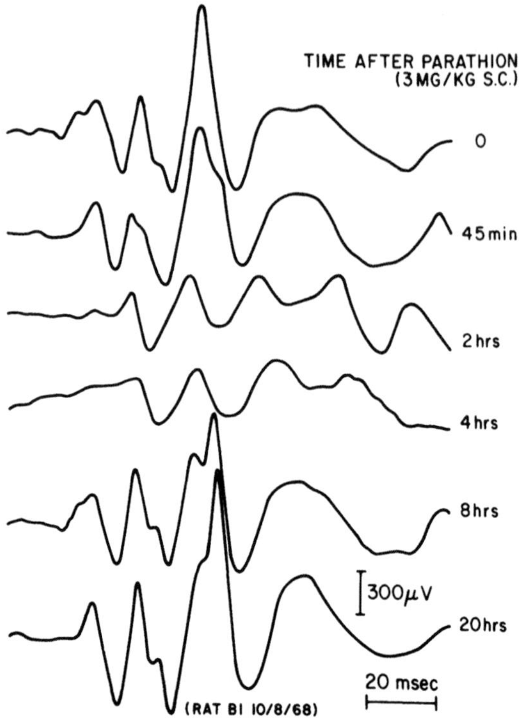

Figure 5. Effects of sc administration of parathion on averaged flash-evoked potentials in the superior colliculus in an awake unrestrained rat with chronically implanted electrodes. Note the increase in latency and decrease in amplitude of the components of the response at 2 and 4 hours and marked recovery at 8 hours.

colliculus. Maximal effects on amplitude and latency usually occurred 2 and 4 hours after parathion administration, and had usually returned to pretreatment values by 8 hours after administration.

In separate studies it was shown that tremorine, which is converted in vivo to oxotremorine, a potent cholinergic agonist, has similar effects on evoked potentials. Therefore, the observed effects of parathion administration would appear to be typical of cholinergic agents. Although it has been demonstrated that there is no cholinergic transmission in cortical and subcortical visual centers in the rat (Bigl and Schober, 1977), Shute and Lewis (1967) have described heavy cholinergic input from midbrain areas to important visual centers. Cholinergic neurons with cell bodies in the nucleus cuneiformis send major inputs to the superior col-

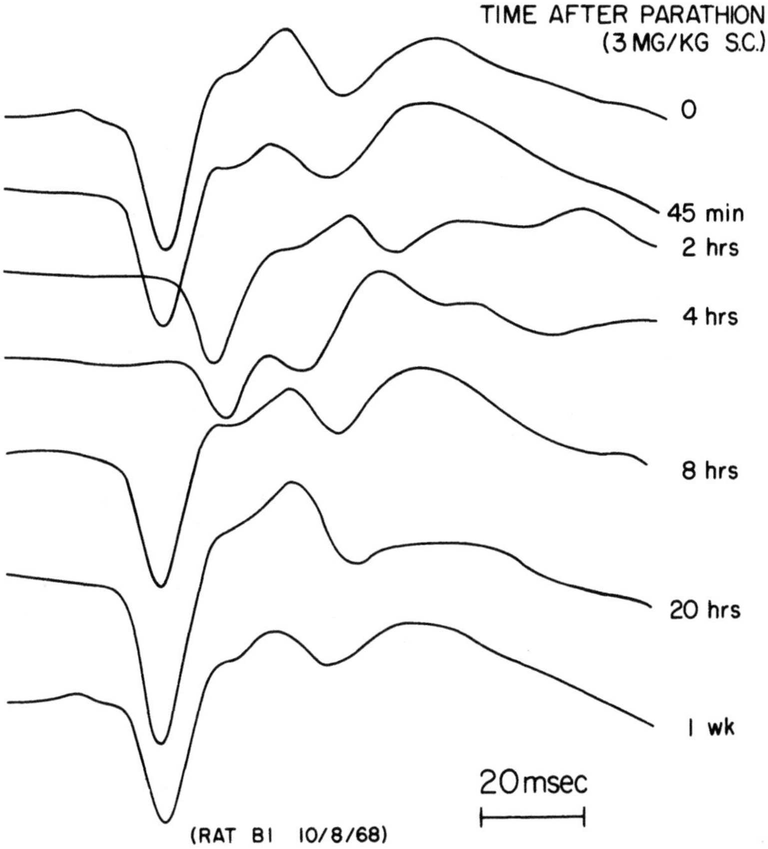

Figure 6. Effects of sc parathion administration on averaged, flash-evoked potentials recorded in the visual cortex of an awake unrestrained rat with chronically implanted brain electrodes. Latencies of the waves of the response were increased and amplitudes were decreased 2 and 4 hours after parathion administration. The effects had disappeared and recovery was complete by 8 hours after administration.

liculus, dorsal and deep pretectal nuclei, and the ventral nucleus of the lateral geniculate body. The cerebral cortex receives a cholinergic input from neurons in the globus pallidus and entopeduncular nucleus. Thus, transmission through these major central visual areas may be modified by cholinergic drugs and toxicants. In addition, AChE has been localized histochemically in the retina of cats (Koelle <u>et al.</u>, 1952) and ACh applied iontophoretically has an excitatory effect on retinal ganglion cells in the cat

(Straschill and Perwein, 1973). Also, activation of cholinergic parasympathetic fibers to the pupil may cause constriction of the pupil, which would affect flash-evoked potentials.

Thus, the observed effects of parathion administration on visual evoked potentials may result from activation of cholinergic synapses at the level of the pupil or retina, or at central visual stations in the lateral geniculate, tectum, cortex or other brain areas. The present findings on visual evoked potentials do not tell us the sites at which these effects occurred. In the pentobarbital anesthetized mouse ip administration of parathion gradually abolished the waves of the electroretinogram (Carricaburu and Lacroix, 1973). These effects would appear to be explainable only by an effect of paraoxon on either the pupil or retina. On the other hand, another anticholinesterase organophosphate insecticide, mevinphos (Phosdrin), altered the directional sensitivity of single neurons in the thalamus by a central action, i.e., by an action on brain cells, rather than by peripheral effects, i.e., effects on the pupil (Revzin, 1976). This was demonstrated by blocking the effects of mevinphos with atropine, a cholinergic blocking drug which acts both centrally and peripherally, but not with methyl atropine, which does not penetrate the blood-brain barrier and so does not act centrally. Similar studies with atropine and methyl atropine need to be conducted to localize better the effects of parathion administration on the visual evoked potentials reported here.

When the time course of the effects of parathion administration in the rat on brain AChE inhibition (Figures 1-3) are compared with the time course of effects on visual evoked potentials (Figures 5-6), it is evident that brain evoked potentials recovered much sooner than did the inhibited brain AChE activity. In order to determine if yet another measure of central nervous system (CNS) function in the rat recovered as rapidly after parathion administration as did visual evoked potentials, we determined the effects of 3 doses of parathion on the durations of the various phases of the maximal electroshock seizure (MES). The procedures for producing the MES have been described previously (Woolley, 1970). Briefly, a pair of silver disc electrodes with a small amount of electrode paste are placed on the scalp on top of the head just in front of the ears. A sine wave (60 cycles/sec) stimulus of 250 mA was delivered by a constant current stimulator. This stimulus normally produces a sequence of stereotyped motor responses as follows: tonic hindlimb flexion, followed by tonic full extension of hindlimbs, and finally a partial tonic extension of the hindlimbs, in which they are in a position intermediate between flexion and full extension (Figure 7). Administration of a single dose of 1.5, 2.0 or 2.5 mg/kg parathion ip significantly altered the MES pattern 1 hour later. The major effect was an increase in the duration of the first phase, i.e., hindlimb flexion,

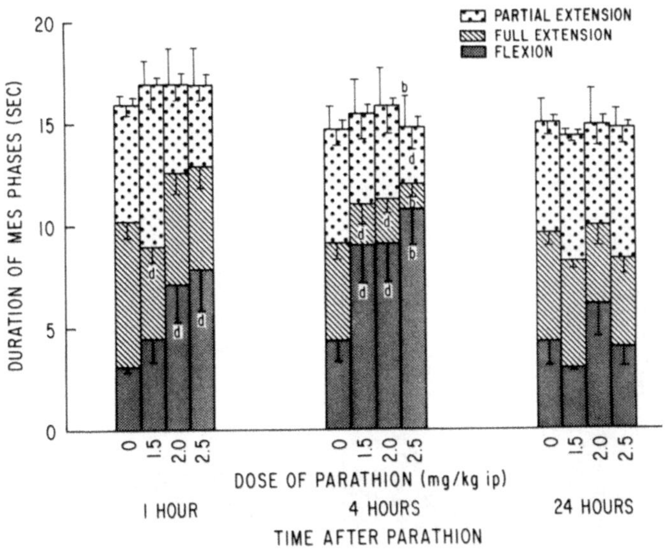

Figure 7. Time course of the effects of increasing doses of parathion administered ip as a single injection on duration of phases of the maximal electroshock seizure (MES) in the rat. The MES was produced as described in the text. In this and the following figure the vertical bracketed line within each portion of the barogram represents the standard error of the mean for the duration of that phase. At the top of each barogram the left vertical bracketed line represents the standard error of the mean for the sum of the extensor phases, and the line on the right represents the standard error of the mean of the total duration of the seizure. The letters indicate the level of statistical significance (*t* test) for the difference between the control and parathion-treated group run at the same time, as follows: a < 0.001, b < 0.01, c < 0.02, d < 0.05. The effects of parathion treatment were already evident one hour later in the two highest treatment groups, were marked in all treatment groups at four hours, and had disappeared by 24 hours after parathion administration.

in animals treated with the two highest doses. The changes were more marked 4 hours after parathion administration, when each of the treated groups exhibited a marked increase in duration of hindlimb flexion and marked decrease in duration of tonic hindlimb extension (Figure 7). Recovery had occurred at 24 hours.

Thus, it is evident that the effects of parathion treatment on the MES, as well as on visual evoked potentials, appear within a few hours after administration and disappear 24 hours after exposure. In other studies (results not shown) it was observed that

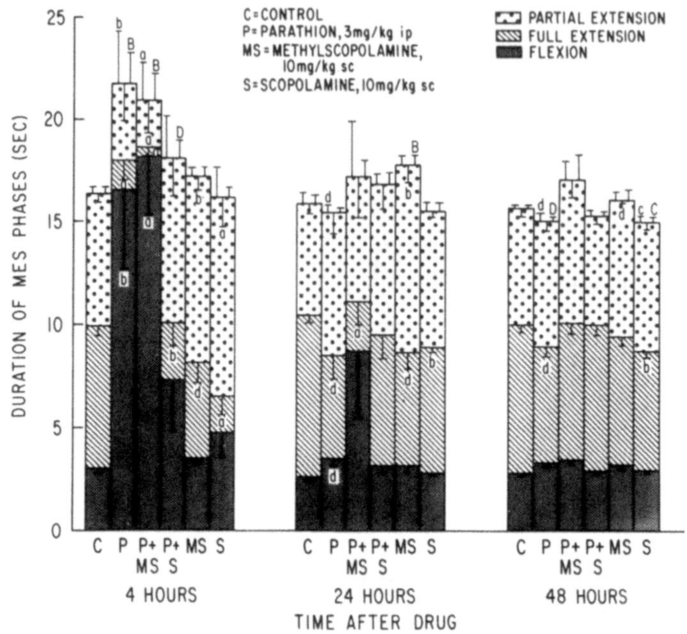

Figure 8. Effects of parathion, methyl scopolamine, scopolamine, alone and combined, on the durations of the MES phases in the rat. Presentation of the data is the same as in Figure 7. Parathion alone markedly increased duration of tonic hindlimb flexion and decreased duration of tonic hindlimb full extension at 4 hours post treatment, but at 24 hours these effects had mostly disappeared. At four hours post parathion treatment scopolamine was effective in blocking the major effects of parathion, whereas methyl scopolamine was not.

administration of tremorine produced changes in the MES similar to those produced by parathion. This may be interpreted to mean that the observed changes in the MES may be typical of at least some cholinergic agents.

In order to determine if the observed effects of parathion administration on the MES were due to effects within the CNS or to peripheral effects, the interaction between parathion, scopolamine and methyl scopolamine were determined. Scopolamine has anticholinergic effects at both central and peripheral synapses and junctions, whereas methyl scopolamine acts at peripheral sites only. The results (Figure 8) demonstrated that at equal doses scopolamine reversed the major effects of parathion administration whereas methyl scopolamine did not. Again, the effects of parathion administration had mostly disappeared by 24 hours.

The major conclusion which may be drawn from the latter study (Figure 8) is that the principal effects of parathion administration on the MES in the rat are exerted on central synapses rather than at peripheral sites.

When the effects of parathion on visual evoked potentials, the MES pattern, and brain and spinal cord AChE activities in the rat are compared (Figures 1-3, 5-8), it is evident that the effects are maximal when CNS AChE activities are falling and that these effects disappear after AChE inhibition has reached its maximum and stabilized.

Studies in the Monkey

The use of the monkey as an animal model to study the effects of parathion would appear to have more obvious relevance for human health than does the use of the rat. Therefore, studies were undertaken to determine the effects of parathion on performance of a learned task and on visual evoked potentials. These were correlated with levels of blood AChE activity.

For the behavioral testing three male rhesus monkeys were conditioned on a visual discrimination task (Reiter et al., 1975). During the testing the monkeys were placed in restraining chairs inside a behavioral chamber with the monkey facing the work panel (Figure 9). The task was to choose the correct one of two patterns. Each pattern consisted of a 3 x 3 array of lucite squares lighted in some combination of 9 light or dark squares. The animals learned to initiate the visual discrmination task by activating a toggle switch which in turn presented the two patterns. If the monkey then touched the correct panel, a banana pellet reward was automatically received. In each behavioral session four problems were presented and each problem was repeated 20 times for a total of 80 responses per session. Correct responses were automatically recorded. The correct one of the two patterns was randomly switched between left and right sides.

Four levels of parathion administration (0.5, 1.0, 1.5, and 2.0 mg/kg) were administered as a single dose (Reiter et al., 1975). The lowest dose had no effect on performance, but the 3 higher doses each disrupted performance for varying periods of time. The effects on 2 animals of 2.0 mg/kg parathion administered per os (po) hidden in fruit are shown in Figure 10. No effect was seen one hour following the administration of 2.0 mg/kg of parathion. At 5 hours, however, performance was completely abolished. The performance of animal MMU 6448 was especially affected by parathion. On the first problem, shown in the upper left-hand quadrant of Figure 10, performance waxed and waned below control levels for 3 weeks. On the last two problems shown on the

Figure 9. Schematic drawing showing monkey in behavioral chamber reaching for the toggle switch to initiate the presentation of two patterns. A pattern is formed by lighting some combination of squares in a 3 x 3 array with the projector shown behind the chamber. If the animal presses the panel with the correct pattern, a banana pellet is automatically dispensed.

right of the figure, the same animal refused to perform for one week.

These observations demonstrated that relatively low doses of parathion (1-2 mg/kg po) had marked effects on performance. Performance was abolished (the animals would not attempt to perform the discrmination task) for periods of time ranging from 1 day to 1 week and performance became irregular in some cases for several weeks.

In other monkeys 16 electrodes were chronically implanted in the visual, parietal and frontal cortex, and the effects of 2-4 mg/kg parathion on visual evoked potentials were determined. The procedures were essentially the same as those described above except that evoked potentials from 6 electrode positions were averaged simultaneously by a Data General Nova 1200 programmable computer. In some studies averaged evoked potentials, plus and minus the standard errors of the mean values were determined and plotted

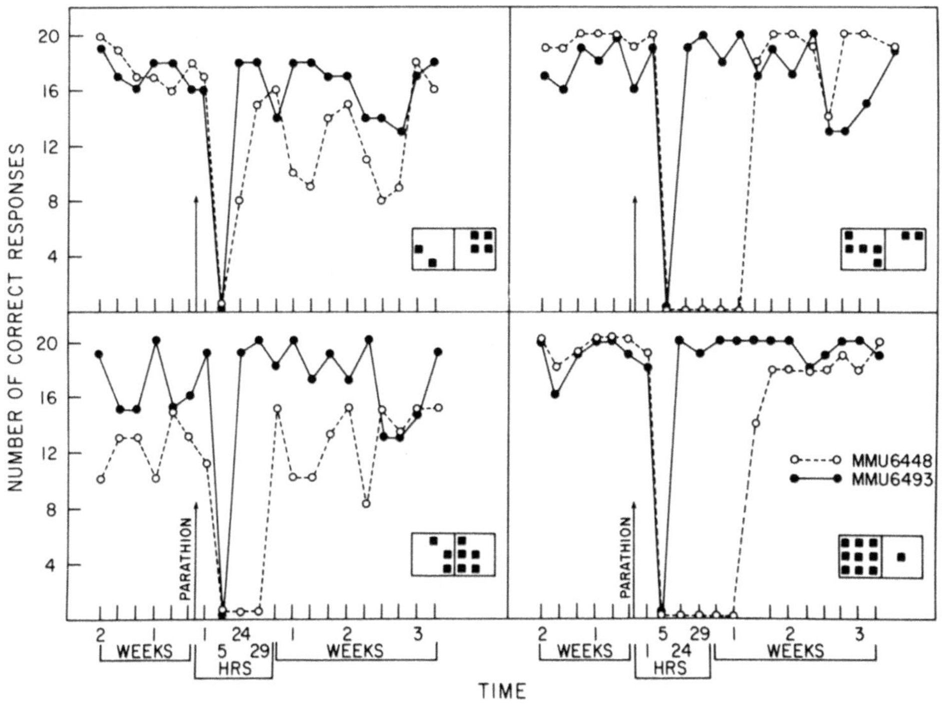

Figure 10. The performance of two monkeys on four different visual discrimination patterns after po administration of 2.0 mg/kg parathion. The two patterns to be discriminated are shown in the lower right portion of each quadrant. When the number of correct responses was zero, this resulted because the animals did not perform the tasks at all. After administration of parathion the major effect was that the animals would not perform the tasks for varying periods of time.

(Figure 11). A dose of 2 mg/kg affected the visual evoked potentials little if at all. A dose of 4 mg/kg parathion po usually had marked effects. Particularly marked effects are shown in Figure 11, and consisted of an increase in latency of the response and a decrease in amplitude. This effect was most marked 4 and 7 hours after parathion. Some recovery was evident at 24 hours and was nearly complete at 48 hours.

When the effects of parathion on visual evoked potentials and on visual discrimination performance are compared and evaluated, it was evident to us that the behavioral task was a more sensitive

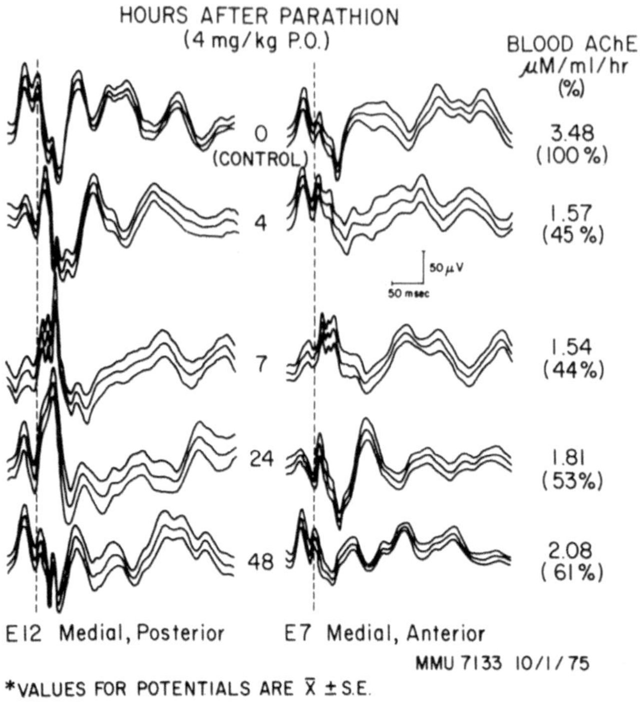

Figure 11. Flash-evoked averaged potentials with ± SE recorded from an anterior and a posterior position in the medial portion of the visual cortex of a rhesus monkey. The effects were most marked at 7 hours and had almost completely recovered by 48 hours. Blood AChE activities are presented on the right.

index of toxic action. Effects on visual evoked potentials were evident only at higher doses of parathion and recovered sooner than did the effects on behavior.

RELATION BETWEEN BRAIN AChE INHIBITION AND NEUROLOGICAL EFFECTS

When the data presented here on the time course of effects of parathion administration on inhibition and recovery of CNS AChE activity and on visual evoked potentials and maximal seizure responses are compared, it is evident that these neurological effects recover sooner than does the inhibited brain AChE. An exception may be the visual discrimination performance in monkeys which is disrupted for a longer period of time and at lower doses than are the visual evoked potentials.

This apparent dissociation between the administration of parathion or other anticholinesterase agents on brain AChE activity and on behavioral or other neurological effects has intrigued investigators for a long time and has been reviewed recently (Bignami _et al._, 1975; Hobbiger, 1976). The following possible explanations for this dissociation should be considered.

1. The rapid recovery of neurological effects may be related more to effects on brain AChE isozymes which recover rapidly but contribute little to the total AChE activity as measured biochemically, for example with the Ellman _et al._ (1961) procedure. This possibility has been discussed above. It goes against the majority opinion which holds that paraoxon produces an irreversible inhibition of brain AChE which is restored primarily by resynthesis. Resynthesis occurs in neuronal cell bodies and the resynthesized AChE travels via slow axoplasmic transport to the axon terminals. Depending on the distance traveled, restoration of AChE activity could require a few days to a couple of weeks. Further work with isozymes, their differential sensitivities to inhibitors, recovery times, and especially physiological significance, is needed to resolve this problem.

2. The neurological effects of parathion administration are related to the elevated ACh levels at cholinergic synapses and junctions, and with time these levels decline to normal despite continued inhibition of AChE activity. In fact, recent work has demonstrated that this is exactly what happens (Wecker _et al._, 1977). After ip administration of paraoxon to adult male rats AChE is rapidly inhibited so that in 15 minutes AChE activity in brain is only 20% that of control activity. After 24 hours AChE activity is only 50% of control levels and 12 days are required for full recovery. At the same time brain ACh levels double in 15 minutes after paraoxon administration and are back to normal levels 24 hours later. Thus, there is a dissociation between the inhibited AChE activity and ACh levels, and it is reasonable to assume that the cholinergic effects of parathion or paraoxon administration parallel ACh levels, not AChE inhibition _per se_.

The effects of paraoxon may be mediated via actions on the presynaptic release of ACh. This was demonstrated in both _in vivo_ and _in vitro_ experiments with the phrenic nerve-diaphragm nerve-muscle preparation in the rat (Laskowski and Dettbarn, 1975). Thirty minutes after exposure to paraoxon the frequency of miniature end-plate potentials (MEPPs) was increased 38-fold. MEPP frequency is proportional to the rate of quantal release of ACh. Therefore, this observation means that paraoxon rapidly increased ACh release from the presynaptic terminals. The exact mechanism of this increased release is not known, but it was blocked by the administration of tetrodotoxin which also blocked the increased spontaneous and stimulated antidromic firing in the presynaptic neuron. The increased MEPP frequency after paraoxon administration gradually declined to control rates within 6 hours after injection despite continued marked inhibition of AChE activity at the junction. In other words, the increased release of ACh at the neuromuscular junction declined to normal within hours after paraoxon administration (Laskowski and Dettbarn, 1975).

Thus, in two different preparations--brain and neuromuscular junction--there is recent evidence that elevated ACh levels return to normal long before the inhibited AChE activity recovers. Perhaps the high levels of ACh in the synapse or junction exert a negative feedback effect on the presynaptic terminals to reduce synthesis and/or release. It has been suggested that high levels of ACh inhibit its synthesis via inhibition of choline acetyltransferase (Kaito and Goldberg, 1969), but this requires further study. Although the mechanism may not be clear, it is evident that elevated ACh levels may recover before AChE activity recovers and that this must be an important factor in adapting to the effects of parathion or paraoxon administration.

3. The postsynaptic or postjunctional membrane may adapt to elevated ACh levels and so cholinergic function may return to normal. This would represent a true tolerance or adaptation of the receptor membrane to elevated neurotransmitter. The best evidence for this comes from work with the neuromuscular junction which demonstrated that the postjunctional membrane adapts to constant high levels of ACh _in vitro_ (Kim and Karczmar, 1967). Unfortunately, similar evidence is not available for central synapses.

4. Paraoxon may combine with the AChR and alter the symptomatology of parathion poisoning in this way. In several neuromuscular preparations in which cholinesterases had been totally inhibited, the addition of a number of anticholinesterase agents, including parathion, acted as receptor agonists (reviewed by Eldefrawi, 1976). In one study the organophosphates--DFP, phospholine and paraoxon, all of which are known to be potent inhibitors of AChE--were applied to monocellular electroplax preparation (Bartels and Nachmansohn, 1969). The electroplax contains very high concentrations of nicotinic AChR. At high concentrations the compounds acted as inhibitors of the AChR. The effects were rapidly reversible. At low concentrations the organophosphates potentiated the action of ACh, as would be expected from their AChE inhibiting effects.

The fact that high concentrations of organophosphate anticholinesterases may combine with the AChR should not be surprising in view of the similarity of some structural features and the distribution of AChE and AChR (Csillik, 1975; Eldefrawi, 1976). However, as may be expected, the affinities of the organophosphates for AChR is lower than for AChE (Eldefrawi _et al._, 1971).

The most relevant observation is that high levels of organophosphates may reversibly inhibit the effects of ACh on the AChR (Bartels and Nachmansohn, 1969). Thus, when AChE becomes inhibited after parathion administration and ACh levels rise, paraoxon may combine with the AChR to reduce postsynaptic sensitivity to the excess ACh. This then could be a mechanism for compensating for the excess ACh during parathion poisoning.

At present there is some evidence supporting each of these possible explanations for the "dissociation" between AChE inhibition and neurotoxicological effects of parathion administration. It is likely that one or the other is relatively more important at different synapses and junctions, at different times after poisoning, or in different species. Although work remains to be done, our knowledge of the mechanisms of action of paraoxon, the active metabolite of parathion, is increasing.

ACKNOWLEDGMENTS

Figures have been reprinted with permission as follows: Figures 1, 2, 4, 6-8 and 11 from Woolley, D. E. Some aspects of the neurophysiological basis of insecticide action. Fed. Proc., 34: 2610-2617, 1976; figure 5 from Woolley, D. E. Electrophysiological techniques in toxicology. In: L. Reiter and H. Zenick, eds. Behavioral Toxicology: An Emerging Discipline. U.S. Gov't. Printing Office, DHEW, in press, 1978; figure 9 from Reiter, L. W., G. M. Talens and D. E. Woolley. Parathion admnistration in the monkey: Time course of inhibition and recovery of blood cholinesterases and visual discrimination performance. Toxicol. Appl. Pharmacol., 33:1-13, 1975. Figure 10 was redrawn from the latter reference. Figure 3 represents unpublished observations by Chernobieff and Woolley. Original research in the authors' laboratory was supported by NIH grant ES-00163.

The expert assistance of Ms. Mary Lou Rodriguez in typing this manuscript is gratefully acknowledged.

REFERENCES

Bartels, E. and D. Nachmansohn. Organophosphate inhibitors of acetylcholine-receptor and -esterase tested on the electroplax. Arch. Biochem. Biophys., 133:1-10, 1969.

Benke, G. M. and S. D. Murphy. Anticholinesterase action of methyl parathion, parathion and azinphosmethyl in mice and fish: Onset and recovery of inhibition. Bulletin of Environmental Contamination and Toxicology, 12:116-122, 1974.

Bigl, V. and W. Schober. Cholinergic transmission in subcortical and cortical visual centers of rats: no evidence for the involvement of primary optic system. Exp. Brain Res., 27:211-219, 1977.

Bignami, G., N. Rosic, H. Michalek, M. Milosevic, and G. L. Gatti. Behavioral toxicity of anticholinesterase agents: Methodological, neurochemical, and neuropsychological aspects. In: B. Weiss and V. G. Laties, eds. Behavioral Toxicology. Plenum Press, New York and London, pp. 155-215, 1975.

Brownson, R. H., H. D. McDougal, D. B. Suter, and V. K. Vijayan. Parathion-induced alterations in acetylcholinesterase of the rat nervous system: A histochemical, biochemical and isozyme study. In: L. Roizin, H. Shiraki, and N. Grcevic, eds. Neurotoxicology, Raven Press, New York, pp. 457-468, 1977.

Carricaburu, P. and R. Lacroix. Effet du parathion sur L'Electroretinogramme de la souris blanche. Vision Res., 13:793-796, 1973.

Chippendale, T. J., C. W. Cotman, M. D. Kozar, and G. S. Lynch. Analysis of acetylcholinesterase synthesis and transport in the rat hippocampus: Recovery of acetylcholinesterase activity in the septum and hippocampus after administration of diisopropylfluorophosphate. Brain Res., 81:485-496, 1974.

Corbett, J. R. Insecticides inhibiting acetylcholinesterase. In: Biochemical Mode of Action of Pesticides. Academic Press, New York, pp. 107-163, 1974.

Csillik, B. Synaptochemistry of acetylcholine metabolism in a cholinergic neuron. In: C. C. Pfeiffer and J. R. Smythies, eds. Intern. Rev. Neurobiol., 18:69-140, 1975.

Davis, G. A. and B. W. Agranoff. Metabolic behaviour of isozymes of acetylcholinesterase. Nature, 220:277-280, 1968.

Davison, A. N. Return of Cholinesterase activity in the rat after inhibition by organophosphorus compounds. Biochem. J., 54:583-590, 1953.

Davison, A. N. Return of cholinesterase activity in the rat after inhibition by organophosphorus compounds. 2. A comparative study of true and pseudocholinesterase. Biochem. J., 60:339-346, 1955.

DeCandole, C. A. and M. K. McPhail. Sarin and paraoxon antagonism in different species. Can. J. Biochem. Physiol., 35:1071-1083, 1957.

Dolivia, M. and G. B. Koelle. Properties of nicotinic and muscarinic receptors in isolated rat ganglia. Experientia, 26:679, 1970.

Doull, J. The treatment of insecticide poisoning. In: C. F. Wilkinson, ed. Insecticide Biochemistry and Physiology. Plenum Press, New York, pp. 649-667, 1976.

DuBois, K. P., J. Doull, P. R. Salerno, and J. M. Coon. Studies on the toxicity and mechanism of action of p-nitrophenyl diethyl thionophosphate (parathion). J. Pharmacol. Exp. Therap., 95:79-91, 1949.

Eldefrawi, A. T. The acetylcholine receptor and its interactions with insecticides. In: C. F. Wilkinson, ed. Insecticide Biochemistry and Physiology. Plenum Publishing Corp., New York, New York, 297-326, 1976.

Eldefrawi, M. E., A. G. Britten, and R. D. O'Brien. Action of organophosphates on binding of cholinergic ligands. Pest. Biochem. Physiol., 1:101-108, 1971.

Ellman, G. L., K. D. Courtney, V. Andres, Jr., and R. M. Featherstone. A new and rapid colorimetric determination of acetylcholinesterase activity. Biochem. Pharmacol., 7:88-95, 1961.

Hitchcock, M. and S. D. Murphy. Activation of parathion and guthion by mammalian, avian, and piscine liver homogenates and cell fractions. Toxicol. Appl. Pharmacol., 19:37-45, 1971.

Hobbiger, F. Pharmacology of anticholinesterase drugs. In: Eleanor Zaimes, ed. Neuromuscular Junction. Springer-Verlag, New York, 487-581, 1976.

Kaito, A. A. and A. M. Goldberg. Control of acetylcholine synthesis. The inhibition of choline acetyltransferase by acetylcholine. J. Neurochem., 16:1185-1191, 1969.

Kewitz, H. and D. Nachmansohn. A specific antidote against lethal alkyl phosphate intoxication. IV. Effects in brain. Arch. Biochem. and Biophys., 66:271-283, 1957.

Kim, K. C. and A. G. Karczmar. Adaptation of the neuromuscular junction to constant concentration of ACh. Int. J. Neuropharmacol., 6:51-61, 1967.

Koelle, G. B., L. Wolfand, J. S. Friedenwald, and R. A. Allen. Localization of specific cholinesterase in ocular tissues of the cat. Amer. J. Ophthalmol., 35:1580-1584, 1952.

Laskowski, M. B. and W.-D. Dettbarn. Presynaptic effects of neuromuscular cholinesterase inhibition. J. Pharmacol. Exper. Therap., 194:351-361, 1975.

Lewis, P. R. and C. C. D. Shute. The cholinergic limbic system: Projections to hippocampal formation, medial cortex, nuclei of the ascending cholinergic reticular system, and the subfornical organ and supra-optic crest. Brain, 90:521-542, 1967.

Main, A. R. Structure and inhibitors of cholinesterase. In: A. M. Goldberg and I. Hanin, eds. Biology of Cholinergic Function. Haven Press, New York, pp. 269, 1976.

O'Brien, R. D. Insecticides action and metabolism. Academic Press, New York, 1967.

Poore, R. E. and R. A. Neal. Evidence for extrahepatic metabolism of parathion. Toxicol. Appl. Pharmacol., 23:759-768, 1972.

Reiter, L. W., G. M. Talens, and D. E. Woolley. Acute and subacute parathion treatment: Effects on cholinesterase activities and learning in mice. Toxicol. Appl. Pharmacol., 25:582-588, 1973.

Reiter, L. W., G. M. Talens, and D. E. Woolley. Parathion administration in the monkey: Time course of inhibition and recovery of blood cholinesterases and visual discrmination performance. Toxicol. Appl. Pharmacol., 33:1-13, 1975.

Revzin, A. M. Effects of organophosphate pesticides and other drugs on subcortical mechanisms of visual integration. Aviation, Space, and Environ. Med., 47:627-629, 1976.

Schaumann, W. Maximal inhibition of cholinesterase in the central nervous system. Br. J. Pharmacol., 15:432-435, 1960.

Schleifer, L. S. and M. E. Eldefrawi. Identification of the nicotinic and muscarinic acetylcholine receptors in subcellular fractions of mouse brain. Neuropharm., 13:53-63, 1974.

Serrone, D. M., A. A. Stein, and F. Coulston. Cholinesterase inhibition by parathion in rhesus monkeys and the limited effect of 2-PAM. Exptl. Molec. Pathol., 11:99-111, 1969.

Shute, C. C. D. and P. R. Lewis. The ascending cholinergic reticular system: Neocortical, olfactory and subcortical projections. Brain, 90:497-522, 1967.

Snyder, S. H., K. J. Chang, M. J. Kuhar, and H. I. Yamamura. Biochemical identification of the mammalian muscarinic cholinergic receptor. Fed. Proceedings, 34:1915-1921, 1975.

Stewart, W. C. Accumulation of acetylcholine in brain and blood of animals poisoned with cholinesterase inhibitors. Brit. J. Pharmacol., 7:270-276, 1952.

Straschill, M. and J. Perwein. The effect of iontophoretically applied acetylcholine upon the cat's retinal ganglion cells. Pflügers Arch., 339:289-298, 1973.

Vijayan, V. K. and R. H. Brownson. Polyacrylamide gel electrophoresis of rat brain acetylcholinesterase: Isoenzyme changes following parathion poisoning. J. Neurochem., 24:105-110, 1975.

Wecker, L., P. L. Mobley, and W.-D. Dettbarn. Central cholinergic mechanisms underlying adaptation to reduced cholinesterase activity. Biochemical Pharmacology, 26:633-637, 1977.

Whitehouse, L. W. and D. J. Ecobichon. Paraoxon formation and hydrolysis by mammalian liver. Pesticide Biochemistry and Physiology, 5:314-322, 1975.

Woolley, D. E. Effects of DDT and of drug-DDT interactions on electroshock seizures in the rat. Appl. Pharmacol., 16:521-532, 1970.

Woolley, D. E. Some aspects of the neurophysiological basis of insecticide action. Fed. Proc., 35:2610-2617, 1976.

Woolley, D. E. Electrophysiological techniques in toxicology. In: L. Reiter and H. Zenick, eds. Behavioral Toxicology: An Emerging Discipline. U.S. Gov't. Printing Office, DHEW, in press, 1978.

EFFECTS OF INSECTICIDES ON THE SENSORY NERVOUS SYSTEM

Joep van den Bercken, Alfons B.A. Kroese and
Louis M.A. Akkermans
Institute of Veterinary Pharmacology and Toxicology
University of Utrecht
Biltstraat 172, Utrecht, The Netherlands

INTRODUCTION

The majority of insecticidal chemicals owe their toxicity to the ability to interfere with the function of the nervous system. For organophosphates and carbamates it is well established that their action on the nervous system is primarily due to inhibition of the enzyme acetylcholinesterase, resulting in a disturbance of cholinergic synaptic transmission. For other insecticides, notably DDT and its derivatives and pyrethroids, such a biochemical mode of action has not been established. It is now generally accepted that these latter compounds act directly on excitable tissues and interfere with the changes in ionic permeability of the nerve membrane that underlie the generation and conduction of the nervous impulse (1, 2).

The most characteristic effect of DDT in the nervous system of insects and other invertebrates is the induction of repetitive activity. After treatment with DDT a brief electrical stimulus applied to the nerve induces a train of action potentials, also called a repetitive discharge, instead of a single action potential. Trains of nerve impulses may also appear, without stimulation, in preparations which are spontaneously active, or upon stimulation of sense organs. Since the early work of Roeder and Weiant it has been known that in the cockroach much lower concentrations of DDT are required to induce repetitive activity in sense organs and in sensory nerve fibres as compared to motor nerve fibres (3-5). The reason for this difference in sensivity has not been clarified. One of the first signs of DDT-poisoning in the axonal nerve membrane is the appearance of a prolonged depolarizing (negative) after-potential following the action potential (6, 7). The induction of repetitive activity and the development of a depolarizing after-potential is probably directly related to the DDT-induced prolongation of the transient increase in sodium conductance of the nerve membrane during excitation (8, 9). A number of DDT-analogues and derivatives also cause repetitive nerve activity, but some of them completely block axonal conduction, even without inducing repetitive activity, presumably by a suppression of the ionic conductance changes of the nerve membrane (10, 11).

DDT-induced repetitive activity has also been observed in peripheral nerves of the rat (12, 13) and of the frog (14), and there are several indications for an increased afferent activity in DDT-treated rats (13, 15, 16). In myelinated nerve fibres of the frog DDT causes a prolongation of the transient increase in membrane sodium conductance very similar to that in nerve fibres of invertebrates (17, 18).

Natural pyrethrins and pyrethroids are also known for their ability to induce repetitive activity in the nervous system of arthropods, notably in the sensory nervous system. In addition, these compounds block axonal conduction and transmission across synapses. Allethrin, one of the early synthetic pyrethroids, induces a depolarizing after-potential and causes a prolongation of the transient increase in sodium conductance in nerve fibres of invertebrates (19-24). As with DDT, the allethrin-induced repetitive activity is probably associated with the prolonged sodium conductance. The blocking of axonal conduction by allethrin has been shown to be due to a suppressive action of this compound on the ionic conductance changes of the nerve membrane (25).

Until recently very little information was available on the action of pyrethroids in the nervous system of vertebrates. It has been reported that peripheral nerves of the rat and of the frog, in contrast to crayfish nerve fibres, were fairly insensitive to the blocking action of pyrethrins, and it was tentatively suggested that nerve membranes of vertebrates were less susceptible to the neurotoxic action of these compounds than those of invertebrates (26, 27).

Although chlorinated hydrocarbon insecticides other than DDT, e.g. dieldrin (HEOD) and other bicyclodienes, are generally considered to act specifically on the nervous system, their mechanism of action is far from understood. In the cockroach dieldrin produces repetitive activity in peripheral sensory neurones, although much less intense than caused by DDT (28, 29). Recent evidence indicates that dieldrin, albeit after a long latency, interferes with synaptic transmission in the central nervous system of the cockroach and causes an increase in spontaneous activity and a prolongation of the postsynaptic response to presynaptic stimulation. This probably results from spontaneous and excessive release of transmitter from presynaptic stores (29, 30). It has been proposed that in the cockroach dieldrin must first be converted to aldrin-transdiol and possibly other metabolites before it can exert its neurotoxic action (1, 29). Recently, however, this hypothesis has been seriously questioned (31).

There are strong indications that in mammals and other vertebrates dieldrin probably acts primarily on the central nervous system (32-35). When the present study was started information about the action of dieldrin on the nervous system of vertebrates at the cellular level was lacking.

Of particular interest is the observation that DDT and pyrethroids are more toxic to insects at lower than at higher temperatures. Furthermore, the symptoms of poisoning by these compounds can be reversed by changing the temperature (1, 36). Such a negative temperature coefficient of toxicity has not been found for any other group of insecticidal compounds. In the past several attempts have been made to clarify this peculiar phenomenon. It has been reported that after treatment with DDT the repetitive activity from the sensory neurons in the cockroach leg increased as the temperature was lowered. In addition, DDT was found to be more effective in inducing trains of impulses at a low than at a high temperature (37, 38). However, this could not be confirmed by other investigators who concluded that events occurring in the central nervous system of the cockroach were responsible for the negative temperature coefficient of DDT-toxicity (36, 39). The allethrin-induced repetitive activity in nerve fibres of invertebrates shows a positive rather than a negative temperature coefficient (19, 24). On the other hand, the nerve blocking action of allethrin increases markedly upon lowering the temperature. This was shown to be due to an increased suppressive action of allethrin on membrane ionic conductances at lower temperature (22).

Our research of recent years has been directed towards a better understanding of the neurotoxic action of insecticides in higher animals. We have studied the effects of DDT, allethrin and dieldrin on various parts of the peripheral nervous system of the frog, with special interest in the sensory nervous system. The effects of temperature on the action of these compounds have also been investigated.

MATERIALS AND METHODS

The experiments were performed on the common frog (Rana esculenta and Rana temporaria) and on the clawed frog (Xenopus laevis). The effects of DDT, allethrin, dieldrin and the dieldrin metabolite aldrin-transdiol on impulse conduction in peripheral nerves, impulse transmission in the neuromuscular junction and impulse generation in sense organs were studied by means of standard electrophysiological techniques. The different preparations and methods are mentioned briefly. More detailed information can be found elsewhere (40-46). Unless otherwise noted, the experiments were conducted at a room temperature of 20-24°C.

Peripheral nerves. The excised sciatic nerve and other nerve branches were mounted on silver wire electrodes in a moist chamber and compound action potentials evoked by a brief electrical stimulus were recorded. Action potentials of single nerve fibres were recorded by means of a suction electrode. For measuring membrane potentials of the isolated node of Ranvier a single nerve fibre was mounted across two partitions in a small Perspex cell fitted with two pairs of Ag-AgCl electrodes.

Synaptic transmission. Synaptic transmission in the neuromuscular junction was studied in the isolated sartorius nerve-muscle preparation of the common frog. Postsynaptic potentials (end-plate potentials) evoked by a brief electrical stimulus applied to the nerve, were recorded intracellularly by means of conventional glass microelectrodes.

Sense organs. Two different sense organ preparations were used: the cutaneous touch receptor and the lateral-line organ of the clawed frog. In both cases a small piece of skin together with the appropriate nerve branches was removed from the animal and afferent nervous activity was recorded by means of a silver wire electrode. In the touch receptor preparation a single spike in one of the afferent nerve fibres was elicited by applying a brief mechanical stimulus to the surface of the skin. In the case of the lateral-line organ activity was recorded from a single stitch, which is innervated by only two afferent nerve fibres. Mechanical stimulation was supplied by local water motions which were produced by a small glass sphere.

Chemicals. DDT (purified p,p'-DDT; Orgachemia), allethrin (K & K Laboratories, Inc.), dieldrin (99.4% HEOD; Shell Chemical Company) and aldrin-transdiol (> 99% trans-6,7-dihydroxy-dihydroaldrin; Shell Research Ltd) were dissolved in ethanol to make up stock solutions. Small amounts of these solutions were added to standard frog Ringer solution with the aid of a hypodermic syringe. The final ethanol concentration was less than 0.5% (v/v) and was found to be without effect in control experiments.

Additional experiments were carried out with preparations isolated from animals which showed severe symptoms of poisoning, either after injection of the insecticides into the dorsal lymph sac (25-35 mg/kg), or after addition of the compounds to the water in which the animals were kept (1-5 ppm; clawed frog only).

RESULTS

Effects on Peripheral Nerves

DDT. Exposure of excised frog nerves to DDT in a concentration of 5×10^{-4}M for 1 hr produced marked repetitive activity (41). Such activity was also observed in sciatic nerves excised from animals which showed severe symptoms after poisoning in vivo. Recording from different nerve branches and from the sciatic nerve with intact spinal roots made it possible to demonstrate that this repetitive activity occurred in sensory fibres only. Afferent nerve fibres innervating skin, muscle and the lateral-line organ were all affected by DDT. The effect of DDT on single myelinated nerve fibres is illustrated in Fig. 1. The first sign of DDT-poisoning was the development of a depolarizing after-potential following

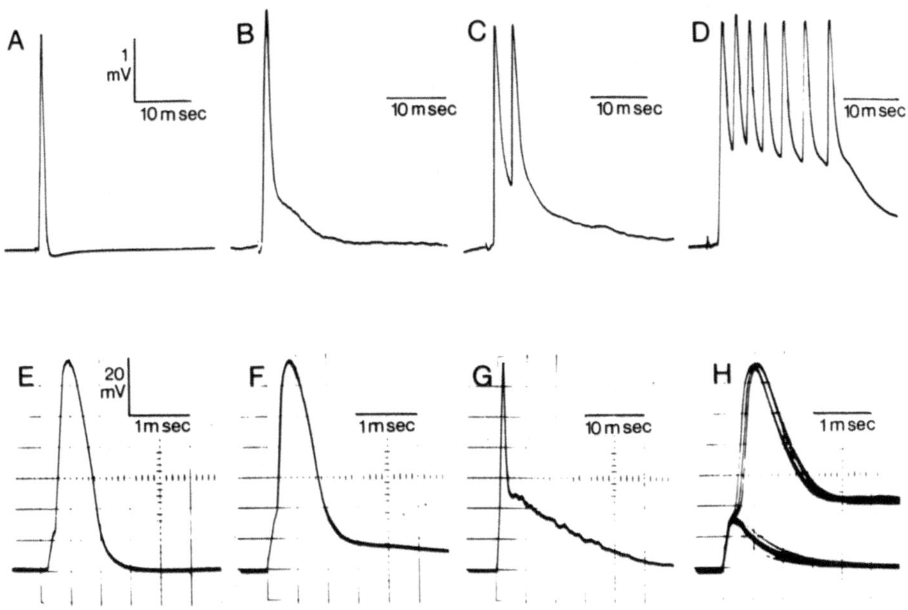

Figure 1. Effects of DDT on single nerve fibres (A-D) and on isolated nodes of Ranvier (E-H). (A) Control. (B) Action potential with depolarizing after-potential, 65 min after application of 5×10^{-4}M DDT. (C-D) Repetitive action potentials, 85 min and 135 min after application of DDT respectively. (E) Control. (F-H) Action potentials with depolarizing after-potential, 6 min, 45 min and 50 min after application of 5×10^{-4}M DDT respectively. Note that in (H) threshold voltage, rate of rise and amplitude of the action potential are not affected by DDT. From van den Bercken (41).

the action potential (Fig. 1B). When the after-potential reached a certain amplitude, in some fibres a second action potential appeared superimposed on the depolarizing after-potential (Fig. 1C). With the advance of time the amplitude of the after-potential and the number of repetitive action potentials gradually increased (Fig. 1D). After prolonged exposure a single stimulus can evoke a train of up to 13 action potentials. In other fibres, however, no repetitive activity was observed, despite the presence of a large depolarizing after-potential. Similarly, in isolated nodes of Ranvier treated with DDT a large depolarizing after-potential was observed, but no repetitive firing occurred (Fig. 1E-G). Resting membrane potential, threshold voltage, rate of rise and amplitude of the action potential were not affected by DDT (Fig. 1H). In other nodes DDT merely caused a prolongation of the whole falling phase of the action potential. Thus, the effect of DDT is exclusively on the falling phase of the action potential. Suppression of the action potential attributable to the effect of DDT was not observed in any of the experiments. These experiments clearly show that a depolarizing after-potential is not the only precondition for a repetitive response and that other important factors must be involved. The development of a depolarizing after-potential and the induction of repetitive activity is probably due to the prolongation of the transient increase in membrane sodium conductance by DDT (17).

Allethrin. Treatment with allethrin (3.3×10^{-7}M to 3.3×10^{-6}M for 15-30 min) also caused pronounced repetitive activity in excised peripheral nerves, which lasted for more than 24 hr without any significant suppression of the action potential (42). As with DDT, the allethrin-induced repetitive activity occurred only in sensory fibres. This was demonstrated by recording compound action potentials from the sensory nerve branch which innervates the skin over the gastrocnemius muscle and from the motor branch which innervates the head of the muscle (Fig. 2A-C). The slight repetitive activity in this latter branch apparently originates from sensory fibres innervating the muscle spindles. In addition, Fig. 2D shows that upon stimulation of the sciatic nerve pronounced repetitive activity occurred in the dorsal spinal root (consisting of sensory fibres), whereas no repetitive activity was observed in the ventral spinal root (which is made up of motor fibres) (46). Higher concentrations of allethrin (3.3×10^{-5}M) also produced repetitive discharges, but the period in which the repetitive firing occurred was transient. After the disappearance of repetitive discharges the action potential was followed by a large depolarizing after-potential. This shows that, as in the case of DDT, a large depolarizing after-potential may occur without repetitive activity. Even after prolonged treatment no blocking effect of allethrin was observed at these concentrations. Only concentrations as high as 3.3×10^{-4}M caused a partial suppression of the compound action potential.

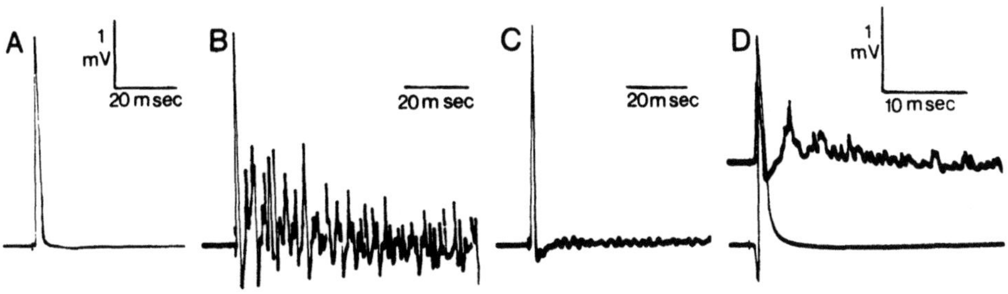

Figure 2. Effects of allethrin on peripheral nerves. (A) Normal action potential of a sensory nerve branch. (B) Same nerve branch, after 40 min of exposure to 3.3 x 10^{-7}M allethrin. (C) Action potential of a motor branch after 120 min of exposure to 3.3 x 10^{-6}M allethrin. (D) Action potentials recorded from the Xth spinal root upon stimulation of the sciatic nerve after 120 min of exposure to 1 x 10^{-6}M allethrin. Upper trace: dorsal root; lower trace: ventral root. Note the absence of repetitive activity in the ventral root. (A-C) from van den Bercken et al (42), (D) from Wouters et al (46).

The effect of allethrin on single myelinated nerve fibres remains to be investigated. It seems, however, reasonable to suggest that the allethrin-induced repetitive activity in myelinated nerve fibres results from a prolongation of the sodium conductance similar to that observed in the nerve membrane of invertebrates (23, 24).

Dieldrin. Exposure of frog nerve preparations to dieldrin in concentrations of up to 1 x 10^{-3}M produced neither repetitive activity, nor any significant effect in single nerve fibres or isolated nodes of Ranvier (42). After prolonged exposure only a slight prolongation of the falling phase of the action potential was observed in some cases. Recently, it has been shown that the dieldrin metabolite aldrin-transdiol at a concentration of 5 x 10^{-5}M caused a blockade of the action potential in squid giant axons, mainly by a suppression of the increase in sodium conductance of the membrane, while dieldrin itself was without any effect (47). Aldrin-cisdiol, which is also a major metabolite of dieldrin (48, 49), caused very similar effects in this preparation (50). Preliminary experiments revealed no blocking action of aldrin-transdiol in frog sciatic nerve (43), but these experiments should be repeated with desheathed nerve preparations.

Effects on Synaptic Transmission

DDT. Treatment of the sartorius nerve-muscle preparation with 1 x 10^{-4}M DDT for 1 hr caused pronounced repetitive activity in the motor end-plate due to repetitive firing of the presynaptic nerve fibres (Fig. 3A, B). This is in sharp contrast with the absence of repetitive discharges in motor nerve fibres as described above. Presumably the repetitive activity arises in the terminal parts of the motor nerve fibres only, as demonstrated with allethrin (see below).

Allethrin. Nerve-muscle preparations treated with allethrin (1 x 10^{-7}M to 1 x 10^{-6}M) also showed repetitive end-plate potentials in response to a single nerve stimulus (46). However, the action potential recorded simultaneously from the sciatic nerve upon stimulation of the ventral spinal root failed to show any sign of repetitive discharges (see Fig. 3C). With higher concentrations of allethrin similar repetitive activity was observed, but later on the number of repetitive end-plate potentials gradually declined until one stimulus again evoked a single response despite the continuous application of allethrin. By recording nervous activity along the whole nerve innervating the sartotius muscle, - from the spinal roots down to the smallest nerve fibres in the muscle -, it was demonstrated that the motor fibres showed no repetitive discharges except in the vicinity of

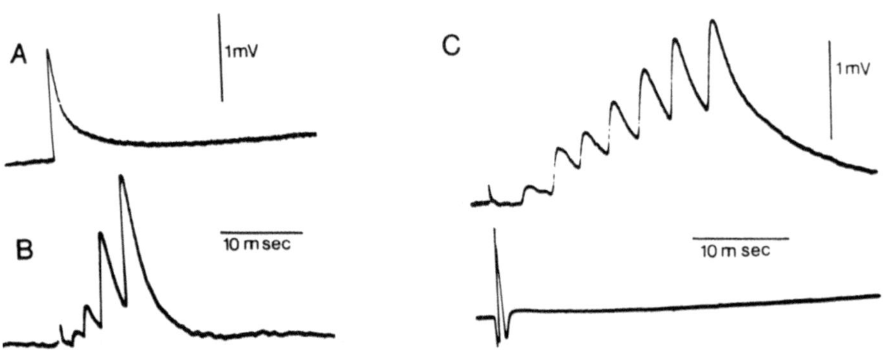

Figure 3. Effects of DDT and allethrin on synaptic transmission in the frog motor end-plate. (A) Intracellulary recorded end-plate potential in a control preparation. (B) Repetitive end-plate potentials after 60 min of exposure to 1 x 10^{-4}M DDT. (C) Repetitive end-plate potentials (upper trace) together with the action potential of the sciatic nerve (lower trace) recorded simultaneously after 50 min of exposure to 1 x 10^{-6}M allethrin. Note the absence of repetitive activity in the sciatic nerve. (C) from Wouters et al (46).

the muscle. Apparently, only the terminal part of the motor fibres becomes repetitive after treatment with allethrin. No further effects of allethrin were observed, either on the postsynaptic membrane, or on the muscle fibre membrane. Therefore, the effect of allethrin is exclusively on the presynaptic nerve membrane resulting in pronounced repetitive activity. Presynaptic repetitive firing was also observed in the allethrin-treated Xth sympathetic ganglion of the frog by means of intracellular microelectrodes. Recently, Evans (51) has reported the occurrence of repetitive end-plate potentials in frog muscle exposed to cismethrin, another synthetic pyrethroid.

Dieldrin. Exposure of the sartorius muscle to dieldrin in a concentration of 1 x 10^{-4}M for more than 3 hr was without any significant effect on neuromuscular transmission. On the other hand, aldrin-transdiol (2.5 x 10^{-5}M) had a powerful and rather complicated action on synaptic transmission in the frog motor end-plate (43). Within 5-10 min after application this metabolite caused a marked increase in spontaneous transmitter release, as

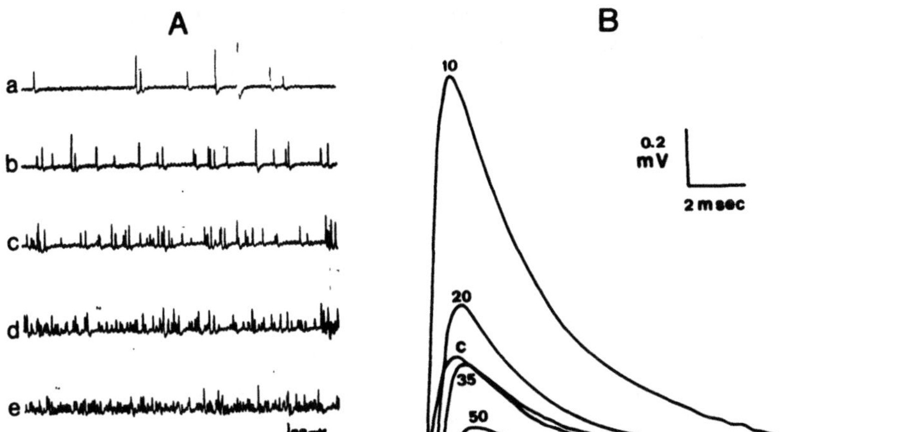

Figure 4. Effects of aldrin-transdiol on spontaneous (A) and evoked (B) transmitter release in frog motor end-plate. (A) Miniature end-plate potentials recorded before (a) and on various times after application of 2.5 x 10^{-5}M aldrin-transdiol (b-f). Note the marked increase in frequency and the decrease in amplitude after application of aldrin-transdiol. (B) Automatically averaged end-plate potentials before (c) and at intervals after the application of 2.5 x 10^{-5}M aldrin-transdiol. Numbers indicate time in minutes. From Akkermans et al (43), with permission of Academic Press, Inc.

revealed by a sharp increase in miniature end-plate potential frequency (Fig. 4A). This was accompanied with a decrease in miniature end-plate potential amplitude. Aldrin-transdiol also affected evoked transmitter release resulting in a dramatic increase in end-plate potential amplitude, followed by a decrease (Fig. 4B). The increase in end-plate potential size was shown to be due to an increase in quantal content, i.e., in the amount of transmitter released by a nerve impulse. Concomitant with the increase in end-plate potential amplitude there was a marked enhancement of facilitation during short trains of stimuli. Both these effects can be explained by assuming that aldrin-transdiol increases the amount of calcium which enters the nerve terminal during the nerve impulse (44). After longer exposure to aldrin-transdiol end-plate potential amplitude and quantal content declined and ultimately neuromuscular transmission was almost completely blocked. This latter effect, together with the fall in miniature end-plate potential amplitude, could be partly explained by a reduction of the chemical sensitivity of the postsynaptic membrane by aldrin-transdiol, as demonstrated by the iontophoretic application of acetylcholine. A presynaptic blocking action may also be involved.Aldrin-transdiol also markedly affected reflex activity of the frog spinal cord, producing both excitatory and inhibitory actions, while again application of dieldrin itself to the isolated preparation remained without effect (52).

Effects on Sense Organs

Cutaneous touch receptors

The cutaneous touch receptors consist of free nerve endings in the skin, without specialized accessory structures. These receptors show no spontaneous afferent nerve activity. By adjusting location and intensity of a brief mechanical stimulus to the surface of the skin it was possible to elicit a single impulse in one of the afferent nerve fibres (Fig. 5A, C).

DDT. After exposure of the isolated preparation to 1×10^{-5}M DDT for 30-90 min, the same stimulus invariably evoked a repetitive response, as illustrated in Fig. 5B (45). The number of repetitive impulses gradually increased with time to a maximum of 8-10 impulses after 120-180 min of exposure. Preparations from animals which showed severe symptoms of poisoning after treatment with DDT in vivo always showed a train of impulses in response to a brief mechanical stimulus.

Allethrin. Allethrin was more effective than DDT in producing repetitive activity in the touch receptor preparation (45). Short trains of impulses were observed as early as 5 min after exposure

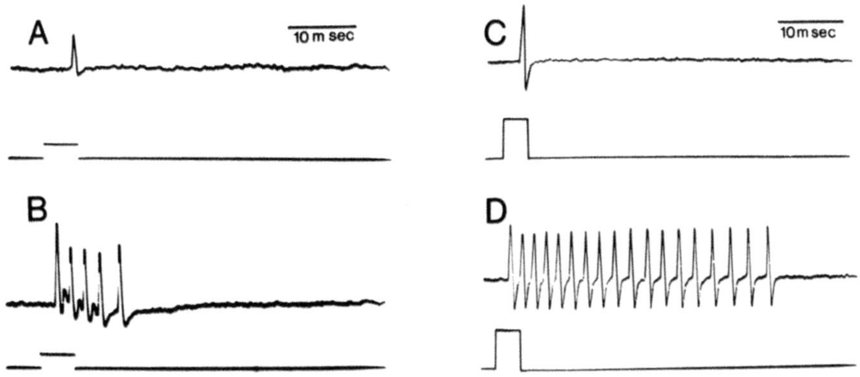

Figure 5. Effects of DDT (A, B) and allethrin (C, D) on cutaneous touch receptors. Response of a cutaneous touch receptor to a brief mechanical stimulus (lower trace) (A, C) before, (B) after 80 min of exposure to 1 x 10^{-5}M DDT and (C) after 60 min of exposure to 1 x 10^{-5}M allethrin. From Akkermans et al (45), with permission of Academic Press, Inc.

to 1 x 10^{-5}M or 15 min to 1 x 10^{-6}M allethrin. After longer exposure the number of repetitive impulses was usually much higher than with DDT (Fig. 5D) and occasionally trains of more than 25 impulses in response to a single stimulus were observed.

Dieldrin. Treatment with dieldrin, either in vivo or in vitro, did not cause any repetitive activity in the cutaneous touch receptors. Exposure of the preparation to aldrin-transdiol (1 x 10^{-5}M) also failed to induce repetitive nerve impulses in response to a brief mechanical stimulus (45). Since the touch receptors are not spontaneously active, a possible blocking effect of aldrin-transdiol cannot be ruled out.

Lateral-line organ

The lateral-line organ contains specialized receptor cells, the sensory hair cells. At their apical surface the receptor cells possess cilium-like processes, the sensory hairs; at their base they form chemical synapses with branches of the afferent nerve fibres (Fig. 6A). The lateral-line organ of Xenopus is located superficially in the skin and serves as a detector of water motions (53, 54). The sensory hair cells are grouped together in organs called neuromasts. The sensory hairs of each neuromast are embedded in a gelatinous cupula which protrudes from the skin into the surrounding water. A number of neuromasts are arranged in a row to form a stitch, and every stitch is innervated by only two

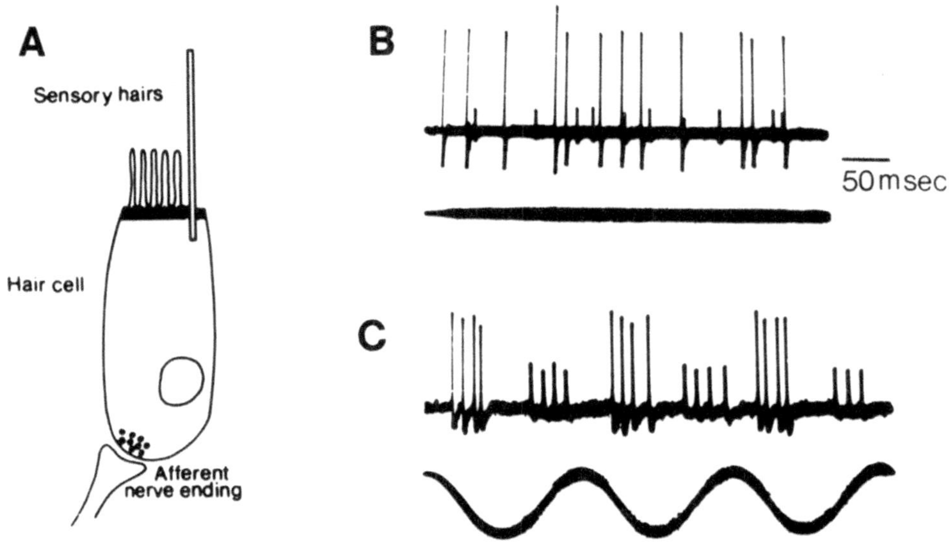

Figure 6. (A) Schematic diagram of a sensory hair cell. For clarity the efferent innervation has been omitted. (B) Spontaneous activity from a single lateral-line organ. Note the two types of impulses, each originating from one of the two afferent nerve fibres. (C) Response of a single lateral-line organ to a sinusoidal water motion.

branching, afferent nerve fibres. A typical characteristic of the lateral-line organ is its irregular, spontaneous activity, even in the absence of any mechanical stimulation, which is thought to be brought about by the spontaneous release of transmitter from the base of the sensory hair cells (55). In most preparations the two types of impulses, each originating from one of the two afferent nerve fibres innervating the stitch, can be easily distinguished (Fig. 6B). A stitch is directionally sensitive and a water displacement evokes opposite reactions in the two afferent nerve fibres; when activity is enhanced in one fibre, it decreases in the other fibre. A sinusoidal stimulus of sufficient strength produces trains of impulses alternating between the two afferent nerve fibres (Fig. 6C). The interval between the impulses depends on stimulus intensity and stimulus frequency.

DDT. Lateral-line preparations from animals which had been exposed to 2-5 ppm DDT for about 18 hr (1 ppm is equivalent to 2.8×10^{-6}M) showed a pronounced repetitive activity, as illustrated in Fig. 7A. Single impulses no longer occurred; all impulses were grouped together in short trains separated by irregular intervals, which were, however, much longer than in the untreated preparation. The number of repetitive impulses per train increased

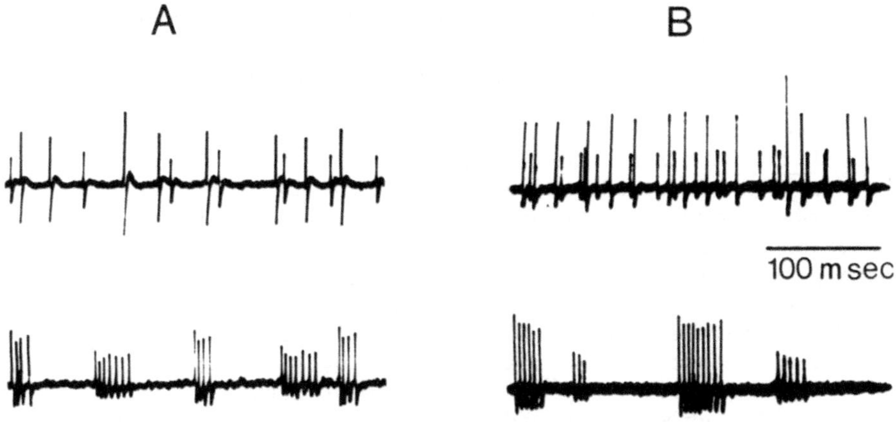

Figure 7. Effect of DDT (A) and allethrin (B) on spontaneous activity from a single lateral-line organ. Upper trace: normal spontaneous activity. Lower trace: (A) Repetitive activity from a lateral-line preparation isolated from an animal exposed to 3 ppm DDT for 18 hr. (B) Repetitive activity from a lateral-line organ 60 min after application of 1 x 10^{-5}M allethrin. (B) from Akkermans et al (45), with permission of Academic Press, Inc.

with increasing DDT-concentrations and continued to increase in the isolated preparation to a maximum of 7-8 impulses at room temperature (40, 42). As a result of this repetitive activity, the relation between stimulus and response of the lateral-line organ was completely upset. It was possible, however, to obtain a synchronization between a low frequency periodic water motion and the impulse trains, but the number of impulses per train and the interval between the impulses remained constant and no longer depended on the stimulus. Attempts to induce repetitive activity by application of DDT (5 x 10^{-4}M) to the isolated lateral-line preparation were unsuccessful, probably because the preparation did not survive long enough to allow sufficient accumulation of the insecticide.

Allethrin. Exposure of a clawed frog to 1-3 ppm allethrin for 20-30 min (1 ppm is equivalent to 3.3 x 10^{-6}M) also produced repetitive activity in the lateral-line organ, very similar to that caused by DDT (42, 45). Repetitive activity could also be induced by application of allethrin to the isolated preparation (1 x 10^{-6}M to 1 x 10^{-5}M for 30-180 min). Fig. 7B shows the spontaneous activity from the lateral-line organ before and after treatment with allethrin. As with DDT, it was possible to obtain a synchronisation between a periodic water motion and the impulse trains, but again the interval between the impulses and the number of impulses per train remained constant and were independent of the stimulus. With higher concentrations of allethrin (3.3×10^{-4}M) the rate of spontaneous firing gradually declined and finally all activity disappeared. In this case no marked repetitive activity was observed.

Dieldrin. After treatment with dieldrin, either in vivo or in vitro, no sign of repetitive activity in the lateral-line organ was observed, nor was there any other significant effect, even in preparations taken from animals which showed severe symptoms of poisoning (45). Application of aldrin-transdiol to the isolated preparation also failed to induce repetitive activity. However, this metabolite had a marked effect on the rate of spontaneous activity, as illustrated in Fig. 8. With lower concentrations of aldrin-transdiol (2.5×10^{-6}M) a significant increase in spontaneous firing was observed. This was followed by a progressive decline and ultimately a blockade of all spontaneous activity. Higher concentrations of aldrin-transdiol produced no increase in firing but the blockade of spontaneous activity developed much faster. After all spontaneous activity had disappeared impulses of full amplitude could be elicited by mechanical stimulation,

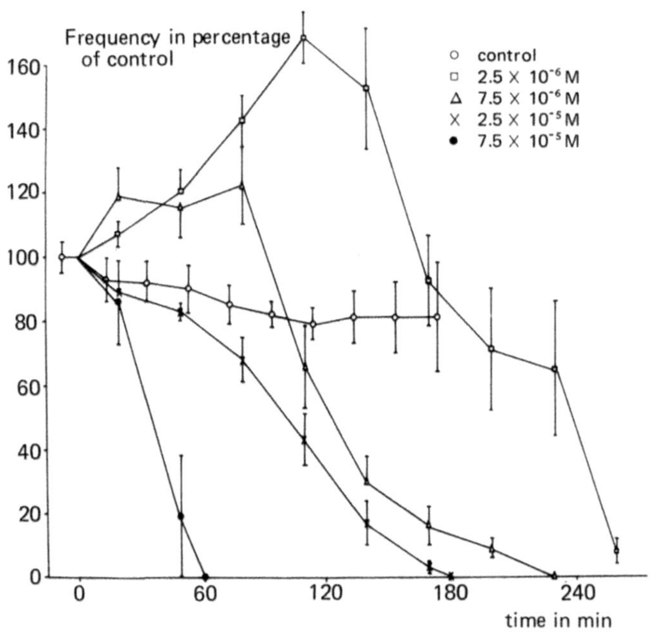

Figure 8. Effect of aldrin-transdiol on the rate of spontaneous firing in single afferent lateral-line nerve fibres. Each point represents the mean frequency ± SD measured over 18 successive periods of 100 sec. From Akkermans et al (45), with permission of Academic Press, Inc.

indicating that the blockade of spontaneous firing was not due to impairment of nervous conduction. After prolonged exposure to aldrin-transdiol the evoked responses were also abolished.

Site of Origin of Repetitive Activity in Sense Organs

The repetitive activity induced by DDT or allethrin in the cutaneous touch receptors and in the lateral-line organ could simply be due to repetitive firing of the afferent nerve fibres as described above. In the cutaneous touch receptor this possibility was tested by electrical stimulation of the nerve branches in the skin in DDT- or allethrin-treated preparations (45). In nearly all cases the compound action potentials elicited in this way were followed by about the same number of repetitive discharges as were observed after mechanical stimulation. No clear difference between either repetitive response could be established. Thus, it was not possible to distinguish between repetitive activity from the touch receptors and repetitive firing of the afferent nerve fibres.

In the lateral-line organ, however, the situation was completely different (40, 42). In several experiments the nerve innervating the lateral-line organ was isolated from preparations which showed pronounced repetitive activity on treatment with DDT or allethrin. In all cases the action potential recorded from this nerve upon electrical stimulation was not followed by repetitive discharges. Furthermore, it was possible to record the electrically-evoked action potential of the lateral-line nerve simultaneously with the spontaneous activity from the lateral-line sense organ. For this purpose the distal end of lateral-line nerve was also dissected free from the skin, leaving intact the small side branch that innervated the selected stitch. The distal end of the nerve was put across a pair of stimulating electrodes and the proximal end across a pair of recording electrodes, as shown in Fig. 9A. The compound action potential of the lateral-line nerve recorded in this way in preparations in which the lateral-line sense organ was highly repetitive after treatment with DDT failed to show any sign of repetition. This is illustrated in Fig. 9B which shows the electrically evoked action potential of the lateral-line nerve together with a train of repetitive impulses from the selected stitch in a preparation from a DDT-treated animal. Similar results were obtained in allethrin-treated preparations. These experiments clearly demonstrate that the repetitive activity of the lateral-line organ does not originate in the afferent nerve fibres, but requires an intact sense organ. On the other hand, it was shown that after treatment with DDT or allethrin the lateral-line organ is still mechanically sensitive and that a synchronization between a periodic stimulus and the

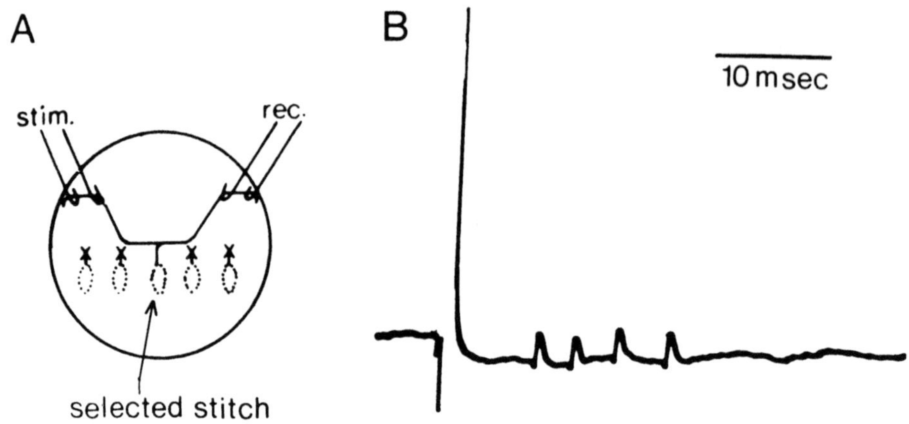

Figure 9. (A) Arrangement of stimulating (stim.) and recording (rec.) electrodes in the lateral-line preparation for recording of the electrically evoked action potential from the lateral-line nerve simultaneously with the spontaneous activity from the selected stitch. (B) Electrically evoked action potential of the lateral-line nerve together with a train of repetitive impulses in a preparation isolated from an animal exposed to 3 ppm DDT for 18 hr. From van den Bercken et al (42).

impulse trains occurred. This indicates that the transducer mechanism in the sensory hair cells is not seriously affected by the insecticides. Further, the probabilistic nature of the spontaneous activity was not significantly altered by the insecticides. Although the intervals between the trains of repetitive impulses were much longer than those between the single impulses in an untreated preparation, their interval distribution and temperature dependency was similar to that of the normal spontaneous activity. It is concluded, therefore, that DDT and allethrin primarily affect the impulse-generating process in the afferent nerve terminals in such a way that the single impulses are converted to trains of impulses.

Mechanism of Repetitive Activity

For a short period of time after the end of an action potential the nerve membrane is in a refractory state, resulting from the development of what is called sodium inactivation (56, 57), i.e., the availability of the sodium system is drastically reduced. The sodium system gradually recovers (inactivation is removed) and about 1 msec after the end of the first action potential a second action potential can be elicited.

In a DDT-treated node of Ranvier the membrane was still in a refractory state during the first part of the depolarizing after-potential, and a short depolarizing pulse produced no active membrane response, as illustrated in Fig. 10A. It was, however, possible to elicit a second action potential by means of a short hyperpolarizing pulse (Fig. 10B). A hyperpolarizing pulse applied during the first part of the depolarizing after-potential induced only a small graded response. The amplitude of this response increased as the delay between the first action potential and the hyperpolarizing pulse was increased, until an almost full size second action potential was induced. During the terminal part of the depolarizing after-potential the hyperpolarizing pulse failed again to induce an action potential. According to the Hodgkin-Huxley nerve membrane model (56, 57) the principal effect of such a short hyperpolarization during the after-potential is removal of sodium inactivation. The prolonged sodium current induced by DDT (or allethrin) can, however, not be terminated by this pulse. In principle, this persistent sodium current exceeds threshold and a second action potential can be evoked because the hyperpolarizing pulse removes sufficient sodium inactivation. Sodium inactivation is still high during the initial phase of the after-potential and only a graded response is induced, while during the terminal phase of the after-potential the persistent sodium current is too small to exceed threshold.

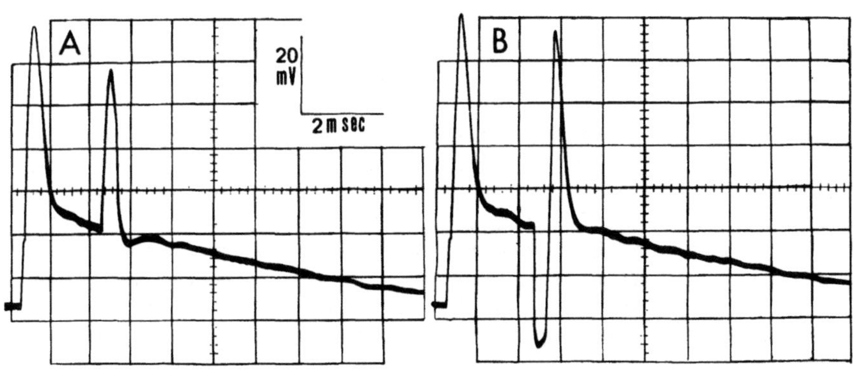

Figure 10. Effects of short current pulses applied during the depolarizing after-potential in a node of Ranvier treated with 5 x 10^{-4}M DDT. (A) A short depolarizing pulse applied during the first part of the depolarizing after-potential producing no active membrane response. (B) A short hyperpolarizing pulse applied at the same instant producing an action potential. From van den Bercken (41).

These results suggest that only those fibres which have a fast rate of recovery from sodium inactivation may produce a repetitive response after treatment with DDT or allethrin. If recovery from inactivation following an action potential is fast enough, even a moderate persistent sodium current may exceed threshold and induce a repetitive response. If, on the other hand, recovery is too slow, a large depolarizing after-potential may occur without repetitive discharges. In point of fact it has been shown that frog sensory fibres tend to have a faster rate of recovery from sodium inactivation than motor fibres (58-60). This would account for the observation that sensory fibres, in contrast to motor fibres, show repetitive activity over their whole length after exposure to DDT or allethrin. Further, it is conceivable that all the membrane conductance parameters do not remain constant over the whole length of the nerve fibres and in particular sodium inactivation kinetics change as the nerve fibres approach their terminals. This would explain why afferent nerve terminals (lateral-line organ) as well as efferent nerve terminals (motor end-plate) are more susceptible to the induction of repetitive activity by DDT and allethrin than the rest of the nerve fibres. Such changes in membrane conductance parameters in favor of a repetitive response are of functional significance in afferent nerve terminals, where the generation of series of impulses is essential for the functioning of the sense organs. Why such changes should also occur in motor nerve terminals is, however, not clear. In addition, certain geometrical factors, e.g. fibre diameter and branching of the nerve fibres, may also contribute to the generation of repetitive discharges in the terminal nerve fibres.

Influence of Temperature

The repetitive activity induced by DDT or allethrin in peripheral nerves was hardly affected by changes in temperature. The only effect that was observed after cooling the nerve was an increase in the intervals between the repetitive discharges (42). In sharp contrast to this, the repetitive activity in the motor nerve terminals in allethrin-treated sartorius muscles increased markedly as the temperature was lowered (46). This effect was readily reversible; raising the temperature caused an almost immediate decrease in the number of repetitive end-plate potentials. Fig. 11 shows the relation between temperature and the number of repetitive end-plate potentials evoked by a single nerve stimulus in 4 different preparations treated with allethrin. Thus, the allethrin induced repetitive activity in the motor nerve terminal exhibits a truly negative temperature coefficient. Preliminary experiments have indicated that this might also be true for the DDT-induced repetitive activity in the motor end-plate.

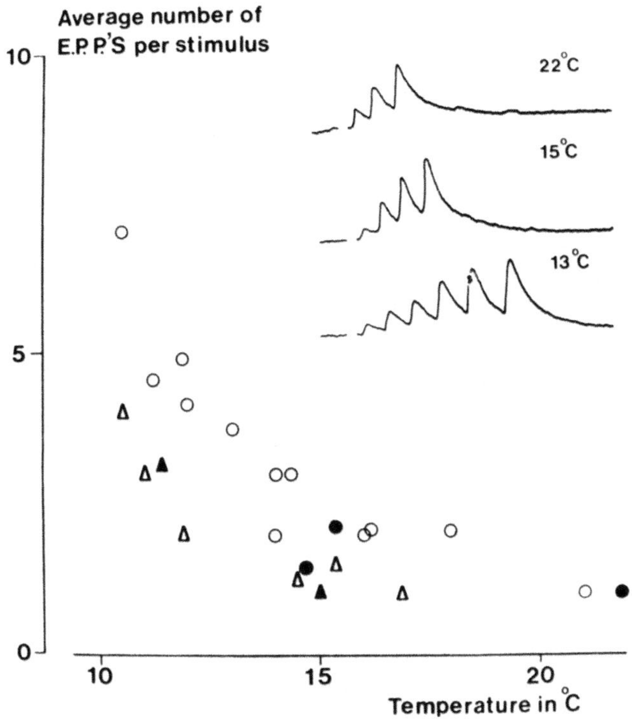

Figure 11. Effect of temperature on repetitive activity in the frog motor end-plate in 4 different preparations exposed to 1 x 10^{-7}M allethrin. Each point represents the mean number of end-plate potentials (E.P.P.'S) evoked by 10 successive stimuli delivered at a frequency of 0.5 Hz. Inset shows three intra-cellular recordings from one end-plate at the temperatures indicated. From Wouters et al (46).

In the cutaneous touch receptor a slight increase in the repetitive activity induced either by DDT or allethrin was sometimes observed when the temperature of the preparation was lowered (45). In most cases, however, there was no effect or even a slight decrease in the number of repetitive impulses after cooling. The situation in the cutaneous touch receptors thus resembles the repetitive firing in peripheral nerves. This is in agreement with the observation that the repetitive activity from the touch receptors cannot be distinguished from repetitive firing of the afferent nerve fibres.

In contrast, the DDT-induced repetitive activity in the lateral-line organ was strongly affected by changes in temperature and lowering the temperature of the preparation caused a dramatic increase in the number of repetitive impulses per train (40, 42). This is illustrated in Fig. 12. In this particular preparation repetitive activity increased from trains of 3-7 impulses at a temperature of 23°C (Fig. 12B) to trains of more than 60 impulses at a temperature of 11°C (Fig. 12E). Contrary to control preparations in which the rate of spontaneous firing always decreases as the temperature is lowered, such preparations even showed an increase in mean firing rate with lowered temperature. It should be noted that the frequency of the impulses during the trains and the frequency of appearance of trains showed a positive temperature coefficient. In other cases the effect of temperature was less dramatic, but there was always a marked increase in the number of impulses per train as the temperature was lowered. Moreover, preparations from animals which were treated with lower concentrations of DDT but did not show repetitive activity at room temperature, might do so after cooling. Raising the temperature caused a decrease in the number of impulses per train, and preparations which were highly repetitive at room temperature

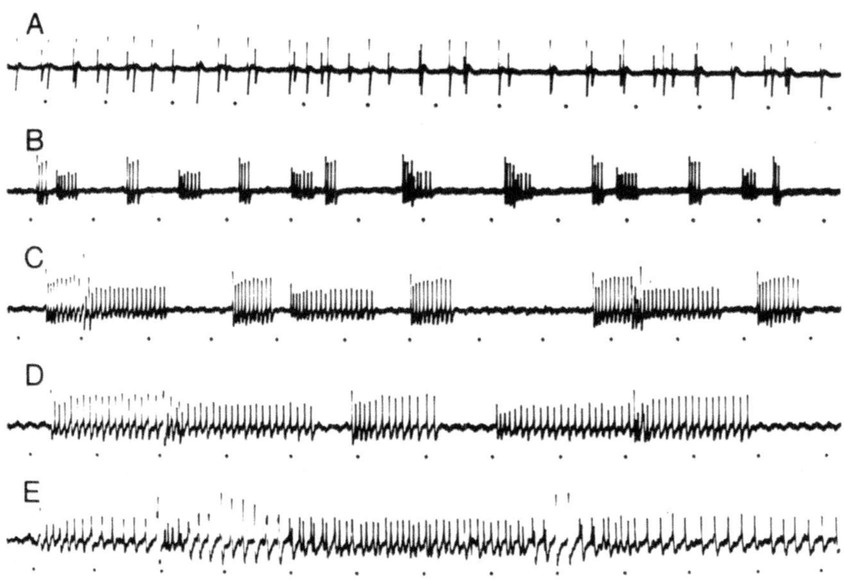

Figure 12. Spontaneous activity from a single lateral-line organ of Xenopus. (A) Control preparation. Temperature 23°C. (B) Preparation isolated from an animal exposed to 3 ppm DDT for 18 hr. Temperature 23°C. (C-E) Same preparation at a temperature of 19, 16 and 11°C, respectively. In (E) two trains of impulses partly coincide. Time mark 100 msec. From van den Bercken and Akkermans (40).

eventually showed almost normal spontaneous activity at a temperature of 28°C.

The effect of temperature was rapidly reversible and could be repeated more than once in the same preparation. Thus, measured by the number of repetitive impulses per train, DDT shows a definite negative temperature coefficient of activity in the lateral-line organ. Strikingly similar results were obtained in preparations treated with allethrin. As in the case of DDT, the number of repetitive impulses per train increased tremendously as the temperature was lowered (Fig. 13), whereas raising the temperature caused the number of impulses per train to fall. Just as in the DDT-treated lateral-line organ the frequency of the impulses during the trains and the frequency of appearance of trains showed a positive temperature coefficient. The action of allethrin, as measured by the number of repetitive impulses per train, also showed a negative temperature coefficient very similar to that following DDT-treatment.

The nerve blocking action of allethrin in squid giant axons also depends highly on temperature and increases markedly upon cooling. This was shown to be mainly due to an enhanced suppressive effect of this compound on membrane ionic conductance at lower temperature (22).

There was no indication for an influence of temperature on the action of the dieldrin metabolite aldrin-transdiol.

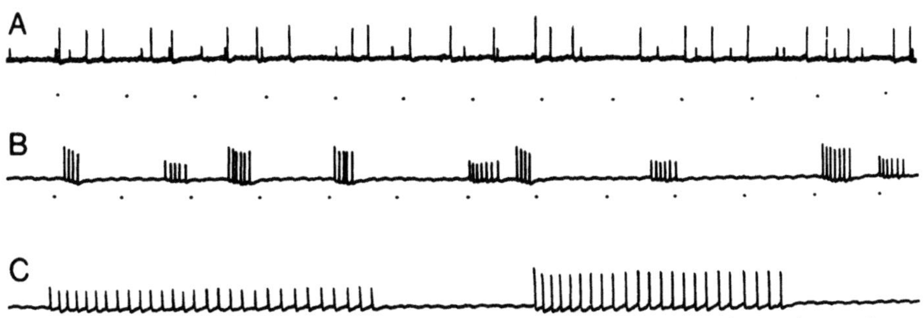

Figure 13. Spontaneous activity from a single lateral-line organ of Xenopus. (A) Control preparation. Temperature 20°C. (B) Same preparation, 70 min after the application of 6.6 x 10^{-6} M allethrin. Temperature 20°C. (C) Same preparation, 5 min later. Temperature 10°C. Time mark 100 msec. From van den Bercken et al (42).

DISCUSSION

The experiments reported here demonstrate a striking similarity between the action of DDT and allethrin in the peripheral nervous system of the frog, despite the marked difference in their molecular structure. Both insecticides strongly affect the sensory nervous system, producing pronounced repetitive activity in sense organs and in sensory nerve fibres. In the frog motor end-plate both compounds also cause marked repetitive firing of the presynaptic motor nerve terminals, but without inducing repetitive activity in the more proximal part of the motor fibres. A genuine synaptic effect of allethrin was not observed, either pre- or postsynaptically, and the same probably holds for DDT. The action of DDT and allethrin in the lateral-line sense organ as well as in the motor nerve terminal shows a negative temperature coefficient, and in both preparations lowering the temperature causes a dramatic increase in repetitive activity. There are only minor differences between the action of both insecticides. Allethrin in high concentrations produces a moderate blocking action which was not observed after treatment with DDT. In addition, allethrin induces repetitive activity at a lower concentration and more rapidly than DDT, but this is probably related to the different physical properties of the molecule.

Voltage clamp experiments by Hille (17) have revealed that in frog myelinated nerve fibres DDT specifically affects the sodium channels of the membrane and causes a prolongation of the transient increase in the sodium conductance of the membrane during excitation. Hille suggests that after treatment with DDT a fraction of the sodium channels that open upon depolarization remain open for a much longer time than normal. Further, DDT would not affect the opening of the sodium channels. A similar prolongation of the transient increase in membrane sodium conductance has been found in the DDT-treated arthropod nerve membrane (8,9).

The action of allethrin on membrane ionic conductances of myelinated nerve fibres has yet to be investigated. In the nervous system of insects and other invertebrates allethrin produces a strong blocking action which was shown to be due to a suppressive effect of this compound on both the sodium and potassium conductance of the nerve membrane during depolarization (25).
In addition, allethrin also produces a prolongation of the transient increase in sodium conductance similar to that observed with DDT (21-24). In the frog peripheral nervous system, however, virtually no blocking effect of allethrin occurs, except after very high concentrations. Because of the striking similarity in action between both insecticides reported here, it seems reasonable to assume that in myelinated nerve fibres allethrin, like DDT, specifically affects the closing of the sodium channels.

The DDT- and allethrin-induced repetitive activity is probably entirely caused by this prolongation of the transient increase in sodium conductance. Differences in sodium inactivation kinetics may adequately account for the observation that in motor nerve fibres the prolonged sodium current results in a large depolarizing after-potential without repetitive discharges, while in sensory nerve fibres and probably also in nerve terminals a similar prolongation of the sodium current induces pronounced repetitive activity. The difference in sensitivity between motor and sensory structures in insects to the induction of repetitive activity by DDT (3-5) can probably be explained on a similar basis.

Since the introduction of DDT a variety of theories on its mode of action has been proposed (for a survey, see ref. 61,62). Yet, nothing definite can be said about its action at the molecular level. Recently it has been found that DDT and a number of other neurotoxic agents, including allethrin, are capable of inhibiting ATPases from nervous tissue (63-65). It is not clear if this has pharmacological implications and is causally related to the effects on membrane sodium conductance. It has also been proposed that DDT affects membrane fluidity (66, 67), but this hypothesis was not confirmed by recent experiments (68). Besides, it seems difficult to see how such a general physical interaction would explain the specific action of DDT on the sodium channels of the nerve membrane.

The most promising theory on the mode of action of DDT presently available might be Holan's elaboration (69 - 71) of Mullin's original hypothesis (72) that DDT fits into a hypothetical pore in the nerve membrane. According to Holan DDT-type molecules act as a "molecular wedge". The two phenylrings comprising the base of the wedge lock the molecule to the membrane, probably by forming a charge transfer complex, while the apex (e.g. the CCl_3-group of DDT) fits into a membrane pore so as to prevent the sodium channel from closing. The molecular dimensions of the apex, and not its chemical nature, are of crucial importance for optimum fit. The base must have electron donating properties, but the size limitations of this part of the molecule are less restrictive. The distribution of electronic charge over the base is also important (73).

Molecular conformation is also the most important factor determining the activity of the pyrethroids (74). Originally it was thought that a cyclopropane ring with a <u>gem</u>-dimethyl group was essential for high activity (75). Recently, however, Ohno et al (76, 77) have developed pyrethroid-like substituted phenyl acetic acid esters with high insecticidal activity. These compounds act in a way similar to the dimethyl cyclopropane carboxylate esters (78, 79). Further, it has been shown that pyrethroid-like compounds lacking an ester grouping, e.g. the ketone analogue of allethrin, have an action on the nerve membrane similar to that of allethrin itself (80). This indicates that the site of action of pyrethroids in the nerve membrane can accomodate several different chemical structures. Since allethrin and probably other pyrethroids, as well as DDT and a number of its analogues primarily interfere with the sodium channels of the nerve membrane, it is likely that both classes of insecticides have common structural features for their interaction with the nerve membrane.

Similar effects of temperature on the DDT- and allethrin-induced repetitive activity as reported in the present study, together with the enhanced nerve blocking action of allethrin at lower temperatures described by Wang et al (22), will contribute to the overall negative temperature coefficient of toxicity of these compounds in insects and probably other poikilothermic animals. The molecular mechanisms responsible for the negative temperature coefficient of action of DDT and allethrin have still to be clarified. These mechanisms might be connected with the dissociation of a charge transfer type of complex between DDT (or allethrin) and membrane macromolecules at higher temperatures as suggested by Holan (69-71). This hypothesis, however, seems difficult to bring into line with the finding that the negative temperature coefficient occurs in certain parts of the nervous system only, i.e., the lateral-line sense organ and the motor nerve terminals, and not in other parts, i.e., the touch receptor and the sensory nerve fibre membrane. It is also possible that, at least with respect to the induction of repetitive activity by these insecticides, lowering the temperature causes a further prolongation of the transient sodium conductance which together with appropriate changes in sodium inactivation parameters will result in enhanced repetitive activity (46). These latter effects would also be rapidly reversible by raising the temperature.

In the present investigation no significant effect of the insecticide dieldrin in any of the isolated nerve preparations was found, even after prolonged exposure at high concentrations. In sharp contrast with this the dieldrin metabolite aldrin-transdiol was highly effective. This strongly supports Narahashi's hypothesis that dieldrin must first be converted to aldrin-transdiol and possibly other metabolites (e.g. aldrin-cisdiol) before it can exert its neurotoxic action (1, 29). Aldrin-transdiol probably primarily affects synaptic transmission, not only in the frog motor end-plate, but also in the isolated frog spinal cord (52). In the motor end-plate this metabolite causes a dramatic increase in both spontaneous and evoked transmitter release, followed by a decrease and finally, after prolonged exposure, a complete blocking of the transmission. The chemical sensitivity of the postsynaptic membrane was also suppressed by aldrin-transdiol. A similar action of aldrin-transdiol on synaptic transmission between the sensory hair cells and the afferent nerve fibres would explain the effects of this metabolite on the spontaneous activity of the lateral-line organ.

Shankland and Schroeder (30) have reported similar effects on synaptic transmission in dieldrin-poisoned ganglia of the American cockroach and these authors have concluded that the action of dieldrin or some metabolite is confined to cholinergic junctions. However, this conclusion does not hold for the frog nervous system. Acetylcholine can probably be excluded as the transmitter in the afferent synapse in the lateral-line hair cells (81). Further, the effect of aldrin-transdiol on reflex activity of the frog spinal cord is difficult to explain on the assumption of a specific cholinergic action (52). Recently it has been argued, mainly on the grounds that injected dieldrin is much more toxic to the American cockroach than either aldrin-transdiol or aldrin-cisdiol, that dieldrin itself is the primary insecticidal agent. Injected aldrin-transdiol is much less toxic to the clawed frog than dieldrin (82), but this might be related to the rapid conversion of aldrin-transdiol to the aldrin-derived dicarboxylic acid (hexachlorohexahydromethanoindenedicarboxylic acid), one of the final products in the metabolic pathway of dieldrin (83). The aldrin-derived dicarboxylic acid was actually detected in the water in which the frog was kept after injection of aldrin-transdiol (52). Moreover, it has been shown that this compound has no effect on synaptic transmission in the frog motor end-plate (43). It would be interesting to know whether a rapid conversion of the diols to the aldrin-derived dicarboxylic acid would also occur in the American cockroach.

ACKNOWLEDGEMENTS

We thank Miss C.Z. Troost for excellent secretarial assistance and Mr. J.M. Eyndhoven for photographic art work. This work was supported in part by the Foundation for Medical Research FUNGO and by Shell Corporation.

LITERATURE CITED

1. Narahashi, T., in "Advances in Insect Physiology", Eds. J.W.L. Beament, J.E.Treherne and V.B. Wigglesworth. Acad. Press, London and New York, (1971) 8, 1.
2. Narahashi, T., in "Insecticide Biochemistry and Physiology", Ed. C.F. Wilkinson, p. 327. Plenum Press, New York and London, 1976.
3. Roeder, K.D. and Weiant, E.A., Science, (1946) 103, 304.

4. Roeder, K.D. and Weiant, E.A., J.Cell. Comp. Physiol., (1948) 32, 175.
5. Roeder, K.D. and Weiant, E.A., Ann. Ent. Soc. Am., (1951) 44,372.
6. Narahashi, T. and Yamasaki, T., J. Physiol., (1960a) 152, 122.
7. Narahashi, T. and Yamasaki, T., J. Cell. Comp. Physiol., (1960b) 55, 131.
8. Narahashi, T. and Haas, H.G., J.Gen. Physiol., (1968) 51, 177.
9. Pichon, Y., J. Physiol. (Paris), (1969) 61, (Suppl. 1), 162.
10. van den Bercken, J., European J. Pharmacol., (1969) 9, 146.
11. Wu, Ch.H., van den Bercken, J. and Narahashi, T., Pestic. Biochem. Physiol., (1975) 5, 142.
12. Eyzaguirre, C. and Lilienthal Jr., J.L., Proc. Soc. Exptl. Biol., (1949) 70, 272.
13. Shankland, D.L., Toxicol. Appl. Pharmacol.,(1964) 6, 197.
14. Shanes, A.M., J. Cell. Comp. Physiol., (1951) 38, 17.
15. Woolley, D.E. and Barrow, B.A., Toxicol. Appl. Pharmacol., (1968) 12, 440.
16. Woolley, D.E., in "The Biological Impact of Pesticides in the Environment", Ed. W. Gillet, p. 114. Oregon State University Press, Corvallis, 1970.
17. Hille, B., J. Gen. Physiol., (1968) 51, 199.
18. Århem, P., Frankenhaeuser, B., Göthe, R. and O'Bryan, P., Acta Physiol. Scand., (1974) 91, 130.
19. Narahashi, T., J. Cell. Comp. Physiol., (1962a) 59, 61.
20. Narahashi, T., J. Cell. Comp. Physiol., (1962b) 59, 67.
21. Narahashi, T., Bull. Wld Hlth Org., (1971) 44, 337.
22. Wang, C.M., Narahashi, T. and Scuka, M., J. Pharmacol. Exp. Ther., (1972) 182, 442.
23. Murayama, K., Abott, N.J., Narahashi, T. and Shapiro, B.I., Comp. Gen. Pharmac., (1972) 3, 391.
24. Narahashi, T. Pestic. Sci., (1976) 7, 267.
25. Narahashi, T. and Anderson, N.C., Tox. Appl. Pharmac., (1967) 10, 529.
26. Camougis, G. and Davis, W.M., Pyrethrum Post, (1971) 11, 7.
27. Camougis, G., in "Pyrethrum. The natural insecticide", Ed. I.E. Casida, p. 211. Academic Press Inc., New York, 1973.
28. Lalonde, D.I.V. and Brown, A.W.A., Canad. J. Zool., (1954) 32, 74.
29. Wang, C.M., Narahashi, T. and Yamada, M., Pestic. Biochem. Physiol., (1971) 1, 84.
30. Shankland, D.L. and Schroeder, M.E., Pestic. Biochem. Physiol., (1973) 3, 77.
31. Schroeder, M.E., Shankland, D.L. and Hollingworth, R.M., Pestic. Biochem. Physiol., (1977) 7, 403.
32. Gowdey, C.W., Graham, A.R., Seguin, J.J. and Stavraky, G.W., Can. J. Biochem. Physiol., (1954) 32, 498.
33. Natoff, I.L. and Reiff, B., Brit. J. Pharmacol. Chemother., (1967) 31, 197.

34. Joy, R.M., Neuropharmacol., (1974) 13, 93.
35. Joy, R.M., Tox. Appl. Pharmac., (1976) 38, 357.
36. Eaton, J.L. and Sternburg, J., Insect Physiol., (1964) 10, 471.
37. Yamasaki, T. and Ishii, T., Oyo-Kontyu, (1953) 9, 87. English translation (1957). In "Japanese Contributions to the Study of the Insecticide-Resistance Problem", p. 130. Publ. by Kyoto Univ. for W.H.O.
38. Yamasaki, T. and Ishii, T., Botyu-Kagaku, (1954) 19, 39. English translation (1957). In "Japanese Contributions to the Study of the Insecticide-Resistance Problem", p. 155. Publ. by Kyoto Univ. for W.H.O.
39. Eaton, J.L. and Sternburg, J.G., J. Econ. Ent., (1967) 60, 1358.
40. van den Bercken, J. and Akkermans, L.M.A., European J. Pharmacol., (1971) 16, 241.
41. van den Bercken, J., European J. Pharmacol., (1972) 20, 205.
42. van den Bercken, J., Akkermans, L.M.A. and van der Zalm, J.M., European J. Pharmacol., (1973) 21, 95.
43. Akkermans, L.M.A., van den Bercken, J., van der Zalm, J.M. and van Straaten, H.W.M., Pestic.Biochem. Physiol.,(1974) 4, 313.
44. Akkermans, L.M.A., van den Bercken, J. and van der Zalm, J.M., European J. Pharmacol., (1975) 31, 166.
45. Akkermans, L.M.A., van den Bercken, J. and Versluijs-Helder, M., Pestic. Biochem. Physiol., (1975) 5, 451.
46. Wouters, W., van den Bercken, J. and van Ginneken, A., European J. Pharmacol., (1977) 43, 163.
47. van den Bercken, J. and Narahashi, T., European J. Pharmacol., (1974) 27, 255.
48. Nelson, J.O. and Matsumara, F., Arch. Environ. Contam. Toxicol., (1973) 1, 224.
49. Matthews, H.B. and McKinney, J.D., Drug Metabol. Dispos., (1974) 2, 333.
50. van den Bercken, J. and Narahashi, T., unpublished observations.
51. Evans, M.H., Pestic. Biochem. Physiol., (1976) 6, 547.
52. Akkermans, L.M.A., van den Bercken, J. and Versluijs-Helder, M., European J. Pharmacol., (1975) 34, 133.
53. Görner, P., Z. Vergleich. Physiol., (1963) 47, 316.
54. Harris, G.G. and Milne, D.C., J. Acoust. Soc. Am., (1966) 40, 32.
55. Flock, A., Jørgensen, M. and Russell, I., in "Basic Mechanisms in Hearing", Ed. A.R. Møller, p. 273. Academic Press Inc., New York, 1973.
56. Hodgkin, A.L. and Huxley, A.F., J. Physiol., (1952) 117, 500.
57. Dodge, F.A., Thesis, The Rockefeller Institute, New York (University Microfilms, Inc., Ann Arbor, Michigan, No. 64-7333), 1963.
58. Vallbo, Å.B., Acta Physiol. Scand., (1964) 61, 429.

59. Schmidt, H. and Stämpfli, R., Helv. Physiol. Acta, (1964) 22, C 143.
60. Frankenhaeuser, B. and Vallbo, Å.B., Acta Physiol. Scand., (1965) 63, 1.
61. Metcalf, R.L., J. Agr. Food Chem., (1973) 21, 511.
62. Brooks, G.T., in "Chlorinated Insecticides, vol. II", p. 130. CRC Press Inc., Cleveland, Ohio, 1975.
63. Matsumara, F. and Narahashi, T., Biochem. Pharmacol., (1971) 20, 825.
64. Schneider, R.P., Biochem. Pharmacol., (1975) 24, 939.
65. Desaiah, D., Cutkomp, L.K., Vea, E.V. and Koch, R.B., Gen. Pharmacol., (1975) 6, 31.
66. Hilton, B.D. and O'Brien, R.D., Science, (1970) 168, 841.
67. Hilton, B.D., Bratkowski, T.A., Yamada, M., Narahashi, T. and O'Brien, R.D., Pestic. Biochem. Physiol., (1973) 3, 14.
68. O'Brien, R.D., Herbert, T.J. and Hilton, B.D., Pestic. Biochem. Physiol., (1977) 7, 416.
69. Holan, G., Nature, (1969) 221, 1025.
70. Holan, G., Bull. Wld Hlth Org., (1971a) 44, 335.
71. Holan, G., Nature, (1971b) 232, 644.
72. Mullins, L.J., in "Molecular structure and functional activity of nerve cells", Publ. No. 1, p. 123. Am. Inst. Biol. Sci., Washington, D.C., 1956.
73. Holan, G. and Spurling, T.H., Experientia, (1974) 30, 480.
74. Elliott, M., in "Synthetic pyrethroids, ACS Symposium Series", No. 42, Ed. M. Elliott, p. 1. ACS Washington, D.C., 1977.
75. Elliott, M., Farnham, A.W., Janes, N.F., Needham, P.H. and Pulman, D.A., in "Mechanism of pesticide action, ACS Symposium Series", No. 2, Ed. G.K. Kohn, p. 80. ACS Washington, D.C., 1974.
76. Ohno, N., Fujimoto, K., Okuno, Y., Mizutani, T., Hirano, M., Itaya, N., Honda, T. and Yoshioka, H., Agr. Biol. Chem., (1974) 38, 881.
77. Ohno, N., Fujimoto, K. Okuno, Y., Mizutani, T., Hirano, M., Itaya, N., Honda, T. and Yoshioka, H., Pestic. Sci., (1976) 7, 241.
78. Clements, A.N. and May, T.E., Pestic. Sci., (1977) 8, 661.
79. Narahashi, T., Nishimura, K., Parmentier, J.L., Takeno, K. and Elliott, M., in "Synthetic pyrethroids", Ed. M. Elliott, p. 85. ACS Symposium Series No. 42, Washington, D.C., 1977.
80. Berteau, P.E., Casida, J.E. and Narahashi, T., Science, (1968) 161, 1151.
81. Russell, I.J., J. Exp. Biol., (1971) 54, 643.
82. Akkermans, L.M.A., unpublished observations.
83. Baldwin, M.K., Robinson, J. and Parke, D.V., Food Cosmet. Toxicol., (1972) 10, 333.

NERVE MEMBRANE IONIC CHANNELS AS THE TARGET SITE OF INSECTICIDES

Toshio Narahashi

Department of Pharmacology, Northwestern University

Medical School, Chicago, Illinois 60611

I. INTRODUCTION

Despite the much debated controversy over the continuing use of DDT as insecticide, and despite the fact that its use is now banned or restricted in some countries, there are reasons to believe that DDT will continue to be a subject of intense investigations in the years to come. In particular, the mode of action of DDT still remains a mystery and is a target of scientific curiosity because of its potent insecticidal activity with relatively low mammalian toxicity. The elucidation of the mechanism of action of DDT is urgently and much needed, since it will provide us with a clue to develop better insecticides with many of DDT's advantages preserved while many of its disadvantages are eliminated.

A wealth of information has been accumulated concerning the biodegradation of DDT in insects, mammals and environment. However, the study of its mode of action on the target tissue has progressed at a much slower pace owing to various technical difficulties involved. If the target site could be isolated *in vitro*, the study would have been done more easily. The best such examples are seen with organophosphate and carbamate insecticides whose major action is to inhibit cholinesterases. Fortunately highly purified cholinesterase preparations are available *in vitro*, so that the interactions of those insecticides with the enzyme can be analyzed in well controlled conditions. Furthermore, a large number of experiments can be performed using a variety of analogs and derivatives in order to elucidate the molecular mechanism of action of these insecticides on cholinesterases. Such studies have indeed proved successful, and our knowledge about this subject has been greatly advanced (Fukuto, 1971; Metcalf, 1971).

Unfortunately, the target molecule of DDT has not yet been identified. DDT has no effect on cholinesterases, and does not affect various oxidases and dehydrogenases. The only enzyme system that has been demonstrated to be affected by DDT is ATPase (Cutkomp et al., 1971; Desaiah et al., 1974; Matsumura and Patil, 1969; Cheng and Cutkomp, 1977; Price, 1976; Schneider, 1975). As will be discussed later in this paper, the role of inhibition of ATPase in the alteration of nerve function by DDT still remains to be seen. The only direct action of DDT on the nervous system is hyperexcitability as manifested by repetitive firing and synaptic facilitation. These changes in nerve excitability can be attributed to modifications of membrane ionic permeabilities which are the key factor for excitation. It should be noted that the changes in ionic permeability of nerve membrane are basically physico-chemical phenomena and independent of metabolic energy. Therefore, it is unlikely that disturbance of any enzyme system by DDT is directly involved in the observed hyperexcitability. However, this does not exclude the possibility that the observed inhibition of ATPases by DDT contributes to certain phases of nervous symptom of poisoning through modulation of some parameters associated with nerve excitation.

It then follows that the key experiment for the study of the mechanism of action of DDT is to identify the membrane parameters as the primary target site. First we have to accurately describe the sequence of events occurring in the nervous system as a result of DDT intoxication. Then we can analyze the action of DDT on each sequence using sophisticated electrophysiological techniques. Using some of these key parameters that have been demonstrated to be the target site of DDT, we will be able to compare a variety of analogs and derivatives to establish the true structure-activity relationship.

The present paper is divided into two parts. First, the effects of DDT on synaptic and neuromuscular transmission will be discussed in an attempt to identify the primary target site of DDT in the nervous system. It will be shown that the facilitatory action of DDT is ascribed to repetitive firing in the presynaptic nerve terminals. This gives us the basis for further analysis of the DDT action on nerve membranes in an attempt to identify the more specific target site. Extensive electrophysiological experiments have been performed along this line to demonstrate the sodium inactivation mechanism as the major target site of DDT.

II. EFFECTS OF DDT ON THE NERVOUS SYSTEM

The symptom of DDT poisoning in insects is characterized by

ataxia, hyperexcitability and convulsions. The poisoned insects become paralyzed later and die eventually, but it takes a long time for the death to come depending on the dose of DDT, the species of insect, and the temperature. It should be borne in mind that DDT itself does not quickly paralyze the nervous system. It has indeed been shown that a toxic substance sometimes called autotoxin is released from the nervous system as a result of DDT-induced hyperexcitation, and that the toxin in turn stimulates and paralyzes the nerve (Hawkings and Sternburg, 1964; Shankland and Kearns, 1959; Sternburg, 1960, 1963; Sternburg and Kearns, 1952; Sternburg et al., 1959). Thus the primary action of DDT is to cause hyperexcitation of the nervous system.

Roeder and Weiant (1946) discovered that DDT causes repetitive discharges to appear in the sensory cells of the cockroach trochanter. Repetitive discharges form trains of impulses which appear periodically. This observation was confirmed and extended by several investigators (Becht, 1958; La Londe and Brown, 1954; Roeder and Weiant, 1948, 1951; Yamasaki and Ishii, 1954a,b) (Fig. 1).

The effect is produced by a very low concentration of DDT ranging from 10^{-7} to 10^{-8}M. It also depends on the temperature,

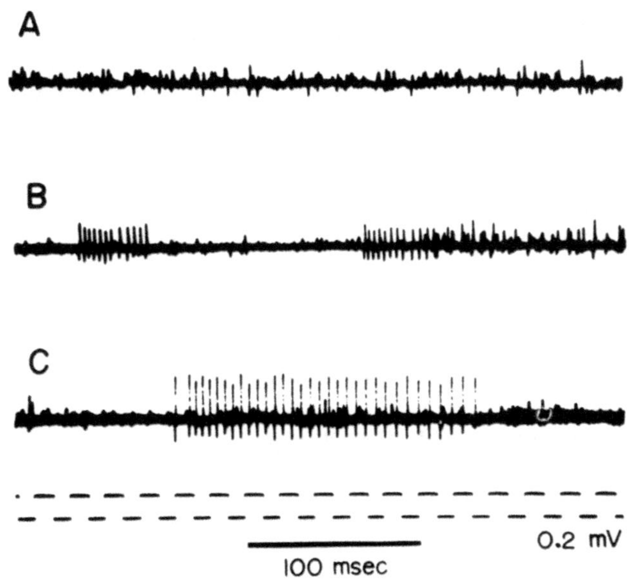

Fig. 1. Trains of impulses from the sensory cells of the cockroach leg after injection of DDT into the leg. A, before injection of DDT; B and C, after injection (Narahashi, 1966).

lowering the temperature greatly enhancing the potency with a Q_{10} of 0.18 (Yamasaki and Ishii, 1953, 1954b). This negative temperature coefficient of DDT potency in inducing trains of impulses is mainly responsible for the increase in killing potency as the temperature is lowered (Yamasaki and Ishii, 1953, 1954b). It should be emphasized that we are talking here about the "potency" of DDT as measured by the effective concentration of DDT to induce trains of impulses. The potency is given by the reciprocal of the concentration of DDT to produce trains in 50% of the population. This potency on the nerve is related to the potency of DDT to induce symptoms of poisoning in insects. On the other hand, the frequency of appearance of trains (as observed by Eaton and Sternburg, 1964, 1967) and the number of impulses in each train represent the "efficacy" of DDT, not the potency. Eaton and Sternburg (1964) have found that the frequency of train appearances in DDT increases as the temperature was raised. These parameters for the efficacy are related to the intensity of the symptoms of poisoning, and not the effective dose of DDT to produce symptoms of poisoning in insects. Therefore, the apparent controversy between Eaton and Sternburg (1964, 1967) and Yamasaki and Ishii (1953, 1954b) with respect to the temperature coefficient of action of DDT on the nerve is due to the different parameters measured (the efficacy by the former group and the potency by the latter), and the data of these two groups agree in essence (see review by Narahashi, 1971).

Although certain sensory organs such as those in the cockroach trochanter are highly sensitive to DDT as described above, other portions of the nervous system are also affected by DDT. These include the chemical sense organs on the tarsus and labellum of the housefly (Smyth and Roys, 1955; Soliman and Cutkomp, 1963), the central nerve cord of the cockroach (Yamasaki and Ishii, 1954a,c; Eaton and Sternburg, 1967), and the synapses in insects and vertebrates (Dresden, 1949; Yamasaki and Ishii, 1952b).Spontaneous discharges in the abdominal nerve cord are increased in frequency, and synaptic transmission is facilitated to produce prolonged after-discharges. Since spontaneous discharges in the insect nerve cord will increase if synaptic facilitation occurs, at least some of the aforementioned observations can be ascribed to the facilitatory effect of DDT on synaptic transmission.

It is surprising to find that very little has been studied on the effect of DDT on the synaptic transmission (see Narahashi, 1971 for references). The only effect of DDT that has been disclosed is the synaptic facilitation as represented by after-discharges induced by a single stimulus. This action is presumably one of the most important aspects of DDT, since it can account for a variety of symptoms of poisoning including atoxia, hyperactivity and convulsions. Therefore, we have conducted detailed electrophysiological experiments in order to elucidate the mechanism of action of DDT on the synaptic transmission. The excitatory neuromuscular junction of the crayfish was used as a model of synapse, because the neurotransmitter is l-glutamate as in insect neuromuscular junctions, and also because a variety of electrophysiological manipulations are feasible. 2,2-(p-Ethoxyphenyl)-3,3-dimethyl oxetane (EDO), a biodegradable DDT analog (Holan, 1971a,b), was used in most experiments because it had essentially the same action as DDT.

III. EFFECTS OF EDO ON EXCITATORY NEUROMUSCULAR TRANSMISSION OF CRAYFISH

When a single stimulus was applied to the excitatory motor nerve of the crayfish, only a small end-plate potential (EPP) was recorded from the muscle using an intracellular microelectrode. The amplitude of EPP was on the order of 0.5 mV. The application of EDO at a concentration of 4×10^{-8} M caused repetitive EPPs to appear in response to a single nerve stimulus, and the EPPs summated to attain a large amplitude (50 mV). Extracellular recording from the motor nerve revealed repetitive action potentials corresponding to each of the repetitive EPPs. Iontophoretic application of l-glutamate to the end-plate caused a transient depolarization. This glutamate potential was not affected by EDO at all. All these observations are in support of the notion that EDO acts on the presynaptic nerve to induce repetitive EPPs. More direct evidence for this hypothesis was obtained using focal recording technique. A microelectrode filled with sodium chloride was brought very close to a junction, and the potential was recorded in response to the nerve stimulation. Under this condition, the recorded potential represents the current flowing near the junction, and therefore, it was possible to record both the action potential from the nerve terminal and the EPP from the muscle. It was clearly shown that each of the repetitive EPPs induced by a single nerve stimulus in the EDO-poisoned preparation is preceded by an action potential of the nerve terminal. Furthermore, it was demonstrated that the quantal content of EPP or the number of transmitter quanta producing an EPP is not affected by EDO. Stimulation of EDO-induced junctional facilitation by applying repetitive nerve stimuli in the normal preparation has revealed that EDO does not alter the degree of facilitation of EPPs. Thus the repetitive firing in the nerve is responsible for inducing repetitive EPPs in the presence of EDO.

IV. EFFECTS OF DDT ON NERVE MEMBRANES

A. Repetitive Discharge and Negative After-potential

The results of the foregoing section have clearly demonstrated that repetitive activity of the nerve is the major cause for synaptic facilitation. In order to elucidate the mechanism of action of DDT or EDO on the nervous system, we have to find out how repetitive discharges are produced in the nerve in the presence of these insecticides.

One of the epoch-making events in the history of the study of DDT action is the discovery of the negative (depolarizing) after-potential as induced by DDT. Shanes (1949) noted this in the crab nerve for the first time, and shortly after that time, Yamasaki and Ishii (1952b) observed it in the action potentials from individual nerve fibers of the cockroach while studying the effect of DDT on the synaptic transmission. The subsequent studies using intracellular microelectrodes clearly demonstrated that DDT greatly increases the negative after-potential in the giant axon of the cockroach (Yamasaki and Narahashi, 1957b; Narahashi and Yamasaki, 1960b,c) (Fig. 2). As a result of the elevation of the negative after-potential by DDT, the axon tends to fire repetitively. This has indeed been observed (Narahashi and Yamasaki, 1960b,c), and as expected, the threshold for excitation decreases during the DDT-induced negative after-potential forming a supernormal phase (Yamasaki and Narahashi, 1957a). Therefore, the basis for repetitive excitation of the DDT-poisoned axon is an augmentation of the negative after-potential. The next step of the study would then be the elucidation of the mechanism whereby DDT increases the negative after-potential.

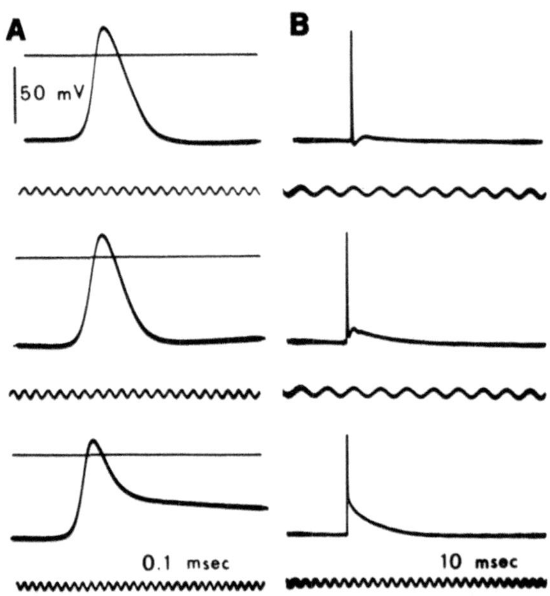

Fig. 2. Changes in intracellularly recorded action potential of the cockroach giant axon after treatment with DDT 10^{-4} M. A, from top to bottom, before, 38 min after, and 90 min after treatment with DDT. The horizontal lines indicate zero potential level. B, as in A, but with slower sweep (Narahashi and Yamasaki, 1960b).

B. Normal After-potentials

Before studying the negative after-potential in the DDT-poisoned axon, we have to know the mechanism whereby the after-potentials are produced in the normal axon. The experiments were carried out using intracellular microelectrodes with the giant axon of the cockroach (Narahashi and Yamasaki, 1960a). In the normal axon, the spike is followed by an undershoot (positive phase) of about 5 mV which in turn is followed by a prolonged but small (about 1.5 mV) negative after-potential (Fig. 3). As in the case of the squid giant axon (Frankenhaeuser and Hodgkin, 1956), the undershoot can be explained in terms of the increase in membrane potassium conductance (Yamasaki and Narahashi, 1959). In other words, when the membrane is repolarized during the falling phase of the action potential as a result of sodium inactivation and potassium conductance increase, the membrane potential approaches the potassium equilibrium potential more closely than at the resting level due to the sustained increase in potassium conductance. The potassium ions that have come out of the axon through membrane potassium channels during the falling phase of the action potential are in the space between the axolemma and the Schwann cell membrane, and diffuse out of this space through the narrow channels between the Schwann cell membranes. Thus a small depolarization is caused by the temporal increase in potassium concentration in the space between the axolemma and the Schwann cell membrane, and the depolarization subsides as potassium ions leak out of the space. This forms a negative after-potential.

C. Negative After-potential in DDT-poisoned Axon

It has been shown that the increased negative after-potential

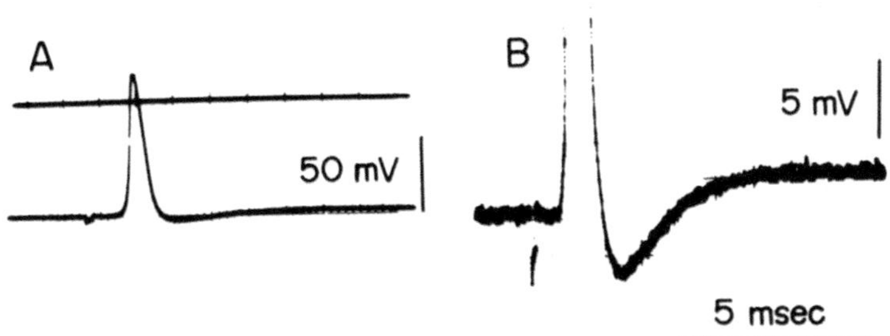

Fig. 3. Action potentials recorded intracellularly from the giant axon of the cockroach. Note that the spike is followed by an undershoot or positive phase which is in turn followed by a slight depolarizing phase or negative after-potential (Narahashi, 1965).

in the DDT-poisoned cockroach giant axon is not produced by a large amount of accumulation of potassium ions in the space just outside the axolemma. In the case of the normal axon, repetitive stimulations cause the negative after-potentials to build up with an exponential time course (Fig. 4A) as is expected from a gradual increase in potassium concentration outside the axolemma. In contrast, the large negative after-potential in the DDT-treated axon does not build up during repetitive stimulation (Fig. 4B). This is contrary to what one might expect from a large accumulation

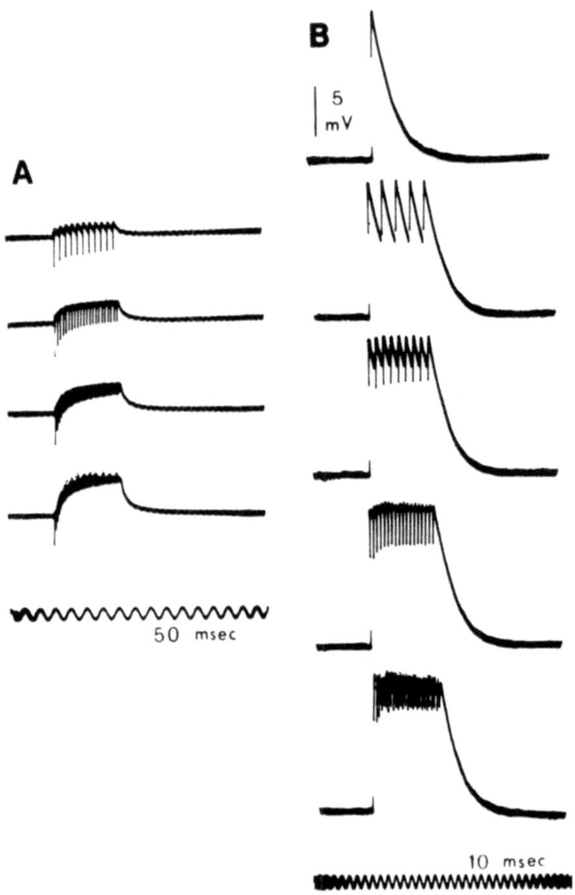

Fig. 4. After-potentials during repetitive stimuli of varying frequencies in the normal (A) and DDT-poisoned (B) cockroach giant axons. The spike potentials are too large to be recorded. The frequencies of stimuli are, from top to bottom in A, 50, 100, 150, and 200 c.p.s., and in B, single stimulus, 50, 100, 200 and 300 c.p.s. (Narahashi and Yamasaki, 1960a,b).

of potassium ions outside the axolemma. Thus we have to seek the cause of the DDT-induced negative after-potential elsewhere.

D. Effects of DDT on Membrane Ionic Conductances

Ionic Mechanism of Nerve Excitation. A series of investigations by Hodgkin, Huxley and Katz in the early 1950's has established the basis for the ionic mechanism of nerve excitation (Hodgkin et al., 1952; Hodgkin and Huxley, 1952a,b,c,d). In short, the nerve membrane is permeable to potassium ions at resting condition but only sparingly so to sodium and other ions. Thus the membrane potential (E_m) approaches the equilibrium potential for potassium (E_K) as defined by the following Nernst equation for potassium:

$$E_m \simeq E_K = \frac{RT}{F} \ln \frac{[K]_o}{[K]_i} \quad (1)$$

where $[K]_o$ and $[K]_i$ refer to the external and internal potassium concentrations, respectively, and R,T and F refer to the gas constant, the absolute temperature and the Faraday constant, respectively. Upon depolarization of the nerve membrane the sodium permeability increases in a regenerative manner making the membrane more permeable to sodium than potassium. This will change the membrane potential from a value close to the potassium equilibrium potential to a value close to the sodium equilibrium potential (E_{Na}) defined by the following Nernst equation for sodium:

$$E_{Na} = \frac{RT}{F} \ln \frac{[Na]_o}{[Na]_i} \quad (2)$$

where $[Na]_o$ and $[Na]_i$ refer to the external and internal sodium concentrations, respectively. Since the concentration gradient for sodium across the nerve membrane is opposite that for potassium, being high outside and low inside, the increase in sodium permeability causes the rising phase of the action potential to appear (Fig. 5). Sodium ions flow across the membrane from outside to inside according to the electrochemical gradient. The increased sodium permeability starts decreasing by a mechanism called sodium inactivation, and the potassium permeability starts increasing, so that the nerve membrane again becomes almost exclusively permeable to potassium bringing back the membrane potential close to the potassium equilibrium potential (Fig. 5). This is the falling phase of the action potential. Potassium efflux occurs during this time according to its electrochemical gradient. The undershoot that follows the falling phase of the action potential in certain nerves such as those of squid and cockroach is explicable in terms of the slow decline of the increased potassium permeability which brings the membrane potential closer to the potassium equili-

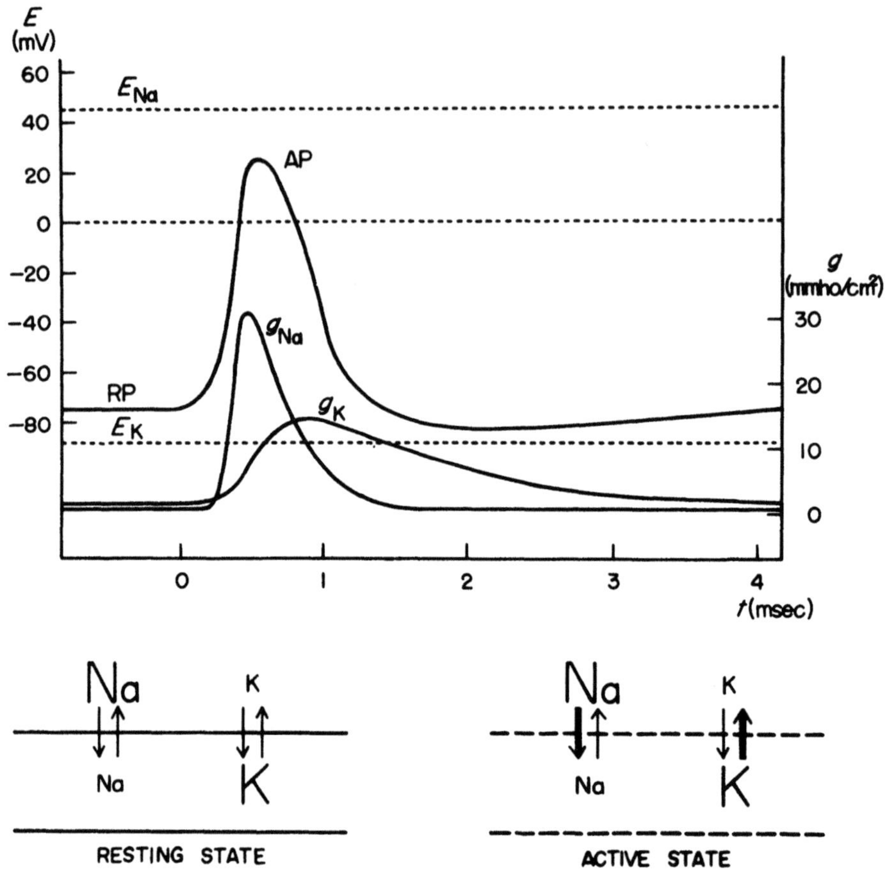

Fig. 5. Diagram of the mechanism of action potential production. RP, resting potential; AP, action potential; E_{Na}, sodium equilibrium potential; E_K, potassium equilibrium potential; g_{Na}, membrane sodium conductance; g_K, membrane potassium conductance. See text for further explanation (Narahashi, 1965).

brium potential than the resting level. There is not the mechanism of potassium inactivation, and the increased potassium permeability declines slowly as a result of repolarization of the membrane.

Voltage Clamp Methods. The ionic permeability changes described above can be measured as membrane ionic conductance changes by the methods called voltage clamp. The voltage clamp is based on the Ohm's law which gives conductance as current/voltage. Thus if the current flowing across the nerve membrane is identified as being carried by an ion and the electromotive force for that ion is known, then the membrane conductance to that ion can easily be calculated. Thus we have

$$g_{Na} = \frac{I_{Na}}{E_m - E_{Na}} \quad (3)$$

$$g_K = \frac{I_K}{E_m - E_K} \quad (4)$$

where g_{Na} and g_K represent sodium and potassium conductances, respectively, and I_{Na} and I_K represent sodium and potassium currents, respectively.

However, in order to measure the current density, the distribution of both membrane current and membrane potential must become uniform along a portion of the axon where measurements are made. Conventional microelectrodes do not serve the purpose, because the current to or from the microelectrode tip spreads along the axon thereby making the membrane current and potential non-uniform. One easy way to achieve "space clamp" is to insert a metal wire longitudinally to short-circuit the axoplasm, and to place a large electrode outside the axon. This makes the current and potential distribution uniform over the entire length of the axon where measurements are to be made. The voltage clamp using the axial wire electrode is applicable to very large nerve fibers such as the squid giant axon whose diameter is around 500 μ. However, it is difficult to use the axial wire method for smaller fibers such as the lobster giant axon with a diameter of about 100 μ. For such preparations, sucrose-gap technique (Julian et al., 1962a,b) has proved quite successful in studying the mechanism of action of various agents including insecticides (Narahashi and Haas, 1967, 1968; Narahashi et al., 1969; Murayama et al., 1972).

Two streams of isotonic sucrose solutions electrically insulate a portion of an isolated giant axon from the rest of the preparation. One end of the axon is perfused with normal physiological saline solution, while the other end is perfused with isotonic KCl solution to depolarize the membrane completely. Thus the membrane potential of the center portion of about 50 μ wide (called artificial node) can be measured in reference to the end of the axon depolarized by KCl, while stimulating currents can be applied to the node through the other end of the preparation perfused with normal saline solution. Since the center portion or node (50 μ width) is much smaller than the length constant of the lobster giant axon which has been estimated to be 1.6 - 1.7 mm (Hodgkin and Rushton, 1946; Tobias, 1960), space clamp condition is established in the node even during the activity when the length constant becomes shorter as a result of the decrease in membrane resistance.

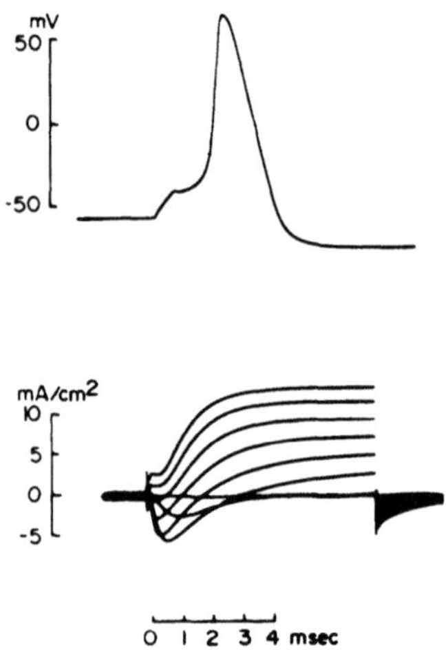

Fig. 6. Action potential and family of membrane currents associated with step depolarizations to various membrane potential levels in squid giant axon (Yeh and Narahashi, 1976).

Figure 6 illustrates an action potential elicited by a brief depolarizing current pulse and a family of membrane currents associated with step depolarizations of various magnitudes under voltage clamp conditions. The membrane current during a depolarizing step is composed of two portions; one is the peak transient current which is inward (downward in the figure) in direction for small to moderate depolarizations and outward for large depolarizations, and the other is the late steady-state current which is always outward in direction. It has been well established that the peak transient current, inward or outward, is carried mostly by sodium ions and that the steady-state current is carried mostly by potassium ions.

When the membrane current is measured at its peak and steady-state and is plotted as a function of the membrane potential, current-voltage relationships such as those shown in Fig. 7 are obtained. The peak sodium current increases in amplitude with depolarization, but decreases with further depolarization reversing the polarity at the sodium equilibrium potential. The steady-state potassium current increases in amplitude with depolarization. The potassium equilibrium potential cannot directly be measured from the current-voltage curve, but can be estimated separately from the measurement of tail potassium currents associated with step repolarizations to various membrane potential levels. Thus the sodium and potassium conductances can be calculated by equations (3) and (4). The effects of any agents on the sodium and potassium conductances can be studied using these analyses.

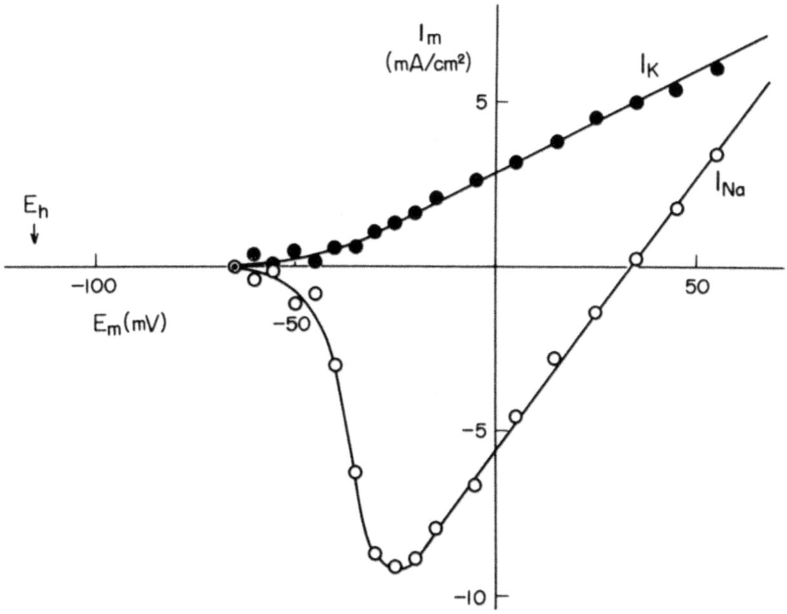

Fig. 7. Current-voltage relations for peak transient sodium current (I_{Na}) and for steady-state potassium current (I_K) in the voltage clamped lobster giant axon. I_m, membrane current; E_m, membrane potential; E_h, holding membrane potential from which the membrane is depolarized to various membrane potential levels (Narahashi, 1964b).

Although the method described above allows us to measure the peak sodium current and the steady-state potassium current separately, it is not easy to measure the kinetics of each current because the two components of the membrane current overlap partially. In order to separate the two currents completely, pharmacological agents have proven to be most convenient and efficient. Tetrodotoxin (TTX) is a poison from the pufferfish, and has been demonstrated to block the sodium current selectively without any effect on the potassium current (Narahashi et al., 1964; Narahashi, 1972,1974). It is effective in blocking the sodium current at a concentration of nanomolar. Thus, in the presence of TTX in the bathing medium, the pure potassium currents can be observed (Fig. 8).

A few potassium current blocking agents are available. The classical agent is tetraethylammonium (TEA) which blocks the potassium current from inside the squid axon membrane (Tasaki and Hagiwara, 1957; Armstrong and Binstock, 1965). In the node of Ranvier of the frog, TEA is effective in blocking the potassium

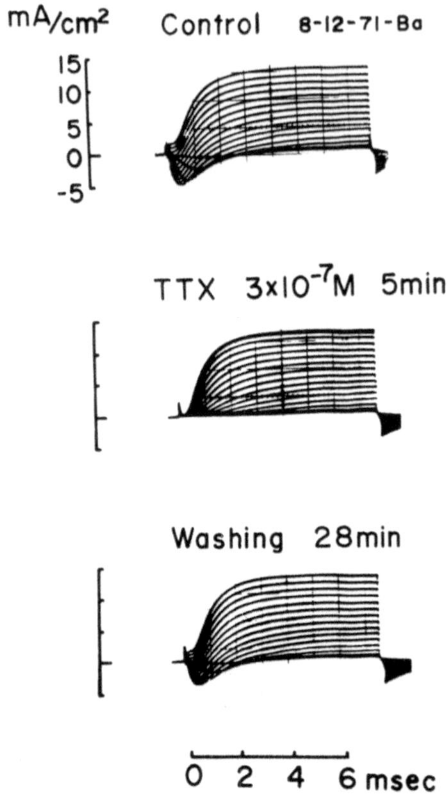

Fig. 8. Families of membrane currents associated with step depolarizations (10 mV steps) in a squid giant axon before and during external application of 3 x 10^{-7} M tetrodotoxin (TTX) and after washing with toxin free medium. TTX blocks peak transient sodium current without any effect on steady-state potassium currents (Narahashi, 1975).

current from either side of the membrane (Hille, 1967). However, TEA has to be applied in a very high concentration. In the squid giant axon, for example, 20-40 mM TEA is required for complete elimination of potassium currents. Another disadvantage is that TEA suppresses the peak sodium current to some extent. Recently, certain aminopyridine compounds have been discovered to block the potassium currents without much effect on the sodium currents (Pelhate and Pichon, 1974). Aminopyridines are effective from outside the membrane, and this constitutes a great advantage when used as tools to eliminate the potassium currents. However, the aminopyridine block of potassium current (Fig. 9) is dependent upon the membrane potential, the duration of depolarization, and the frequency of stimulation (Yeh et al., 1976). In addition to 2-, 3-,

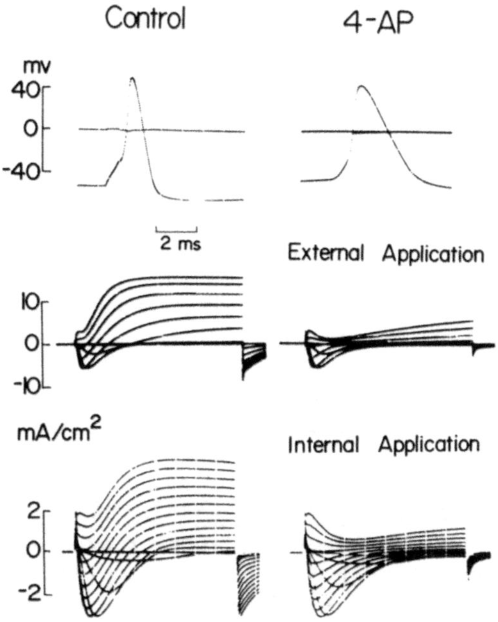

Fig. 9. Effects of 4-aminopyridine on squid giant axons. Upper records are membrane action potentials, and middle records are clamp families before and after external application of 1 mM 4-AP. Lower records are clamp families from another axon before and after internal application of 1 mM 4-AP (Yeh et al., 1976).

and 4-aminopyridine which have commonly been used to eliminate the potassium currents in millimolar concentrations, 3,4-diaminopyridine has recently been discovered to possess a very potent potassium blocking action (Kirsch and Narahashi, 1978). It is effective at micromolar concentrations to block the potassium current, and is expected to become a useful tool to eliminate the potassium currents.

Through the use of these pharmacological agents, it is possible to study the kinetics of sodium or potassium current. For instance, the time course of sodium inactivation can directly be measured from the pure sodium current in the presence of TEA or 4-aminopyridine. It can also be measured from the pure sodium current obtained by subtraction of the current after application of TTX from that before application.

The steady-state sodium inactivation is another important parameter to interpret the action of chemicals on the membrane.

It represents the degree of sodium inactivation as a function of the membrane potential (Fig. 10). In order to obtain this curve, conditioning depolarizing and hyperpolarizing pulses of various magnitudes and of about 50 msec duration are applied to the membrane and are immediately followed by a test depolarizing pulse (e.g. to 0 mV). The amplitude of peak sodium current associated with the test pulse is measured, normalized to the maximum value, and plotted against the membrane potential of the prepulse. At large negative membrane potentials the sodium current attains the maximum, whereas it decreases with depolarization and eventually becomes zero with large depolarizations. Thus the membrane no longer produces action potentials with prolonged depolarizations beyond a certain period of time (e.g. 10 msec). If any agent causes a shift in the steady-state sodium inactivation curve along the potential axis, the availability of the sodium activation system changes thereby altering the excitability.

Effects of DDT on Ionic Conductances. The effects of p,p'-DDT on the ionic conductances of the lobster axon membrane were studied using the sucrose-gap voltage clamp. (Narahashi and Haas, 1967,1968). Before carrying out voltage clamp measurements of membrane currents, the action potential was observed as induced by a single stimulus. The negative after-potential started to

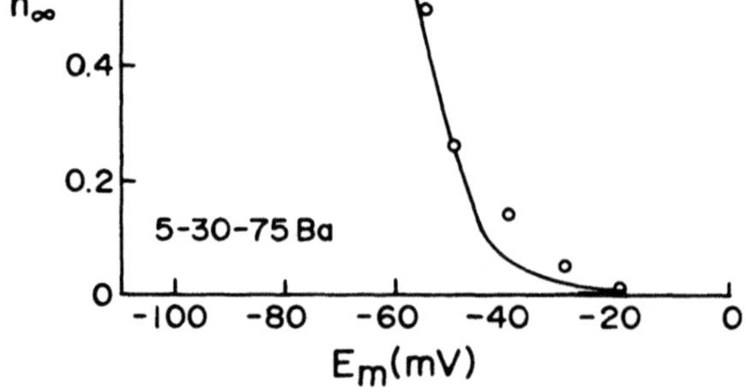

Fig. 10. Steady-state sodium inactivation (h_∞) curve in a squid giant axon.

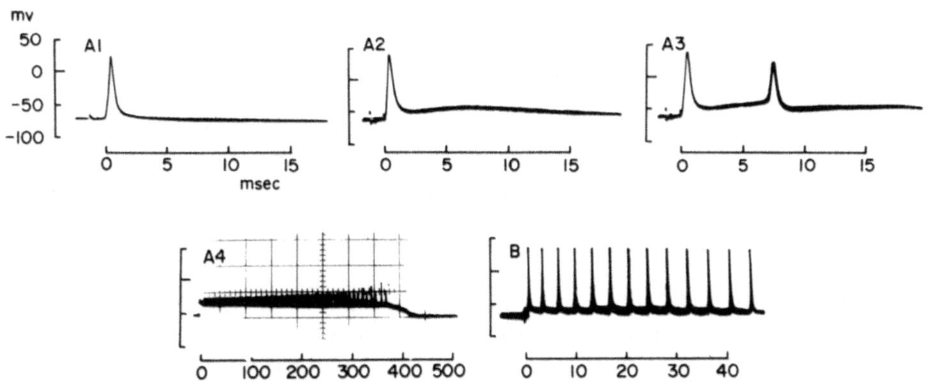

Fig. 11. Action potentials before and after application of 5×10^{-4} M DDT. A1, control record before DDT; A2, 36 min in DDT, increase in negative after-potential; A3, 42 min in DDT, after-discharge superimposed on the increased negative after-potential; A4, 78 min in DDT, repetitive after-discharge superimposed on a prolonged, plateau action potential. The initial spike is too faint to be reproduced. B, another axon, repetitive after-discharge superimposed on a slightly increased negative after-potential, 9 min in DDT (Narahashi and Haas, 1968).

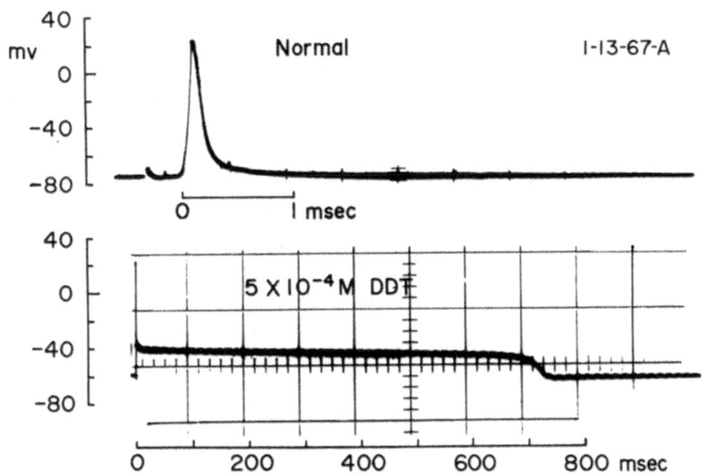

Fig. 12. Action potential recorded from a normal lobster giant axon (upper record) and action potential recorded from a DDT-poisoned axon (lower record). Note different time scales, and the prolonged falling phase of the action potential forming a plateau in DDT.

increase in amplitude and to be prolonged in duration about 20 minutes after application of 5 x 10^{-4} M DDT (Fig. 11, A2), and repetitive discharges were induced superimposing on the elevated negative after-potential (Fig. 11, A3). Prolonged repetitive discharges were often observed (Fig. 11, A4,B). The negative after-potential in DDT was further augmented with time eventually forming a prolonged plateau lasting as long as one second (Fig. 12). These observations are in general agreement with those using the cockroach giant axon (Narahashi and Yamasaki, 1960,b,c).

Slow action of DDT on the lobster giant axon caused some difficulty in performing the voltage clamp analysis in the sucrose-gap chamber. Since the artificial node, once established, could survive only for a short period of time, usually about 20 minutes at most, it was almost impossible to make voltage clamp measurements of membrane currents with the same node before and after full development of DDT action. Therefore, we were forced to use separate preparations for control and test. For measurements of DDT-poisoned axons, the preparations had been soaked in 5 x 10^{-4} M DDT solution for 40 minutes before being mounted in the sucrose-gap chamber. Figure 13 illustrates families of membrane currents associated with various magnitudes of step depolarizations in a normal and a DDT-treated axon. By comparison of these two families of membrane currents, it is clear that DDT poisoning does not much alter the time course of the rising phase of the sodium current but greatly slows the falling phase. The bottom set of records in Fig. 13 shows the membrane currents associated with depolarizations to -20 mV before and 4 minutes after application of 3 x 10^{-7} M TTX to the DDT-poisoned axon. The difference between the two currents represents the sodium current in the DDT-poisoned axon indicating a very prolonged decay.

The effects of DDT on the membrane current are illustrated more clearly in Fig. 14, in which membrane ionic currents are superimposed after corrections for leakage currents. Three points become apparent: 1) The mechanism by which the sodium conductance increases upon depolarization is not markedly affected by DDT. If there is any change, it is a slight slowing. 2) The sodium inactivation is greatly slowed. 3) The potassium current is partially suppressed and slowed. Since the sodium inactivation and the potassium activation are responsible for the falling phase of the action potential, inhibition of these two mechanisms can account for the observed slowing of the falling phase of the action potential. Similar observations were made of the DDT action with the node of Ranvier of the frog (Hille, 1968) and the giant axon of the cockroach (Pichon, 1969a,b), except that the potassium current of the frog was not appreciably affected by DDT. It was suggested that DDT has no effect on the sodium channels that are not open (Hille, 1968).

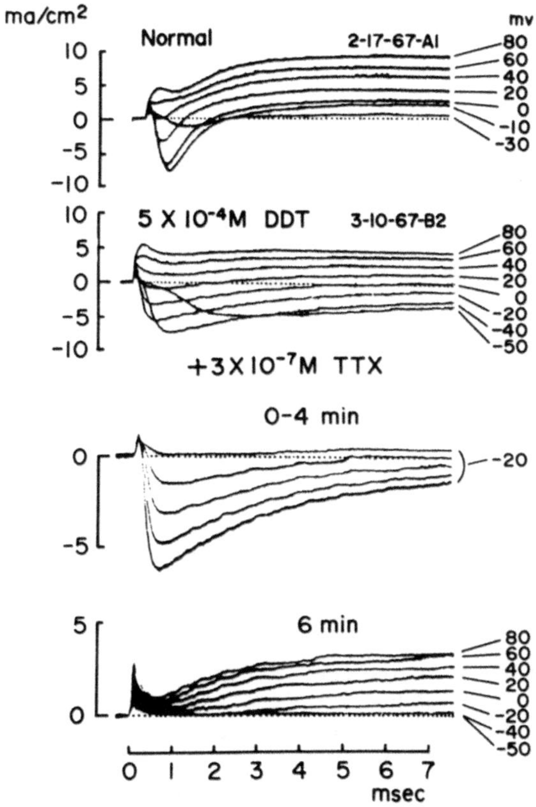

Fig. 13. Families of membrane currents associated with step depolarizations from the holding membrane potential of -100 mV to membrane potentials indicated in a normal axon and in an axon treated with DDT and with DDT plus tetrodotoxin (TTX). The third set of records shows changes in the current during the course of TTX blockage; the largest current was recorded immediately before introducing TTX, and the smallest current 4 min after the introduction of TTX. The dotted lines refer to the zero base lines (Narahashi and Haas, 1967).

Effects of DDT Analogs on Action Potential. In the course of development of a variety of biodegradable DDT derivatives, 2,2-bis(p-ethoxyphenyl) -3,3-dimethyloxetane (EDO or GH149) was discovered to have a potent insecticidal activity (Holan, 1971a,b). It lacks the 2,2,2-trichloroethane moiety, so that it is insensitive to DDT-dehydrochlorinase. Its mammalian toxicity is approximately half of that of DDT.

When applied to the isolated crayfish giant axon, EDO increased the negative after-potential from which repetitive discharges were

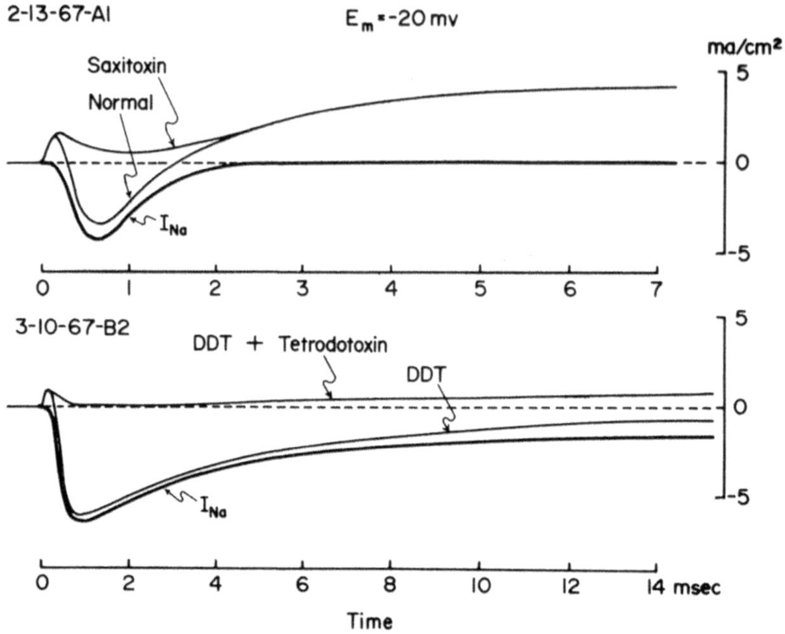

Fig. 14. Separation of membrane current into peak transient (sodium) current and steady-state (potassium) current by the use of saxitoxin or tetrodotoxin at a concentration of 3 x 10^{-7} M in a normal and in a DDT-treated axon. The membrane current in saxitoxin and that in DDT plus tetrodotoxin show the steady-state (potassium) currents. The peak transient (sodium) current (I_{Na}) was obtained by subtraction of the steady-state (potassium) current from the total current (Narahashi and Haas, 1968).

produced (Wu et al., 1975). The action potential was eventually blocked leaving spontaneous oscillations of the membrane potential (Fig. 15). It took only 16 minutes for the action potential to be blocked by 1 x 10^{-4} M EDO. DDT itself never blocked the action potential of the crayfish giant axon completely even after 90 minutes of exposure. Two other DDT derivatives also caused an increase in negative after-potential followed by a block. They are 2,2-bis(p-formylphenyl)-1,1,1-trichloroethane (p,p'-CHO-DDT) and 2,2-bis(p-chlorophenyl)-1,1-dichlorocyclopropane (DCC). These three DDT derivatives may be called dualist groups.

There are many derivatives which behave similarly to p,p'-DDT. They include p,p'-NO_2-DDT,p,p'-CH_3O-DDT (methoxychlor), p,p'-C_2H_5O-DDT (ethoxychlor), p,p'-C_3H_7O-DDT, o,p'-DDT, p,p'-DDD and p,p'-C_2H_5-DDD (Wu et al., 1975). These compounds increased the negative after-potential often to induce repetitive discharges, but never blocked the conduction. This group may be called an excita-

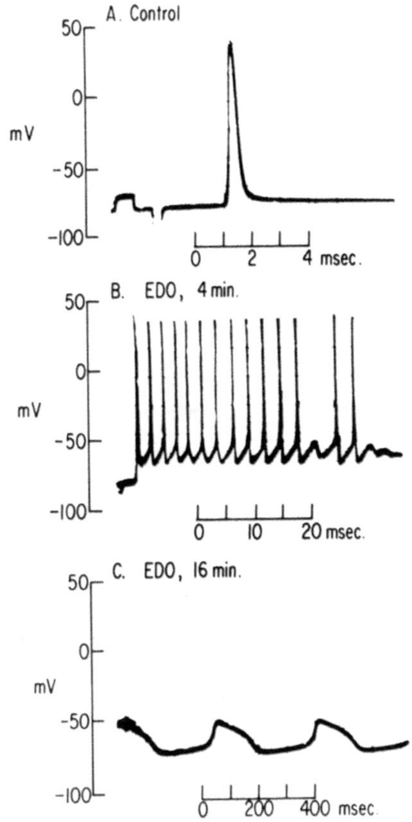

Fig. 15. Dual action of 1 x 10^{-4} M EDO. A. Control. B. Increased depolarizing after-potential and repetitive discharges produced after 4 min of exposure to EDO. C. Spontaneous oscillation in membrane potential after 16-min exposure. Note the slow time scale in C (Wu et al., 1975).

tory group.

On the other hand, at least two DDT derivatives have been found to be blockers. They are p,p'-NH_2-DDT and p,p'-HO-DDT, and blocked conduction without increasing the negative after-potential (Fig. 16) (Wu et al., 1975). This group may be called blocking group.

It is of utmost importance to note that some of the aforementioned DDT derivatives which are active on the isolated nerve are inactive as insecticides. When tested on houseflies and

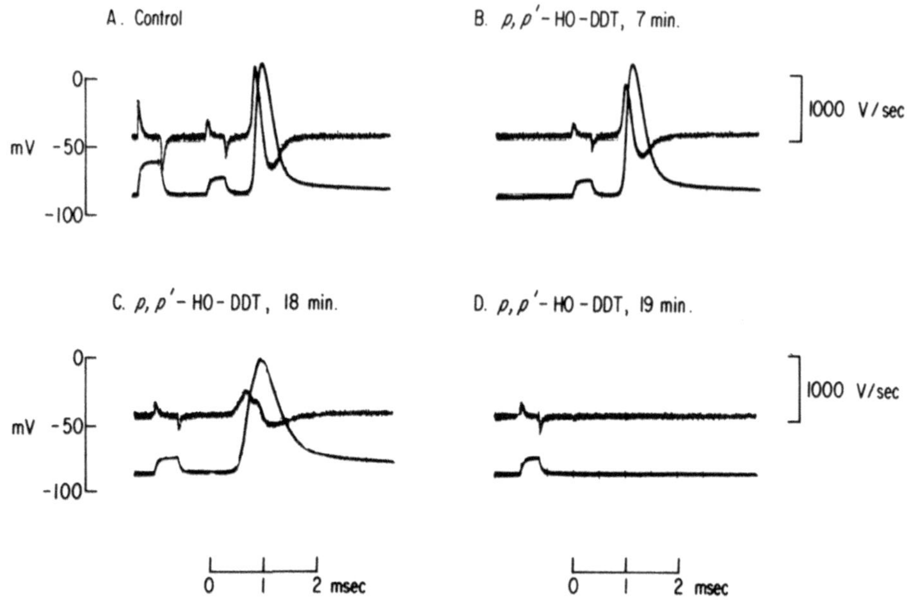

Fig. 16. Blocking action of 1 x 10^{-4} M p,p'-HO-DDT. Upper tracings show the rate of change of membrane potential. Lower tracings membrane potential (Wu et al., 1975).

mosquitos, p,p'-NO_2-DDT, p,p'-NH_2-DDT, and p,p'-HO-DDT were devoid of insecticidal activity (Metcalf and Fukuto, 1968), but they were all effective on the isolated nerve preparation. Thus in order to establish the true structure-activity relationship, one must compare the direct action on the target site, because the killing potency includes other factors such as the penetration of insecticide through the cuticle, and the detoxication in the insect (Narahashi, 1964a, 1971). Study of a large number of synthetic pyrethroids along this line has also demonstrated that this is indeed the case for some derivatives (Nishimura and Narahashi, 1978; Narahashi et al., 1977; Narahashi, 1978).

Effects of EDO on Ionic Conductances. Since EDO is more potent and acts more quickly than DDT on the isolated nerve preparation, it provides us with an excellent opportunity to perform voltage clamp experiments with a sucrose-gap chamber. In the case of the previous study of p,p'-DDT, it was necessary to make separate measurements of the control and DDT-treated lobster giant axons, because the artificial node, once established, could not survive more than 20 minutes, the time period too short for p,p'-DDT to exert its full effect. With EDO, however, it was possible to make measurements of membrane currents before and after application of the test solution. Therefore, we have carried out an extensive study of the effects of EDO on the membrane ionic conductances of the lobster giant axon.

In this study, we were concerned mainly with the effects of EDO on the sodium conductance. The best and easiest way of performing such experiments is to eliminate potassium current. In squid giant axons, this can be achieved by using internal perfusion technique (Baker et al., 1961) and perfusing externally and internally with potassium free media. However, since the squid giant axon in insensitive to DDT (Narahashi, 1971) and EDO, we had to use the lobster giant axon to which the routine internal perfusion method was not applicable. We could not use tetraethylammonium (TEA) to block the potassium, because it had to be applied inside. Therefore, 4-aminopyridine (4-AP) was applied externally to block the potassium current. 4-AP is a specific potassium blocking agent, being able to block the potassium current by external application without affecting the sodium current (Pelhate and Pichon, 1974; Yeh et al., 1976). In squid axons, the 4-AP block of potassium current is dependent on the voltage, time and stimulus frequency but in lobster axons, the block is much less dependent on these parameters making the

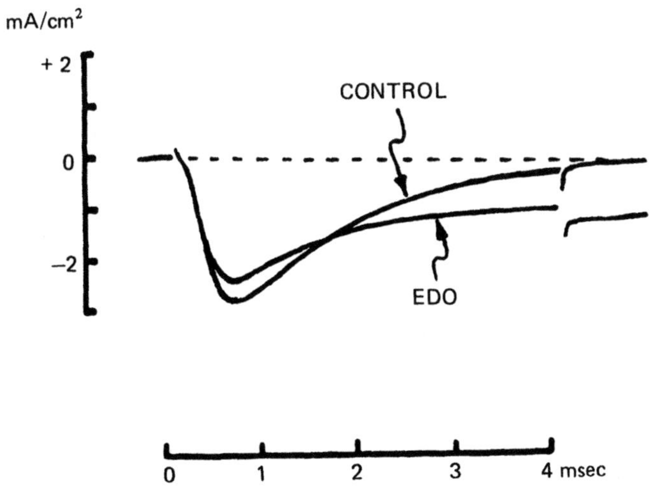

Fig. 17. Effect of 50 μM EDO on the membrane sodium current associated with a step depolarization to 0 mV in a lobster giant axon exposed to 2 mM 4-aminopyridine to block potassium current.

measurements less complicated. Most experiments with EDO were performed in the presence of 2 mM 4-AP in the bathing medium.

The effect of 50 μM EDO on the lobster giant axon was exerted very quickly, the falling phase of the action potential being slowed markedly one minute after introduction of the drug. The duration of the action potential was prolonged from a few milliseconds to several hundred milliseconds. Not only the onset of action but also the recovery from the action after washing were quick, a 13 minute washing being enough to bring back the action potential duration to its original value.

Voltage clamp analyses have clearly demonstrated that the falling phase of the sodium current was markedly slowed. Figure 17 illustrates the superimposed tracings of the sodium currents associated with a step depolarization of the membrane from -80 mV to 0 mV before and 2.5 minutes after application of 50 μM EDO. The potassium current was eliminated by 2 mM 4-AP. The peak amplitude of the sodium current was decreased somewhat, while the falling phase of the sodium current was greatly slowed forming a large residual inward current. The sodium currents associated with various step depolarizations were measured at the peak and at the end of 8-msec pulse, and were plotted as a function of the membrane potential. The two current-voltage curves showed almost the same reversal potentials between 50 mV and 60 mV, supporting the idea that the peak and residual currents are carried by sodium ions.

However, the peak sodium current and the residual sodium current in the EDO-poisoned axon have some different characteristics. The steady-state sodium activation (m_∞) of the Hodgkin-Huxley formulation (Hodgkin and Huxley, 1952d) was computed from the data of the sodium currents. When compared with the m_∞ curve for the peak sodium current, the m_∞ curve for the residual sodium current was shifted by 6 mV in the direction of hyperpolarization.

When the nerve membrane is depolarized and then repolarized at the peak of sodium current, a large inward current (tail current) appears and decays exponentially. The tail current is carried by sodium ions, and results from a sudden increase in the electromotive force for sodium at the moment when the sodium conductance is maintained at a high level. Since the sodium inactivation is much slower in time course than the sodium activation, the time course of the tail sodium current associated with a step repolarization represents mostly that of the sodium activation. Thus the time constant for the sodium activation (τ_m) can directly be measured from the tail current. Figure 18 illustrates the tail currents before and after application of 100 μM EDO when the depolarizing pulse is terminated at the peak of the sodium current. In the control record, the tail current decays exponentially. After exposure to EDO, however, the tail current decays quickly

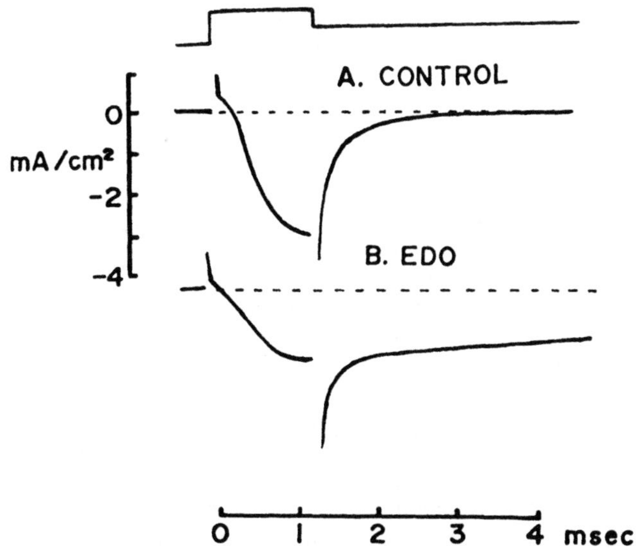

Fig. 18. Tail sodium currents before (A) and during (B) 100 μM EDO treatment of a lobster giant axon in the presence of 2 mM 4-aminopyridine. The membrane was depolarized from -80 mV to +10 mV, and at the peak of the sodium current repolarized to -40 mV.

first, and is followed by a very slow phase. Semilogarithmic plots of the tail current from the EDO-treated axon have revealed two exponential functions, with time constants of 30 msec and 170 msec for the initial and later phases, respectively. The dual phase of the sodium tail current has a very important implication on the interpretation of the mechanism of action of EDO as described later.

Steady-state sodium inactivation (h_∞) in the Hodgkin-Huxley forumlation (Hodgkin and Huxley, 1952d) refers to the degree of the availability of the sodium mechanism at various membrane potentials. Its value can be measured by the conventional double-pulse experiments. The membrane is depolarized or hyperpolarized to various levels for 20-50 msec and is immediately followed by a test step depolarization (e.g. to 0 mV). The sodium current associated with the test depolarization is normalized to the maximum value and is plotted against the membrane potential of the conditioning pulse to obtain a steady-state sodium inactivation curve or h_∞ curve. The h_∞ value attains the unity at high negative potentials, decreases with depolarization, and reaches eventually zero forming a sigmoid curve. It was found that the h_∞ curve for the peak sodium current was shifted by about 20 mV in the direction of hyperpolarization by application of EDO. In other words, the availability of the sodium mechanism is less in EDO than in control at the membrane potential near the resting level. This at least in part accounts for the slight decrease in the peak current by EDO.

The residual sodium current exhibits an interesting h_∞ curve. The h_∞ value never reached zero with depolarization forming a foot. This accounts for the absence of the complete inactivation of the residual current. The time constant of the sodium inactivation (τ_h) can directly be measured from the falling phase of the sodium current recorded in the presence of 4-AP. It was increased by 57% after application of EDO.

Hypothesis of EDO Action on the Sodium Channel. Two observations summarized in the foregoing sections are of particular importance in interpreting the mechanism of action of EDO on the sodium channel. One is the shift of the steady-state sodium activation (m_∞) and inactivation (h_∞) curves along the potential axis by EDO, and the other is the dual phase of the sodium tail current in EDO-treated axons. One possible cause for the shift of m_∞ and h_∞ curves is a change in the surface potential. Since the surface potential is due to ionized groups and oriented dipoles at the membrane, a compound with a large dipole potential could alter the surface potential. The change in surface potential in turn changes the potential field (or gradient) within the membrane without any change in the observed membrane potential which is a sum of the potential field and the surface potentials at the outer and inner membrane edges. In this connection, it is quite interesting to see that bis-(p-chlorophenyl)-acetic acid (DDA) and bis-(p-chlorophenyl)-acetamide (DDAmide) cause a sizable shift (15-20 mV) in n_∞ curve (steady-state potassium activation) and h_∞ curve, respectively (Århem and Frankenhaeuser, 1974). Since DDA, DDAmide and EDO have an asymmetrical portion in the structure, it is reasonable to assume that they cause a change in the surface potential through their contribution of dipole potential. It should be noted that p,p'-DDT, which is a symmetrical molecule, does not cause a shift of the h_∞ and m_∞ curves (Hille, 1968; Narahashi and Haas, 1968).

The sodium tail current associated with a step repolarization in the EDO-treated axon decays with two exponential functions of long time constants. This suggests that the EDO-bound channel is conducting sodium ions without being subjected to inactivation. The time constant for the sodium inactivation is prolonged only by 50%. These changes brought about by EDO can be accounted for in terms of the kinetic model proposed by Moore and Cox (1976) as illustrated in Figure 19. The sodium channel is in a closed configuration P at the resting membrane potential. Upon depolarization, the channel starts opening following a kinetics of the third

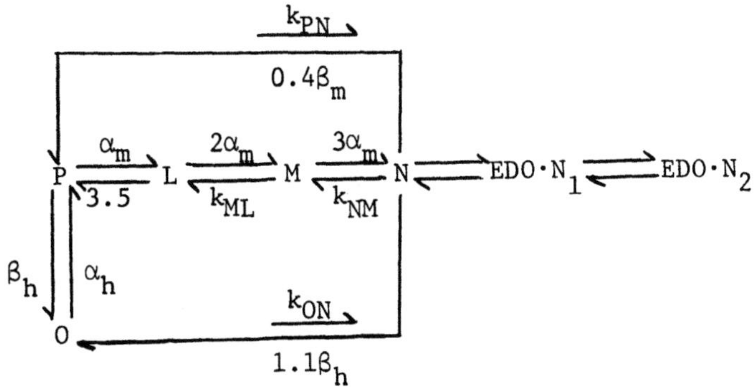

Fig. 19. Modified Moore-Cox model to explain the effect of EDO on sodium conductance.

power finally attaining the open state N via intermediate states L and M. When the membrane is repolarized while the sodium channel is open, the channel closes quickly following the β_m kinetics as can be observed as a tail current. When the membrane is kept depolarized, the sodium channel is closed slowly due to an inactivation mechanism via state O. According to the experimental results described here, EDO does not greatly affect the sodium inactivation time constant, prolonging it only by 50%. However, EDO binds to the open sodium channel to convert it into state EDO·N_1 and then to another state EDO·N_2. These two states are in open configuration, but cannot be inactivated allowing sodium current to keep flowing. The two time constants of the sodium tail current in EDO are explicable on this basis. In other words, dissociations of EDO from EDO·N_1 and EDO·N_2 are slow and represent the rate limiting step.

V. SUMMARY

DDT and its oxetane analog EDO facilitate synaptic and neuromuscular transmission through induction of repetitive firing in the presynaptic nerve terminal. No appreciable effect is observed on the postsynaptic membrane. The basis for repetitive response by DDT and EDO is the increase in negative after-potential which serves as a sustained stimulation. The increase in negative after-potential is due primarily to the slowing of the falling phase of the sodium current and partly to the decrease in the steady-state potassium current. Voltage clamp analyses of the sodium channel kinetics are compatible with the notion that EDO interacts with the open sodium channel in two steps forming the non-inactivating, EDO-bound, conducting sodium channel. The EDO-bound channel allows sodium ions to flow causing a prolonged sodium current which in turn increases the negative after-potential. The time constant of the sodium inactivation is prolonged only by 50% after application of EDO.

This study was supported by NIH grant NS14143.

VI. REFERENCES

Århem, P. and B. Frankenhaeuser (1974) DDT and related substances: Effects on permeability properties of myelinated Xenopus nerve fibre. Potential clamp analysis. Acta Physiol. Scand. 91: 502-511.

Armstrong, C.M. and L. Binstock (1965) Anomalous rectification in the squid giant axon injected with tetraethylammonium chloride. J. Gen. Physiol. 48: 859-872.

Baker, P.F., A.L. Hodgkin and T.I. Shaw (1961) Replacement of the protoplasm of a giant nerve fibre with artificial solutions. Nature 190: 885-887.

Becht, G. (1958) Influence of DDT and lindane on chordotonal organs in the cockroach. Nature 181: 777-779.

Cheng, E.Y. and L.K. Cutkomp (1977) The sensitivity of mitochondrial Mg^{2+} ATPase to DDT and analogs at different temperature. Pesticide Biochem. Physiol. 7: 360-366.

Cutkomp, L.K., H.H. Yap, E.V. Vea and R.B. Koch (1971) Inhibition of oligomycin-sensitive (mitochondrial) Mg^{2+} ATPase by DDT and selected analogs in fish and insect tissues. Life Sci. 10: 1201-1209.

Desaiah, D., L.K. Cutkomp and R.B. Koch (1974) A comparison of DDT and its biodegradable analogs tested on ATPase enzymes in cockroach. Pesticide Biochem. Physiol. 4: 232-238.

Dresden, D. (1949) Physiological Investigations into the Action of DDT. 114 pp. Drukkerij en Uitgeverij G.W. van der Wiel and Co. Arnheim, Netherlands

Eaton, J.L. and J. Sternburg (1964) Temperature and the action of DDT on the nervous system of Periplaneta americana (L). J. Insect Physiol. 10: 471-485.

Eaton, J.L. and J.G. Sternburg (1967) Temperature effects on nerve activity in DDT-treated American cockroaches. J. Econ. Entom. 60: 1358-1364.

Frankenhaeuser, B. and A.L. Hodgkin (1956) The after-effects of impulses in the giant nerve fibres of Loligo. J. Physiol. 131: 341-376.

Fukuto, T.R. (1971) Relationships between the structure of organophosphorus compounds and their activity as acetylcholinesterase inhibitors. Bull. W.H.O. 44: 31-42.

Hawkins, W.B. and J. Sternburg (1964) Some chemical characteristics of a DDT-induced neuroactive substance from cockroaches and crayfish. J. Econ. Entom. 57: 241-247.

Hille, B. (1967) The selective inhibition of delayed potassium currents in nerve by tetraethylammonium ions. J. Gen. Physiol. 50: 1287-1302.

Hille, B. (1968) Pharmacological modification of the sodium channels of frog nerve. J. Gen. Physiol. 51: 199-219.

Hodgkin, A.L. and A.F. Huxley (1952a) Currents carried by sodium and potassium ions through the membrane of the giant axon of Loligo. J. Physiol. 116: 449-472.

Hodgkin, A.L. and A.F. Huxley (1952b) The components of membrane conductance in the giant axon of Loligo. J. Physiol. 116: 473-496.

Hodgkin, A.L. and A.F. Huxley (1952c) The dual effect of membrane potential on sodium conductance in the giant axon of Loligo. J. Physiol. 116: 497-506.

Hodgkin, A.L. and A.F. Huxley (1952d) A quantitative description of membrane current and its application to conduction and excitation in nerve. J. Physiol. 117: 500-544.

Hodgkin, A.L. and W.A.H. Rushton (1946) The electrical constants of a crustacean nerve fibre. Proc. Roy. Soc. B133: 444-479.

Hodgkin, A.L., A.F. Huxley and B. Katz (1952) Measurements of current-voltage relations in the membrane of the giant axon of Loligo. J. Physiol. 116: 424-448.

Holan, G. (1971a) Rational design of degradable insecticides. Nature 232: 664-647.

Holan, G. (1971b) Rational design of insecticides. Bull. W.H.O. 44: 355-362.

Julian, F.J., J.W. Moore and D.E. Goldman (1962a) Membrane potentials of the lobster giant axon obtained by use of the sucrose-gap technique. J. Gen. Physiol. 45: 1195-1216.

Julian, F.J., J.W. Moore and D.E. Goldman (1962b) Current-voltage relations in the lobster giant axon membrane under voltage clamp conditions. J. Gen. Physiol. 45: 1217-1238.

Kirsch, G.E. and T. Narahashi (1978) Effects of aminopyridine derivatives on potassium channel of squid axon. Biophys. J. 21: 165a.

Lalonde, D.I.V. and A.W.A. Brown (1954) The effect of insecticides on the action potentials of insect nerve. Canad. J. Zool. 32: 74-81.

Matsumura, F. and K.C. Patil (1969) Adenosine triphosphatase sensitive to DDT in synapses of rat brain. Science 166: 121-128.

Metcalf, R.L. (1971) Structure-activity relationships for insecticidal carbamates. Bull. W.H.O. 44: 43-78.

Metcalf, R.L. and T.R. Fukuto (1968) The comparative toxicity of DDT and analogues to susceptible and resistant houseflies and mosquitoes. Bull. W.H.O. 38: 633-647.

Moore, J.W. and E.B. Cox (1976) A kinetic model for the sodium conductance system in squid axon. Biophys. J. 16: 171-192.

Murayama, K., N.J. Abbott, T. Narahashi and B.I. Shapiro (1972) Effects of allethrin and Condylactis toxin on the kinetics of sodium conductance of crayfish axon membranes. Comp. Gen. Pharmacol. 3: 391-400.

Narahashi, T. (1964a) Insecticide resistance and nerve sensitivity. Japan. J. Med. Sci. Biol. 17: 46-53.

Narahashi, T. (1964b) Excitable membrane and calcium. Seibutsu-Butsuri (Biophysics) 4: 101-114.

Narahashi, T. (1965) The physiology of insect axons. In: The Physiology of Insect Central Nervous System. ed. J.E. Treherne and J.W.L. Beament, p. 1-20, Academic Press, London and New York.

Narahashi, T. (1966) Mode of action of insecticides. Kagaku To Seibutsu (Chemistry and Biology) 4: 134-140.

Narahashi, T. (1971) Effects of insecticides on excitable tissues. In: Advances in Insect Physiology, Vol. 8, J.W.L. Beament, J.E. Treherne and V.B. Wigglesworth, eds. pp. 1-93, Academic Press, New York.

Narahashi, T. (1972) Mechanism of action of tetrodotoxin and saxitoxin on excitable membranes. Fed. Proc. 31: 1124-1132.

Narahashi, T. (1974) Chemicals as tools in the study of excitable membranes. Physiol. Rev. 54: 813-889.

Narahashi, T. (1975) Mode of action of dinoflagellate toxins on nerve membranes. In: Proceedings of the First International Conference on Toxic Dinoflagellate Blooms, Nov. 4-6, 1974, Mass. Science and Technology Foundation, p. 395-402.

Narahashi, T. (1978) Neurophysiological study of the structure-activity relations of insecticides. In: Pesticides and Venom Neurotoxicity. ed. D.L. Shankland, R.M. Hollingworth and T. Smyth, Jr. p. 119-131. Plenum, New York.

Narahashi, T. and H.G. Haas (1967) DDT: Interaction with nerve membrane conductance changes. Science 157: 1438-1440.

Narahashi, T. and H.G. Haas (1968) Interaction of DDT with components of lobster nerve membrane conductance. J. Gen. Physiol. 51: 177-198.

Narahashi, T. and T. Yamasaki (1960a) Mechanism of the after-potential production in the giant axons of the cockroach. J. Physiol. 151: 75-88.

Narahashi, T. and T. Yamasaki (1960b) Mechanism of increase in negative after-potential by dicophanum (DDT) in the giant axons of the cockroach. J. Physiol. 152: 122-140.

Narahashi, T. and T. Yamasaki (1960c) Behaviors of membrane potential in the cockroach giant axons poisoned by DDT. J. Cell. Comp. Physiol. 55: 131-142.

Narahashi, T., J.W. Moore and W.R. Scott (1964) Tetrodotoxin blockage of sodium conductance increase in lobster giant axons. J. Gen. Physiol. 47: 965-974.

Narahashi, T., J.W. Moore and B.I. Shapiro (1969) *Condylactis* toxin: Interaction with nerve membrane ionic conductances. Science 163: 680-681.

Narahashi, T., K. Nishimura, J.L. Parmentier, K. Takeno and M. Elliott (1977) Neurophysiological study of the structure-activity relation of pyrethroids. In: Synthetic Pyrethroids. ACS Symposium Series, No. 42, Ed. M. Elliott, p. 85-97, Amer. Chem. Soc.

Nishimura, K. and T. Narahashi (1978) Structure-activity relationships of pyrethroids based on direct action on nerve. Pesticide Biochem. Physiol. 8: 53-64.

Pelhate, M. and Y. Pichon (1974) Selective inhibition of potassium current in the giant axon of the cockroach. J. Physiol. 242: 90-91P.

Pichon, Y. (1969a). Effets du D.D.T. sur la fibre nerveuse isolée d'insecte. Étude en courant et en voltage imposés. J. Physiol., Paris 61 (Suppl. 1): 162-163.

Pichon, Y. (1969b) Aspects Électriques et Ioniques du Fonctionnement Nerveux chez les Insectes. Cas Particulier de la Chaine Nerveuse Abdominale d'une Blatte Periplaneta americana. L. Thèse, Univ. Rennes.

Price, N.R. (1976) The effect of two insecticides on the Ca^{2+} + Mg^{2+}-activated ATPase of the sarcoplasmic reticulum of the flounder, Platichthys flesus. Comp. Biochem. Physiol. 55C: 91-94.

Roeder, K.D. and E.A. Weiant (1946) The site of action of DDT in the cockroach. Science 103: 304-306.

Roeder, K.D. and E.A. Weiant (1948) The effect of DDT on sensory and motor structures in the cockroach leg. J. Cell. Comp. Physiol. 32: 175-186.

Roeder, K.D. and E.A. Weiant (1951) The effect of concentration, temperature and washing on the time of appearance of DDT-induced trains in sensory fibers of the cockroach. Ann. Entom. Soc. Amer. 44: 372-380.

Schneider, R.P. (1975) Mechanism of inhibition of rat brain (Na + K)-adenosine triphosphatase by 2,2-bis(p-chlorophenyl)-1,1,1-trichloroethane (DDT). Biochem. Pharmacol. 24: 939-946.

Shanes, A.M. (1949) Electrical phenomena in nerve. II. Crab nerve. J. Gen. Physiol. 33: 75-102.

Shankland, D.L. and C.W. Kearns (1959) Characteristics of blood toxins in DDT-poisoned cockroaches. Ann. Entom. Soc. Amer. 52: 386-394.

Smyth, T., Jr. and C.C. Roys (1955) Chemoreception in insects and the action of DDT. Biol. Bull. 108: 66-76.

Soliman, S.A. and L.K. Cutkomp (1963) A comparison of chemoreceptor and whole-fly responses to DDT and parathion. J. Econ. Entom. 56: 492-494.

Sternburg, J. (1960) Effect on insecticides on neurophysiological activity in insects. J. Agric. Food Chem. 8: 257-261.

Sternburg, J. (1963) Autointoxication and some stress phenomena. Ann. Rev. Entom. 8: 19-38.

Sternburg, J. and C.W. Kearns (1952) The presence of toxins other than DDT in the blood of DDT-poisoned roaches. Science 116: 144-147.

Sternburg, J., S.C. Chang and C.W. Kearns (1959) The release of a neuroactive agent by the American cockroach after exposure to DDT or electrical stimulation. J. Econ. Entom. 52: 1070-1076.

Tasaki, I. and S. Hagiwara (1957) Demonstration of two stable potential states in the squid giant axon under tetraethylammonium chloride. J. Gen. Physiol. 40: 859-885.

Tobias, J.M. (1960) Further studies on the nature of the excitable system in nerve. I. Voltage-induced axoplasm movement in squid axons. II. Penetration of surviving, excitable axons by proteases. III. Effects of proteases and of phospholipases on lobster giant axon resistance and capacity. J. Gen. Physiol. 43: suppl. 57-71.

Wu, C.H., J. van den Bercken and T. Narahashi (1975). The structure-activity relationship of DDT analogs in crayfish giant axons. Pesticide Biochem. Physiol. 5: 142-149.

Yamasaki, T. and T. Ishii* (1952b) Studies on the mechanism of action of insecticides (V). The effects of DDT on the synaptic transmission in the cockroach. Oyo-Kontyu 8: 111-118.

Yamasaki, T. and T. Ishii* (1953) Studies on the mechanism of action of insecticides (VI). DDT symptoms in the cockroach, with special reference to the effect of temperature. Oyo-Kontyu 9: 87-96. English translation (1957). In: Japanese Contributions to the Study of the Insecticide-Resistance Problem. pp. 130-139. Publ. by Kyoto University for W.H.O.

Yamasaki, T. and T. Ishii* (1954a) Studies on the mechanism of action of insecticides (VII). Activity of neuron soma as a factor of development of DDT symptoms in the cockroach. Botyu-Kagaku 19: 1-14. English translation (1957). In: Japanese Contributions to the Study of the Insecticide-Resistance Problem. pp. 140-154. Publ. by Kyoto Univ. for W.H.O.

Yamasaki, T. and T. Ishii * (1954b) Studies on the mechanism of action of insecticides (VIII). Effects of temperature on the nerve susceptibility to DDT in the cockroach. Botyu-Kagaku 19: 39-46. English translation (1957). In: Japanese Contributions to the Study of the Insecticide-Resistance Problem. pp. 155-162. Publ. by Kyoto Univ. for W.H.O.

Yamasaki, T. and T. Ishii* (1954c) Studies on the mechanism of action of insecticides (IX). Repetitive excitation of the insect neuron soma by direct current stimulation and effects of DDT. Japan. J. Appl. Zool. 19: 16-28. English translation (1957). In: Japanese Contributions to the Study of the Insecticide-Resistance Problem. pp. 163-175. Publ. by Kyoto Univ. for W.H.O.

*Former name of T. Narahashi

Yamasaki, T. and T. Narahashi (1957a) Increase in the negative-after-potential of insect nerve by DDT. Studies on the mechanism of action of insecticides (XIII). Botyu-Kagaku 22: 296-304.

Yamasaki, T. and T. Narahashi (1957b) Intracellular microelectrode recordings of resting and action potentials from the insect axon and the effects of DDT on the action potential. Studies on the mechanism of action of insecticides (XIV). Botyu-Kagaku 22: 305-313.

Yamasaki, T. and T. Narahashi (1959) The effects of potassium and sodium ions on the resting and action potentials of the cockroach giant axon. J. Insect Physiol. 3: 146-158.

Yeh, J.Z. and T. Narahashi (1976) Mechanism of action of quinidine on squid axon membranes. J. Pharmacol. Exp. Ther. 196: 62-70.

Yeh, J.Z., G.S. Oxford, C.H. Wu and T. Narahashi (1976) Dynamics of aminopyridine block of potassium channels in squid axon membrane. J. Gen. Physiol. 68: 519-535.

CHARACTERISTICS OF DDT-SENSITIVE Ca-ATPase IN THE AXONIC MEMBRANE[1]

Fumio Matsumura and S. M. Ghiasuddin

Pesticide Research Center
Michigan State University
East Lansing, MI 48824

ABSTRACT

A highly DDT-sensitive Ca-ATPase was found in an axonic nerve preparation from the lobster legs. As a result of several attempts to isolate the enzyme apart from other ATPases, an assay condition was devised. This method gives rise to the ATPases activity which is totally sensitive to DDT. By using such an approach the characteristics of the Ca-ATPase was studied. The I_{50} of the enzyme toward DDT was found to be in the neighborhood of 10^{-9} M. It has a rather high Ca^{++} optimum (0.3 mM) suggesting that it is likely an ecto Ca-ATPase located at the surface of the axons. This was confirmed by using an axonic suspension preparation which gave essentially identical results on Ca-ATPase activities and its characteristics. The inhibition of the Ca-ATPase appears to result in the suppression of Ca^{++} uptake process at the surface of the membrane as judged by a $^{45}Ca^{++}$ uptake experiment by using both the membrane and the axonic suspension preparations in the presence of ATP. These results help to explain why DDT poisoned nerve membranes are prone to destabilization and subsequent excitation.

---

[1] A major portion of the experiment presented in this manuscript was conducted in the Department of Entomology, University of Wisconsin - Madison.

INTRODUCTION

It has been more than 35 years since DDT was first marketed as an insecticide. In the history of man's quest for control of insect pests, no other single chemical has ever been used so extensively. DDT is economical, reliable as an insect controlling agent, and its effectiveness is long lasting. On the other hand, DDT has been found to accumulate in many biological systems and was found to cause varieties of biological effects which no doubt led to its eventual downfall in this country.

From the viewpoint of biochemical basis of its mode of action, DDT is a very mysterious compound. It causes a violent excitatory neurotoxic symptom in most of insects and other arthropod species, and yet it is not a reactive chemical as attested to by its long residual property in the environment. More curious is the phenomenon that DDT's neurotoxic action unlike other insecticides, is enhanced at low temperature and in a low Ca^{++} environment *in situ* (Matsumura and Narahashi 1971). There have been many efforts to study and understand the reason for such DDT effects, and many hypotheses have been advanced. However, no adequate experimental data have ever been produced to explain the biochemical cause for DDT intoxication to date.

In 1969 it was shown independently by two research groups (i.e. by Koch and Matsumura and his colleagues) that DDT characteristically affects the Na-K ATPase and Mg-ATPase in rabbit and rat brain. Subsequent studies from a variety of tissues and animal species have confirmed this finding, and that DDT inhibits not only Na-K ATPase but many other defined and undefined ATPases as well. A careful study of the literature shows that DDT inhibition of these ATPases does not usually reach 100% no matter how high its concentration is raised suggesting thereby that only a part of the total ATPase complex was sensitive to DDT.

In this paper we report the discovery that there is a highly DDT-sensitive ATPase in the peripheral nervous system of a crustacea, and that such a system has many qualifications to be the actual target of DDT attack.

MATERIALS AND METHODS

The materials consisted of peripheral nerves from walking legs of the american lobster, *Homarus americanus*. The nerves were dissected out from live lobsters and glass-glass homogenized in 0.725 M sucrose containing 1 mM EDTA and subjected to fractional ultracentrifugation as described by Doherty and Matsumura (1974). The membrane fraction obtained at 90,000 g was used as the main enzyme source.

In another set of experiments lobster nerves were cut into 0.2 mm sections by tissue chopper, suspended in 0.725 M sucrose and 1 mM EDTA by using a loose fit teflon-glass homogenizer (clearance 1 mm), centrifuged at 800 g and the pellet thus obtained was used as enzyme source of "axonic suspensions".

To measure the ATPase activity in lobster nerves, the inorganic phosphate (or Pi) formed as a result of the ^{32}P ATP hydrolysis in the supernatant of the reaction mixture was measured using activated charcoal as described by Crane and Lipmann (1953). Unless otherwise specified a fixed final concentration of 8 x 10^{-8} M ^{32}P ATP (tris-salt from New England Nuclear) was always employed. The ATPase reaction mixture consisted of 0.1 ml enzyme preparation (containing 4 mg equivalent of wet nerves unless mentioned otherwise) and 0.9 ml assay buffer containing usually 60 mM Na^+, 60 mM K^+ and 30 mM tris buffer, pH adjusted by HCl to 7.3; and 0.3 mM Ca^{++}. In addition to these 0.1 mM (final concentrations) of ouabain was present in each tube to inhibit any Na-K ATPase activity. DDT 1,1,1-trichloro-2,2-bis(p-chlorophenyl) ethane was added in 10 µl ethanol solution and the reaction was carried out for 10 minutes at 30°C. After the incubation, the reaction was stopped by adding 0.2 ml of 10% trichloroacetic acid (TCA) to each tube and the proteins were co-precipitated with 1 mg bovine serum albumin added in 0.1 ml of water. The precipitated proteins were removed by centrifugation and the clear supernate transferred to a similar set of clean tubes. About 100 mg activated charcoal (Nuchar-CN) was then added to each tube to absorb unreacted ATP and other adenosine-containing phosphate compounds leaving behind the inorganic ^{32}P inorganic phosphate (or Pi) in the supernate. A 0.5 ml aliquot of the clear supernate was used for radioassay. Zero minute controls were prepared by adding ^{32}P ATP at the end of the incubation period followed by the immediate addition of TCA. They accompanied each experiment and their values were subtracted from controls and controls treated with DDT to estimate the exact amount of (Pi) formed as a function of the ATPase activity.

The binding of $^{45}Ca^{++}$ (obtained from New England Nuclear) was carried out under identical incubation conditions used for the ATPase assay at 30°C, by a millipore filtration method as described by Nakamaru and Schwartz (1971). For this purpose $^{45}Ca^{++}$ of a high specific activity was diluted in cold $CaCl_2$ solution to get the desired concentration and specific activity of 0.3 mM Ca ions in 1 µl volume; usually the level of radioactivity was 10^6 dpm per assay. $^{45}Ca^{++}$ binding by lobster nerve proteins was studied both in the absence and presence of ATP. After the incubation the reaction was stopped with chilled assay buffer and the reaction mixture was then filtered through 0.45 µm millipore filters and washed for three times. The membranes and axonic suspensions retained on the filters were washed to remove exogenous unbound

$^{45}Ca^{++}$, dried and transferred to scintillation vials for counting. Values for zero minute controls in which $^{45}Ca^{++}$ was added to nerve proteins at 0°C at the end of the incubation period were always subtracted from experimental values to correct for non-metabolic bindings of Ca^{++} to the preparations.

RESULTS

Sensitivity of Ca-ATPase to Cations

Figure 1 shows the effect of different concentrations of Ca^{++} on lobster nerve ATPase activity under the standard assay condition.

It can be seen in Fig. 1 that Ca^{++} stimulates the total ATPase activity in lobster nerves both in controls and controls treated with DDT. The optimum concentration of Ca^{++} which produces maximum stimulation in the ATPase activity was found to be 0.3 mM. The total % increase at this concentration over the base value was found to be around 94%.

It must be noted here that not all of the Ca-ATPase activity was inhibited by DDT under this experimental condition. For the sake of clarity here Ca-ATPase activities are divided into two groups: DDT-sensitive and DDT-insensitive Ca-ATPases. The activity due to the DDT-sensitive ATPase may be obtained by subtracting the value for control by that for DDT-treated (i.e. "difference").

The percent DDT inhibition, shown as the "difference" between control and control + DDT was found to be between 30-40% under the present conditions.

To study the nature of this enzyme, using the standard assay condition, the effects of monovalent cations (Na^+ and K^+) were studied. The optimum concentrations of Na^+ and K^+ for DDT-sensitive Ca-ATPases were found to be 20 mM. The values were lower than those for DDT-insensitive Ca-ATPases. In another set of experiment Na^+ was replaced with Li^+. Under such conditions no appreciable increase in the ATPase activity was observed suggesting thereby that Na ions were essential for the maximum activity of this enzyme.

Temperature and Dose Response Effect of DDT and DDE

The effect of incubation temperature on the levels of DDT inhibition was also studied. It was found that DDT inhibition was more pronounced at low temperature (Fig. 2). The DDT-sensitive ATP-ase has a lower temperature quotient ($Q_{10} = 1.26$) than the insensitive one ($Q_{10} = 1.70$).

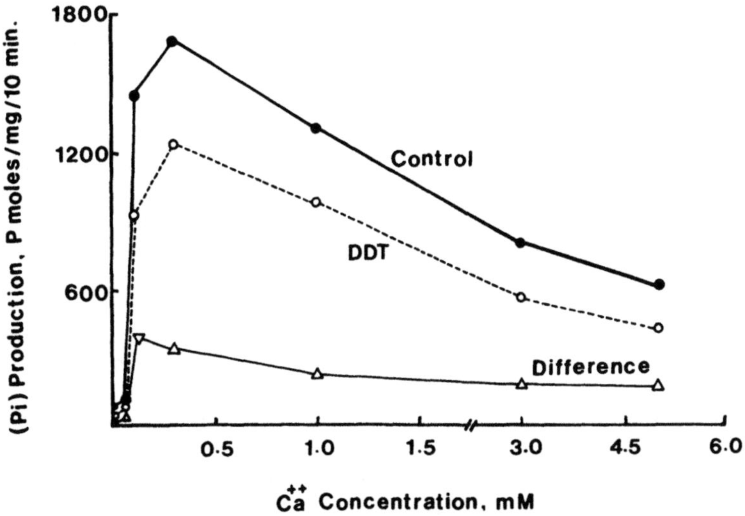

Figure 1. Effect of Ca^{++} concentration on ATPase activity of lobster nerve membrane preparation. The standard assay conditions were: 60 mM Na^{+}, 60 mM K^{+}, pH 7.3, 0.1 mM ouabain, 8 x 10^{-8}M ^{32}P ATP and temperature 30°C. Difference between "control" and "DDT" (10^{-5}M) treated is the indication of the level of the activity of DDT sensitive Ca-ATPase.

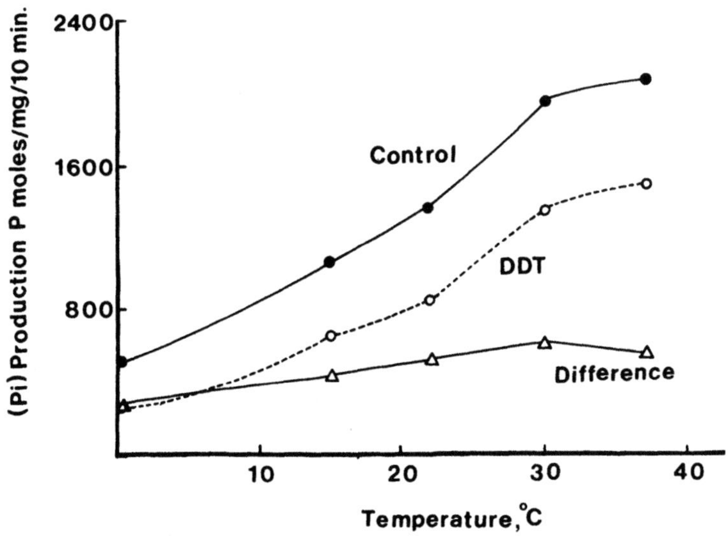

Figure 2. DDT (10^{-5}M) effect on Ca-ATPase activity of the nerve membrane preparation at different temperatures, under the standard assay conditions as in Figure 1.

The effect of the substrate concentration was studied. As shown in Table 1 the change in ATP concentrations was found to have a profound effect on the level of DDT inhibition, and hence the ratio of DDT-sensitive to -insensitive Ca-ATPase.

Under the standard condition, DDT inhibits approximately 30 to 50% of the total ATPase activity at concentrations between 10^{-6} to 10^{-4} M (i.e. plateauing), indicating that approximately one-third to one-half of the enzyme activity comes from a DDT-sensitive ATPase.

Various combinations of the incubation conditions to utilize the above differences in properties between DDT-sensitive and -insensitive Ca-ATPases resulted in varying degrees of success.

The best method to eliminate the DDT-insensitive ATPase activity was, however, found to be the lowering of enzyme concentrations. As pointed out by Bratkowski (1970) and Schneider (1975) the level of DDT inhibition of ATPases increases as the total protein content per assay is decreased. When the protein content was reduced to 1/8 of the standard assay condition, DDT at 10^{-5} M was found to inhibit 100% of the total ATPase activity (Fig. 3) even at relatively high temperature (30°C).

The inhibitory pattern of ATPase by DDT and DDE at 10^{-8} M of ATP at 30°C is shown in Fig. 4. The observed I_{50} for DDT was in the order of 10^{-9} M which is 10 to 10,000 times lower than that reported for other ATPases against DDT (Matsumura and Patil, 1969; Cutkomp _et al._, 1971, 1971; Janicki and Kinter, 1971; Desaiah _et al._, 1974 etc.). Such a sensitivity is comparable to that of tetrodotoxin on invertebrate axons, and thereby makes this case as one of the most sensitive neural inhibitory phenomenon.

TABLE 1. DDT EFFECT AS A FUNCTION OF ATP CONCENTRATION ON Ca-ATPase

ATP Concentration, M	(Pi) Production, P moles/mg/10 min.		% DDT Inhibition
	Control	Control + DDT	
8×10^{-7}	2032.25	1480.41	27.15
8×10^{-8}	2011.62	1446.62	28.08
8×10^{-9}	102.42	51.33	49.88
8×10^{-10}	6.08	2.44	64.14

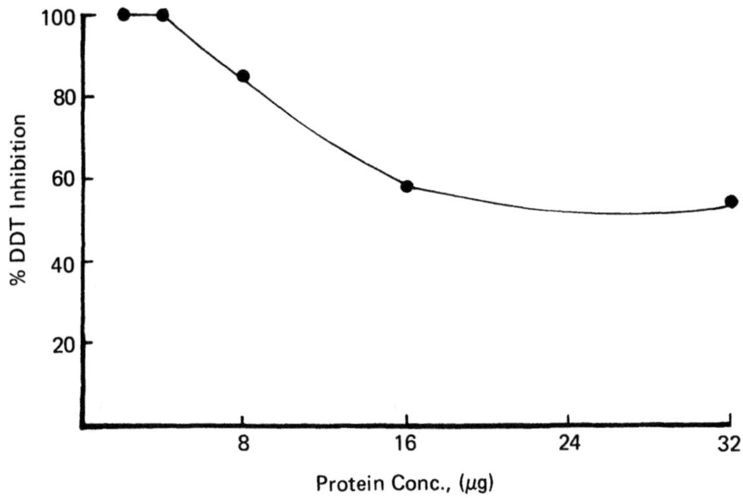

Figure 3. Percent DDT (10^{-5}M) inhibition of Ca-ATPase activity at varied protein concentrations of the nerve membrane preparations. The protein content was assayed by Lowry et al. (1951) method.

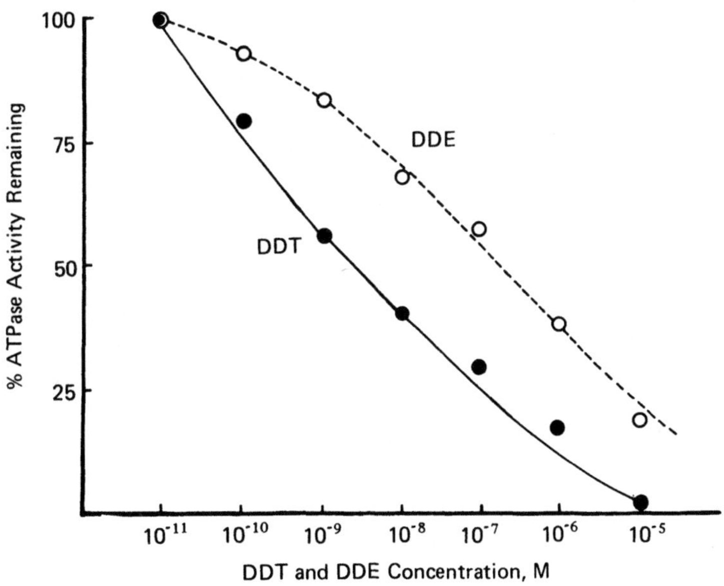

Figure 4. Dose-response effect of DDT and DDE on Ca-ATPase activity of the nerve membrane preparation. A low protein concentration (4 μg/assay) was employed in this experiment under the standard assay condition.

Effect of Various Inhibitors and Neuroactive Agents

To study the nature of this DDT-sensitive ATPase, we have also examined the effect of various inhibitors and neuroactive agents (Table 2). It was found to be sensitive to lanthanum ($I_{90} = 1$ mM) and Ruthenium red ($I_{50} = 10$ μM). The high potency and specificity of Ruthenium red as an inhibitor of Ca-ATPase as opposed to Mg-ATPase is well documented by Watson et al. (1971).

The sensitivities to other neuroactive agents which have similar mode of action as DDT were: veratrine 37% inhibition at 10^{-4} g/ml and D-trans-allethrin 41% inhibition at 10^{-4}M. Cyanide ion had little effect up to 10 mM. Likewise iodide ion had little or no effect. However, fluoride and mersalyl acid seem to have a pronounced effect and appears to inhibit almost total ATPase activity at very low concentrations (Table 2).

$^{45}Ca^{++}$ Binding and Effect of DDT

Because of its ion sensitivities and known roles of ATP in regulating Ca^{++} concentrations in axonic membranes (Dipolo, 1974; Mullins and Brinley, 1975; Martin and Shaw, 1974 etc.) it is likely that this Ca-ATPase plays some role in Ca^{++} regulation in the axonic membrane. To test this possibility binding of $^{45}Ca^{++}$ to the membrane fraction was studied under an identical incubation condition as the above ATPase assay. As shown in Table 3, the extent of Ca^{++} binding was slightly enhanced in the presence of added ATP, and DDT inhibited Ca^{++} binding only when ATP was present in the medium. Judging by the fact that a relatively high Ca^{++} concentration (0.3 mM) was required to activate the enzyme, we have first presumed that the enzyme is working at the outer surface (i.e. ecto-ATPase) of the cell to regulate its Ca^{++} levels (Trams and Lauter, 1974, 1977). To test this possibility an axonic suspension as outlined before was prepared and the Ca^{++} optimum test for the Ca-ATPase activity, and $^{45}Ca^{++}$ binding test as shown in Table 3 were repeated. In both cases it was clearly established that the enzyme characteristics and the DDT-sensitivity of the axonic suspension were identical to the ones previously obtained by using the 90,000 g nerve membrane preparations.

DISCUSSION

The foregoing data have clearly demonstrated that lobster nerves possess a highly DDT-sensitive Ca-ATPase which is cation dependent and is totally inhibited by DDT. The possibility that its inhibition is vitally related to the process of DDT poisoning of the nervous system must be examined first. The strongest supporting evidence to this comes from in situ studies with the lobster nerves wherein

TABLE 2. VARIOUS INHIBITORS AND NEUROACTIVE AGENTS AND PERCENT INHIBITION OF Ca-ATPase

Compound	Concentration	% Inhibition
Lanthanum	1 mM	91.9
Ruthenium Red	10 μM	50.0
Allethrin	0.1 mM	41.2
Veratrine	0.1 mg/ml	36.6
KCN	2 mM	5.6
NaI	10 mM	18.8
Na/KF	10 mM	80.1
Mersalyl Acid	2 μM	82.5

TABLE 3. DDT INHIBITION OF $^{45}Ca^{++}$ BINDING BY NERVE ATPase

ATP	$^{45}Ca^{++}$ binding, n moles/mg/10 min.		% DDT
Concentration, M	Control	Control + DDT	Inhibition
Nerve membrane (90,000 g precipitate)			
0	1.81	3.08	—
1×10^{-3}	1.77	0.00	100
5×10^{-4}	2.00	0.44	78.0
8×10^{-6}	2.20	0.32	85.6
8×10^{-8}	2.08	0.38	81.7
Axonic suspension (800 g precipitate)			
0	2.95	3.11	—
1×10^{-3}	2.59	0.00	100
5×10^{-4}	2.72	0.59	78.3
8×10^{-6}	2.41	0.21	91.3
8×10^{-8}	2.51	0.00	100

Matsumura and Narahashi (1971) have demonstrated that neurophysiological symptoms of DDT poisoning (i.e. prolonged falling phase or increased negative after potential) can be markedly enhanced at low (0-20 mM), and reversed to normal at high external Ca^{++} concentration, (to 100 mM) in the bathing medium.

Clearly then the external Ca^{++} concentration has something to do with DDT-intoxication mechanism at the axonal level.

The effect of changes in Ca^{++} concentrations on the axonic excitability has been studied in detail by Frankenhaeuser and Hodgkin (1957). In short, the absence of sufficient Ca^{++} causes membrane destabilization by decreasing membrane threhsold. Here the surface charge of the membrane is decreased by the lack of Ca^{++}. The result is that the potential relative to the inner to the outer surface of the membrane decreases. The axons lacking sufficient external Ca^{++}, therefore, are easy to depolarize and prone to elicit a series of firing of action potentials. Ecto Ca-ATPases are envisioned as the enzyme situated at the surface of the membrane and regulating Ca^{++} levels at the surface at the expense of ATP's energy. Here the enzyme is expected to be activated when the external Ca^{++} concentration becomes low such as 10^{-4} to 10^{-3} M; whereas whenever the external concentration exceeds the necessary level it either should cease to function, or should actively expel Ca^{++}. It has been known that in alive cell membranes there must be a gradient of Ca^{++} concentrations; externally the order of concentration being 10^{-2} to 10^{-3} M and internally 10^{-6} to 10^{-8} M in terms of the presence of free Ca^{++} ions (Fig. 5). Ca-ATPases which act at the inside or have Ca^{++} affinity site on the inner surface of a cell are expected to have low Ca^{++} optimum (e.g. Ca-Mg ATPase has usually the optimum Ca^{++} concentration of 10^{-6} to 10^{-7} M), whereas the ones working outside of the membrane should accordingly have high Ca^{++} optimum values. The high Ca-optimum of the DDT sensitive Ca-ATPase here, and its characteristics and activity in the axonic suspension preparation strongly support that the ATPase in question is an *ecto* Ca-ATPase.

More convincing is the recent demonstration that the level of this type of *ecto*-Ca-ATPase is lower in the brain of the seizure prone mice than that in the normal mice (Rosenblatt *et al.*, 1976), indicating that the lack of the enzyme activity to maintain a proper level of the surface Ca^{++} in the brain cells can indeed lead to the increase in the chance for epileptic seizures as predicted by the above Ca^{++} theory.

Indeed our experiment with $^{45}Ca^{++}$ clearly shows that DDT inhibits Ca^{++} uptake by the membrane in the presence of ATP.

In summary the DDT sensitive Ca-ATPase found in the lobster nerve has many qualifications to be an important enzyme in the process of DDT poisoning. It is highly sensitive to DDT, acts at

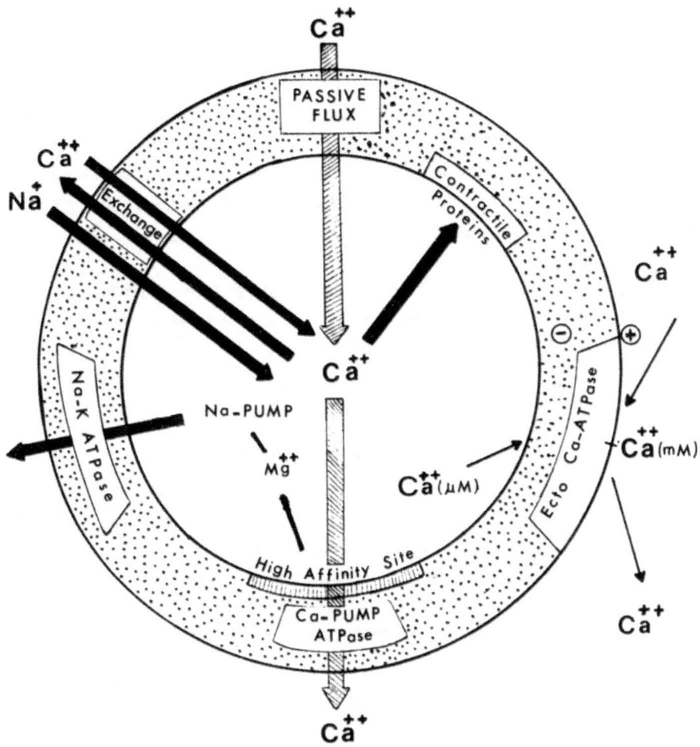

Figure 5. Diagramatic sketch showing the regulation of Ca^{++} and the inter-relationships among the levels of <u>extra</u> and <u>intra</u>-cellular Ca^{++}, the Ca-pump and Na-pump of nerves. Ca^{++} may enter the nerve cells and axons <u>via</u> passive influx or Na^{+} - Ca^{++} exchange diffusion. The concentration of Ca^{++} bound to the outer surface of the membrane is probably controled by <u>ecto</u>-ATPase and its optimum concentration is in mM range (10^{-2} - 10^{-3} M). The levels of free intracellular Ca^{++} coming to the inner surface of the membrane is in the range of μM levels (10^{-6} - 10^{-7} M). The levels of free intracellular Ca^{++} are dependent on the balance of influx and efflux, and equilibrium reached by the interaction of Ca^{++} binding proteins. (The role of mitochondria is omitted from this figure for simplification). The Ca-pump and Na-pump are the function of the membrane bound ATPase while the <u>ecto</u>-Ca-ATPase work to keep the level of Ca^{++} at the outer surface of the membrane. Inhibition of this enzyme leads to a decrease in the level of surface bound Ca^{++} and results in destabilization of the nerve cell. [Sketch modified from, Vincenzi and Hinds (1976)].

low temperatures, and low ATP concentrations, and is sensitive to other neuroactive agents with similar actions as DDT. Above all the most important aspect of this study is that the phenomenon of Ca-ATPase inhibition fits very well with the overall picture of DDT induced nerve excitation processes. There is a wealth of information on the role of Ca^{++} on the axonic functions to back up the theory as well as the presence of experimental evidence that the external Ca^{++} concentration changes indeed greatly affect the expression of nerve poisoning by DDT.

Thus, even if the Ca-ATPase was not the prime target of DDT, it is certain that inhibition of such an enzyme should play a significant supporting role in the process of the nerve poisoning by DDT.

ACKNOWLEDGEMENTS

Supported by Division of Research, College of Agricultural and Life Sciences, University of Wisconsin - Madison, College of Agriculture and Natural Resources, and College of Natural Science, Michigan State University, and by a research grant ES-00857 from the National Institute of Environmental Health Sciences, Research Triangle Park, N.C. We thank Dr. Toshio Narahashi, Northwestern University, for his valuable discussion relating to the role of Ca^{++} on nerve function.

REFERENCES

Bratkowski, T. A. 1970. Ph.D. Thesis, Dept. Entomol., University of Wisconsin.

Crane, R. K. and F. Lipmann. 1953. J. Biol. Chem. 201, 235.

Cutkomp, L. K., H. H. Yap, E. Y. Cheng and R. B. Koch. 1971. Chem.-Biol. Interact. 3, 439.

Cutkomp, L. K., H. H. Yap, E. V. Vea and R. B. Koch. 1971. Life Sci. 10, 1201.

Desaiah, D., L. K. Cutkomp and R. B. Koch. 1974. Pest. Biochem. Physiol. 4, 232.

Dipolo, R. 1974. J. Gen. Physiol. 64, 503.

Doherty, J. D. and F. Matsumura. 1974. J. Neurochem. 22, 765.

Frankenhaeuser, B. and A. L. Hodgkin. 1957. J. Physiol. 137, 218.

Janicki, R. H. and W. B. Kinter. 1971. Nature 233, 148.

Koch, R. B. 1969. J. Neurochem. 16, 269.

Lowry, O. H., N. J. Rosenbrough, A. L. Farr and R. J. Randall. 1951. J. Biol. Chem. 193, 265.

Martin, K. and T. I. Shaw. 1970. J. Physiol. 208, 171.

Matsumura, F. and T. Narahashi. 1971. Biochem. Pharmacol. 20, 825.

Matsumura, F. and K. C. Patil. 1969. Science 166, 121.

Mullins, L. J. and F. J. Brinley, Jr. 1975. J. Gen. Physiol. 65, 135.

Nakamaru, Y. and A. Schwartz. 1971. Arch. Biochem. Biophys. 144, 16.

Narahashi, T. 1976. "Insecticide Biochemistry and Physiology", p. 327, edited by Wilkinson, C. F. Plenum Press, N. Y.

Rosenblatt, D. E., C. J. Lauter and E. G. Trams. 1976. J. Neurochem. 27, 1277.

Schneider, R. P. 1975. Biochem. Pharmacol. 24, 939.

Trams, E. G. and C. J. Lauter. 1974. Biochim. Biophys. Acta. 345, 180.

Trams, E. G. and C. J. Lauter. 1977. Amer. Society Neurochem. 8th Annual meeting, p. 173.

Vincenzi, F. F. and T. R. Hinds. 1976. "The Enzymes of Biological Membranes", p. 261, edited by Mortonsi, A. Plenum Press, N.Y.

Watson, E. L., F. F. Vincenzi and P. W. Davis. 1971. Biochim. Biophys. Acta 249, 606.

INTERACTION OF INSECTICIDES WITH ACETYLCHOLINE RECEPTORS

S. W. Jones, P. Sudershan, R. D. O'Brien

Section of Neurobiology and Behavior

Cornell University, Ithaca, New York 14853

The rational design of insecticides depends on the discovery of biochemical differences between target insects and organisms (including man) that should not be affected. This permits the selection of agents that act on a site that is not present in vertebrates, or at least has lower sensitivity to the agent. Our comparative study of vertebrate and insect acetylcholine receptors, which is just beginning, is in part an attempt to identify compounds with selective blocking ability on the acetylcholine receptors of insects.

Acetylcholine receptors are potentially important targets for insecticide action for several reasons. First, the insecticidal action of anticholinesterases demonstrates that interference with cholinergic synaptic transmission can be fatal to insects. More direct interference, by blocking the acetylcholine receptor itself, should be equally effective. In fact, there are insecticides in use that are believed to act in this way, by binding to insect acetylcholine receptors. The classic one is nicotine. Recently, another insecticide has been developed that has a cholinergic blocking activity in insects and vertebrates--cartap (1). Furthermore, there are several antihelminthic drugs that are thought to act on acetylcholine receptors. Finally, as we will discuss in detail, there are differences in drug sensitivity between vertebrate acetylcholine receptors and some putative receptors of invertebrates.

Acetylcholine is one of several chemical messengers in the nervous system. After release from a nerve cell, it can cause a neighboring nerve, gland or muscle to respond. This response is triggered by a combination of the acetylcholine with the acetylcholine receptor, a protein complex which is built into the membrane of certain cells. The receptor is therefore the key macromolecule involved in the response of nerve cells to acetylcholine. It should be stressed that it is entirely separate in function, composition and location from acetylcholinesterase, which is the enzyme responsible for breaking down acetylcholine once the message has been delivered. The current view (Fig. 1) is that in insects and vertebrates, the acetylcholinesterase is not built into the limiting membrane of the cell, but occupies a non-cellular layer which is adjacent to it. By contrast, the receptor is, at least in vertebrates, an intrinsic part of the cell's limiting membrane.

The next point to make is that acetylcholine receptors from different vertebrate tissues have different properties. A 50-year-old classification provides the principal division into [1] nicotinic receptors, which are readily stimulated by nicotine; and [2] muscarinic receptors, which are readily stimulated by muscarine. The receptor of the skeletal muscle is nicotinic, and that of smooth muscle is muscarinic. Quite apart from the drugs nicotine and muscarine, a whole series of agents exists which is "typically nicotinic" and another one which is "typically muscarinic." The situation in the vertebrate is complicated by the fact that the receptor in the autonomic ganglia is nicotinic, but differs in its response to some drugs from the nicotinic receptor of the neuromuscular junction. Furthermore, mammalian brain appears to contain both muscarinic and nicotinic receptors (2,3,4,5), and possibly a third kind with some of the properties of each (6,7,8). In this short paper we shall turn to relatively new information about the acetylcholine receptor of insects.

Biochemical work on vertebrate acetylcholine receptors has produced two quite valuable agents for studying receptor binding: α-bungarotoxin, a protein from a snake venom that binds very tightly to neuromuscular junction nicotinic receptors; and quinuclidinyl

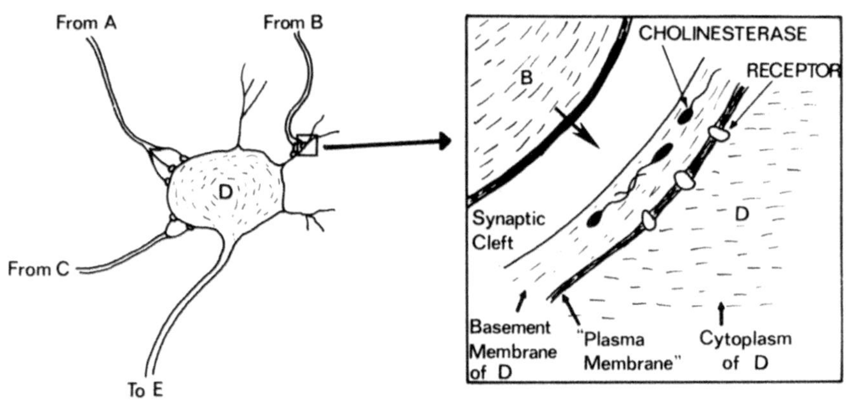

Figure 1. Schematic diagram of cholinergic synaptic transmission.

benzilate (QNB), which binds to muscarinic receptors. Both of these have recently been used to identify acetylcholine receptors in insects, especially in *Drosophila*.

The α-bungarotoxin binding component of insects is in many ways quite similar to the vertebrate nicotinic receptor, as might be expected. Both are found in heavy membrane fractions. Autoradiography demonstrates that the toxin binds only to synaptic regions of the *Drosophila* brain, as expected for a neurotransmitter receptor (9,10). The *Drosophila* heads bind α-bungarotoxin with high affinity--binding reaches a maximum at nanomolar concentrations--but unlike the vertebrate receptor, the binding is slowly reversible (11). The pharmacology of toxin binding is rather close to that of the vertebrate nicotinic receptor (Table 1); it is quite sensitive to nicotine and curare, but not to muscarinic agents. The affinity for acetylcholine is rather low, however, and the powerful nicotinic drug decamethonium is quite inactive. There is evidence that a similar toxin binding site is present in moths (12), house flies (13; our unpublished results), and in the mollusc *Aplysia* (14).

Table 1. Inhibition of α-bungarotoxin binding to *Drosophila* heads.

	μM Concentration for 50% block
Nicotinic Agents	
Nicotine	0.8
Curare	2
Decamethonium	900
Muscarinic Agents	
Atropine	50
Pilocarpine	80
Muscarine	> 10^{-3}
Dexetimide	500
Others	
Acetylcholine	20
Carbamylcholine	90

Inhibition of the initial binding rate of 15 nM toxin was studied by Millipore filtration. A 500 g supernatant of homogenized heads was used. From Dudai (13).

Table 2. Blockade of QNB binding to *Drosophila* heads.

	I_{50} (μM)
Nicotinic Agents	
Decamethonium	> 1000
Nicotine	> 1000
Muscarinic Agents	
Dexetimide	0.004
Scopolamine	0.007
Atropine	0.03
Muscarine	0.06

Inhibition of 5 nM [^{3}H]QNB binding at equilibrium was followed by filtration through glass fiber filters. A 500 g supernatant of homogenized heads was used. From Dudai and Ben-Barak (15).

Dudai and Ben-Barak (15) have also identified a putative muscarinic acetylcholine receptor in *Drosophila* by the binding of QNB. That receptor was detected by a filter assay, suggesting that it also is particulate. Its pharmacology is that of a typical muscarinic receptor (Table 2), blocked by muscarinic but not nicotinic agents.

In addition to these fairly classical nicotinic and muscarinic receptors, there is an unusual receptor that has been found in insects, that is quite unlike any known vertebrate acetylcholine receptor. This was in fact the first demonstration of what might be a receptor in insects (16). The agent muscarone (which is equally effective on muscarinic and nicotinic receptors) bound strongly to a supernatant fraction derived from house fly heads. The binding was blocked by a variety of drugs expected to block acetylcholine receptors; but Table 3 shows that the binding was blocked both by nicotinic drugs (such as curare, succinylcholine and decamethonium) and also by muscarinic drugs (such as atropine and pilocarpine). By contrast, the well-charactierzed nicotinic receptor from *Torpedo* electroplax is relatively insensitive to muscarinic drugs. It should be noted that several agents that are not cholinergic (i.e. do not normally block acetylcholine receptors) are quite effective at blocking muscarone binding in house fly, e.g. amphetamine, tyramine, and eserine. We next explored the binding of other tritiated cholinergic ligands to extracts of house fly heads: nicotine, decamethonium and atropine (17,18). All three bound to essentially the same number of sites as did muscarone, and all three had the same pharmacology. That is, binding of the classical muscarinic agent atropine was inhibited by both nicotinic and muscarinic agents; and

Table 3. % Blockade of 10^{-6} M muscarone binding to house fly brain and Torpedo electroplax.

		House Fly	Torpedo
Cholinergic (Nicotinic)			
Nicotine	10^{-5} M	39	59
Curare	10^{-5} M	19	79
Decamethonium	10^{-5} M	65	77
Cholinergic (Muscarinic)			
Atropine	10^{-4} M	72	21
Pilocarpine	10^{-4} M	84	0
Noncholinergic			
Eserine	10^{-4} M	78	0
Amphetamine	10^{-4} M	51	0
Tyramine	10^{-4} M	60	0

The data for Torpedo is with lyophilized powder prepared from a 12,000 g pellet of electroplax homogenized in water. For house fly heads, the supernatant of a 100,000 g centrifugation for one hour of a 20% homogenate in water was used in experiments given in Tables 3-7, and Fig. 3. Equilibrium dialysis was used to study binding throughout. From Eldefrawi *et al.* (16)

binding of the classical nicotinics nicotine and decamethonium were also inhibited by both classes of drug. This is strong evidence that all of these agents bind to one site with a mixed pharmacology, not to separate nicotinic and muscarinic receptors. These agents bind with high affinity (0.1-3.2 μM), and to a rather large number of sites, 2-3 nmoles per gram of heads. This is actually a higher concentration of receptor than in Torpedo electroplax, which is almost exclusively synaptic tissue. The separate nicotinic and muscarinic receptors described earlier are present in at least 40-fold lower concentration, comparable to levels in vertebrate brain.

One important aspect of this early work is the action of nicotine. It has long been assumed, by analogy with the vertebrate, that nicotine owed its toxic action on insects to an effect on the acetylcholine receptor. In confirmation, we found (Table 4) a rough correlation betwen the toxicity of some nicotine analogs to insects and their effectiveness on the mixed pharmacology receptor of the supernatant (18).

Several groups have attempted to purify this mixed receptor. Jewess *et al.* (19) and Mansour *et al.* (20) have achieved

Table 4. Nicotine and analogs on house flies.

	% Block of Muscarone Binding	LD_{50} mg/kg
Nicotine	100	250
Anabasine	98	200
3-Pyridylmethyl-dimethylamine	92	800
3-Pyridylmethyl-diethylamine	97	550
N-Diethylnicotinamide	0	5000
N-(3-Pyridylmethyl)-morpholine	0	5000

[^{3}H] muscarone 10^{-6} M; inhibitors 10^{-4} M. From Eldefrawi et al. (18).

approximately 25-fold purification by gel filtration on Sepharose or Sephadex gels. This partial purification does not affect the receptor's binding properties. Further purification attempts have been disappointing, however. Ion exchange chromatography gives less than 2-fold further purification (19) and preparative disc gel electrophoresis results in loss of binding activity (20). A very powerful purification technique, which has been very successful for purification of enzymes and receptors, is affinity chromatography. In the case of the electric skate receptor, good purification can be achieved with an affinity column containing phenyltrimethylammonium bound to Sepharose. Mansour et al. (20) found that for the house fly, a similar column eluted with the nicotinic agent decamethonium gave a peak with 3 components revealed by electrophoresis (Fig. 2). Recently R. K. Tripathi (unpublished) obtained a single peak in a similar system. As the figure shows, elution with the muscarinic agent QNB also elutes the same peak, reinforcing the view that the receptor is of mixed character. The two preparations also have closely similar subunit patterns by SDS gel electrophoresis--two closely spaced bands of 60,000 daltons and one band at 94,000 daltons [Jewess et al. (19) found two bands at 82,000 and 90,000 daltons on SDS gels, after ion exchange chromatography.] Unfortunately, Mansour et al. (20) and Tripathi (unpublished) agree that in the course of affinity purification, nearly all the binding activity is lost, so one has denatured the receptor in the course of purification.

We have undertaken a more thorough pharmacological study of the house fly head mixed receptor, with the goals of comparing it further to other acetylcholine receptors, and of studying the action of insecticides. We have used tritiated decamethonium as our ligand, because of its high affinity for this receptor (20,21,17). One problem with decamethonium is that it is also a rather good inhibitor

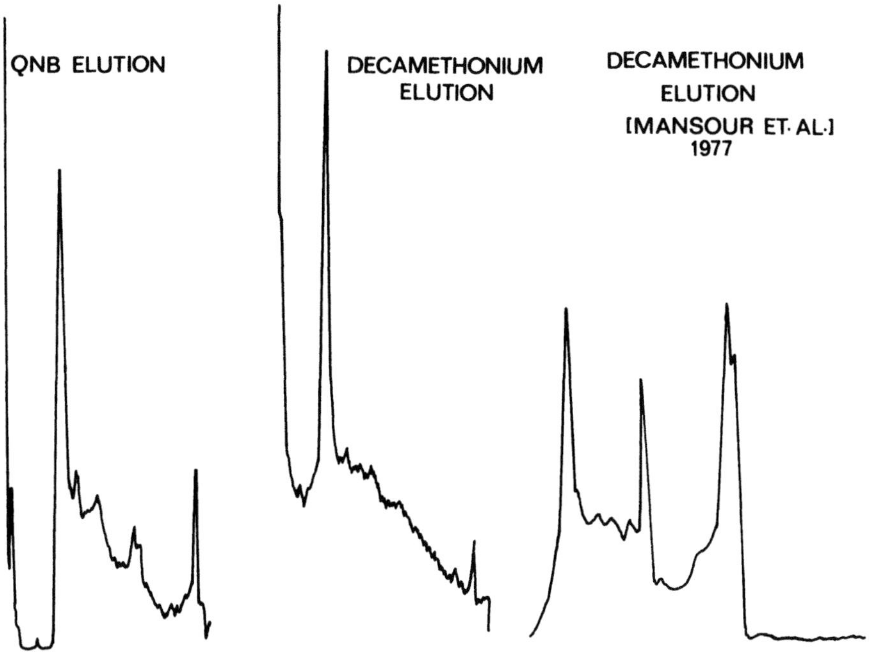

Figure 2. Scans of Coomassie blue stained discontinuous electrophoresis gels on affinity chromatography purified house fly head acetylcholine receptor. The top of the gels is to the left.

of acetylcholinesterase, which is also present in high amounts in house fly head supernatants (22). We have concluded that under the conditions we use for the inhibition studies, essentially all of the binding is in fact to the mixed receptor, by several lines of evidence. First, the affinity of house fly acetylcholinesterase for decamethonium is 30 μM, approximately 200-fold lower than its affinity for receptor (22). Second, the concentration of acetylcholinestrase is approximately 4-fold less than that of receptor in the supernatant [calculated from Tripathi and O'Brien (22)]. Finally, the cholinesterase inhibitor diisopropylfluorophosphate gives essentially complete inhibition of house fly acetylcholinesterase at 100 μM without inhibiting the binding of 0.1 μM decamethonium.

A typical binding curve for decamethonium is shown in Fig. 3. The curvature in the Scatchard plot indicates that there are two binding sites for decamethonium in the preparation. Our inhibition studies are done on the high affinity part of the curve. We are not sure what the low affinity site is. One possibility is that at high concentrations decamethonium does bind to cholinesterase. If

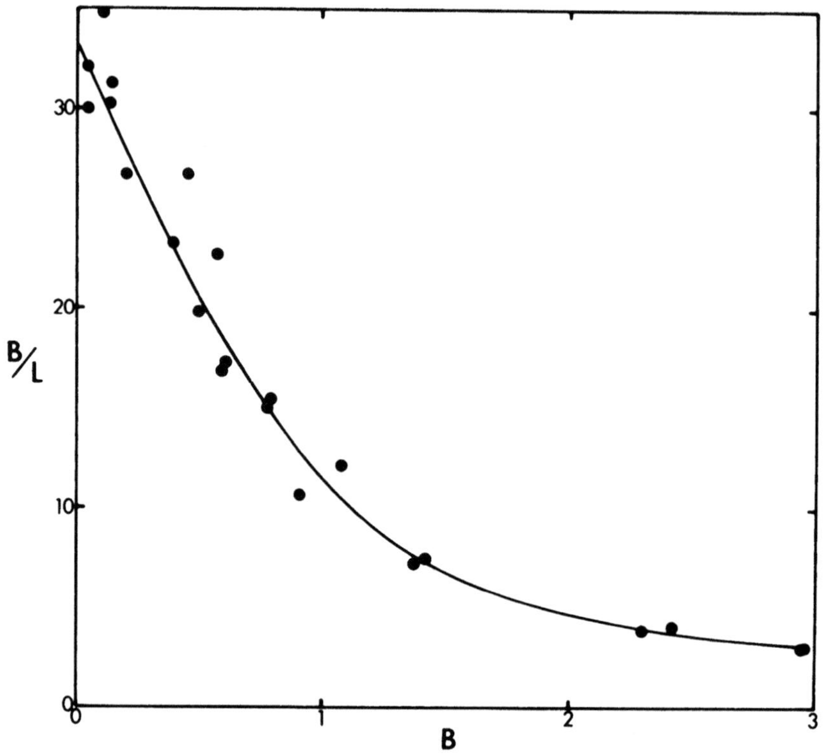

Figure 3. Scatchard plot of [^{3}H] decamethonium binding to house fly head supernatant. B = amount of decamethonium bound, nanomoles per gram heads; L = concentration of free decamethonium, nM.

so, this would not affect our inhibition studies done at lower concentrations of decamethonium. This low affinity site has also been reported by Donnellan _et al_. (21); in the partially purified preparation that site is not present (19,20).

The action of cholinergic agents is shown in Table 5. Most nicotinic agents, whether they are most effective on autonomic ganglia or at the neuromuscular junction, are powerful inhibitors of decamethonium binding. A notable exception is α-bungarotoxin, which has previously been shown not to bind to this receptor (20, 23). Several muscarinic agents and choline esters, including acetylcholine itself, are also effective. It is worth noting that atropine, scopolamine, and QNB have greater than a 1000-fold higher affinity for vertebrate muscarinic receptors, however (5). A new class of cholinergic agent, which appears to interfere with coupling between the acetylcholine binding and the physiological response,

Table 5. % Block of 10^{-7} M decamethonium binding to house fly head.

Nicotinic Agents (10^{-5} M)			
Neuromuscular		Ganglionic	
Decamethonium	94	Hexamethonium	84
Nicotine	43	Tetraethylammonium	36
Curare	16	Lobeline	73
Succinylcholine	66	Dimethylphenylpiper-azinium	89
Gallamine	31		
Phenyltrimethylammonium	69	Mecamylamine	0
α-Bungarotoxin	0		
Muscarinic Agents (10^{-5} M)			
Pilocarpine	72	Scopolamine	27
Atropine	46	QNB (quinuclidinyl benzilate)	17
Agents for Muscarinic and Nicotinic Sites (10^{-5} M)			
Arecoline	13	Acetylcholine + DFP	42
Acetyl-β-methylcholine	84	Butyrylcholine + DFP	63
Carbamylcholine	27		
Agents for Ionic Conductance Mechanism			
Chlorobenzilate	3	Amantadine	10
cis-2-methyl-6-pentadecyl piperidine	10		

is not effective. The piperidine derivative is an analog of the active component of fire ant venom (24), which itself is similar in structure to histrionicotoxin, the first agent of this class to be reported (25). (For chlorobenzilate and amantadine, see Mansour et al. (26) and Albuquerque et al. (27).

Some other agents of interest are shown in Table 6. Several agents that act on axons, and not on acetylcholine receptors, are inactive. Procaine, a local anaesthetic, is a rather good inhibitor here. Although local anaesthetics are best known for their axonal effects, they also inhibit the nicotinic acetylcholine receptor. A number of anticholinesterases are good inhibitors of decamethonium binding. These agents are also effective against vertebrate nicotinic receptors as well (28). It is possible that some anticholinesterase insecticides owe some of their toxicity to interactions with the acetylcholine receptor as well. 9-aminoacridine, an

Table 6. % Block of 10^{-7} M decamethonium binding to house fly head.

Axonic Agents (10^{-5} M)

DDT	2	Ouabain	2
Veratridine	4	Allethrin	5
Procaine	37		

Anticholinesterases (10^{-5} M)

Organophosphates		Carbamates		Quaternary	
DFP	8	Eserine	35	BW 284	97
Tetrachlorvinphos	16	Neostigmine	39	Phenyltrimethylammonium	68

Muscle Agent (10^{-5} M)

9-Aminoacridine	73

inhibitor of the action potential K^+ gate in muscle (29), is quite effective. However, two other K^+ gate agents, tetramethylammonium and 4-aminopyridine, are not effective.

Table 7 shows the effects of four antihelminthic drugs (worming agents). All of them have been proposed to act at acetylcholine receptors (30,31,32). The only one of the four that is not effective, piperazine, has been found more recently to have quite a different mechanism of action [rather than inhibiting an excitatory cholinergic response, it mimics an inhibitory response normally caused by γ-aminobutyric acid (33)]. To our knowledge, none of these compounds has been tested for insecticidal activity.

Having described the data on the so-called mixed receptor in some detail, we must nevertheless point to some facts that raise questions as to whether this binding site is in fact a receptor. First, it has a rather low affinity for acetylcholine, 16 μM by inhibition of nicotine binding (34), or 5 μM by direct binding (19). This is a lower affinity than is found for vertebrate nicotinic receptors. However, it is not clear what affinity a macromolecule must have to be able to function as a receptor for acetylcholine. There are indications that at the neuromuscular junction of vertebrates, the concentration of acetylcholine during synaptic transmission is over 3000 μM, so that a rather low affinity receptor would be quite sufficient (35). There is no information about the acetylcholine concentration at other synapses. Second, the proposed receptor is present in the supernatant of high-speed centrifugations, unlike other acetylcholine receptors. Proteins found in such supernatants are generally considered to be soluble, i.e. cytoplasmic,

Table 7. Percent block of 10^{-7}M decamethonium binding to house fly head by antihelminthics (10^{-5}M)

HYCANTHONE 57

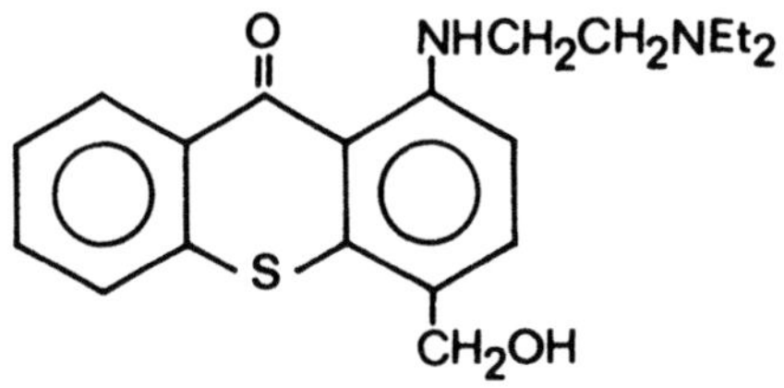

LEVAMISOLE 63

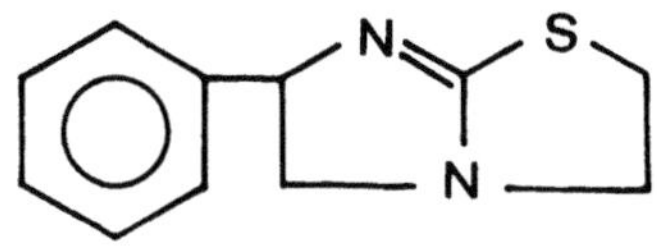

PYRANTEL 75

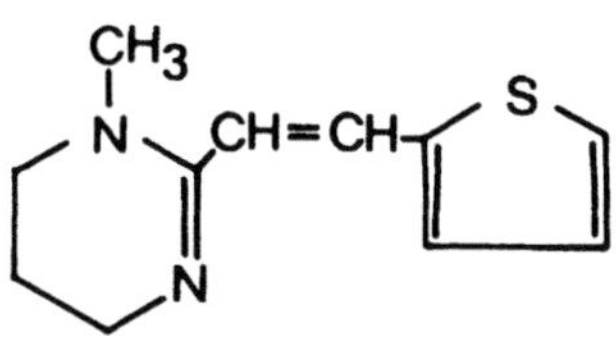

PIPERAZINE 0

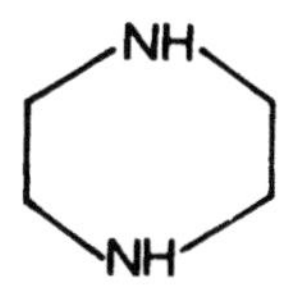

whereas neurotransmitter receptors must be membrane bound to function as such. This criterion for solubility, however, was established for vertebrate tissues. There is no guarantee that house fly head membrane and rat liver membrane will behave similarly upon centrifugation. There is in fact some evidence that the mixed receptor is actually membrane bound: centrifuging 5 times as long (34) or twice as long and twice as hard (19) has been reported to put the mixed receptor binding activity in the pellet. Treatment with Ringer's then appears to solubilize the receptor, in that it behaves in gel filtration like a protein rather than like a piece of membrane (19, 20). If this interpretation is correct, then the mixed receptor is present in very small pieces of membrane that can be solubilized by relatively high salt concentrations, as is true for acetylcholinesterase.

Another difficulty with the mixed pharmacology receptor is its similarity to a macromolecule in the nerve of some arthropods (36, 37,38). Figure 4 shows that there are macromolecules in lobster and horseshoe crab walking leg nerve that bind cholinergic ligands with high affinity. We call them ACBM, for axonal cholinergic binding macromolecules. They are almost certainly not acetylcholine receptors, because there are no synapses present. These molecules of arthropod nerve have a mixed cholinergic pharmacology, more like that of house fly head than like typical nicotinic or muscarinic

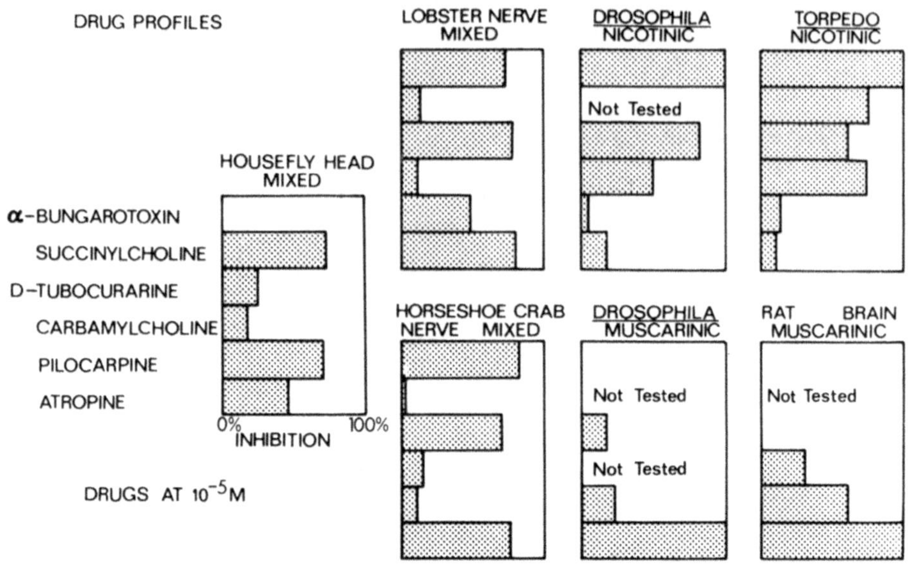

Figure 4. The data in this figure are from the following sources: lobster and horseshoe crab (37); house fly head mixed receptor (this paper); Drosophila nicotinic receptor (13); Drosophila muscarinic receptor (15); Torpedo nicotinic receptor (61); and rat brain muscarinic receptor (5). The Torpedo data is on Lubrol solubilized receptor. Data for the Drosophila and rat brain receptors was converted from I_{50} data to predicted percent inhibition at 10^{-5} M drug. Ligands used were nicotinic, 10^{-7} M (horseshoe and lobster); acetylcholine, 10^{-7} M (Torpedo); decamethonium, 10^{-7} M (house fly); QNB, concentration not given (rat brain); QNB, 5 nM (Drosophila muscarinic); α-bungarotoxin, 15 nM (Drosophila nicotinic).

sites. They differ in some ways from the house fly mixed receptor, most notably their sensitivity to α-bungarotoxin. α-Bungarotoxin binds reversibly to lobster nerve, with much lower affinity than to classic nicotinic receptors, but still quite well in comparison to most drugs (37,38). Acetylcholine binds to ACBM with a dissociation constant of 43 μM, an even lower affinity than for the house fly mixed site (36).

As these considerations demonstrate, the fact that a preparation binds suitable drugs is not adequate proof that it contains an acetylcholine receptor. What data would be needed to prove that a particular macromolecule is a receptor for acetylcholine? One would need to show that acetylcholine is released on neuronal stimulation; also that acetylcholine, whether released neuronally or applied experimentally, can produce a physiological response; also that these responses have a pharmacology that matches those found in binding studies; also that the proposed receptor is located at the appropriate synapses and is not present in nonsynaptic tissue. Ideally, the receptor should be purified and reconstituted into artificial membranes and shown to increase membrane permeability to the same ions to the same extent as the receptor does physiologically, with the same drug sensitivity as the physiological response.

A striking case in which binding data looked very persuasive, but was shown to be misleading, is that of α-bungarotoxin binding to mammalian autonomic ganglia. These bound the toxin with high affinity, and the binding was inhibited by appropriate concentrations of acetylcholine and many nicotinic drugs (39,40,41). However, as other groups pointed out as early as 1969 (42,43,44,45), these neurotoxins do not block the physiological action of acetylcholine at autonomic ganglia, in sharp contrast to their powerful block of neuro-muscular transmission. Two recent studies (46,47) on autonomic ganglion cells in tissue culture demonstrated conclusively both α-bungarotoxin binding and inability of toxin to block the response to acetylcholine. In fact, curare, which blocks both that response and toxin binding, will still block the response to acetylcholine when the toxin binding sites are saturated with α-bungarotoxin. This demonstrates that the toxin binding is not to the site (or sites) in autonomic ganglia that produce a physiological response to acetylcholine that is blocked by curare. Whether the toxin binding has anything at all to do with a macromolecule that actually functions as an acetylcholine receptor is unknown.

These results throw doubt on a number of other studies of α-bungarotoxin binding to tissues where the physiological action of the toxin is not known. This includes the data on house fly heads that we have already mentioned, and studies on mammalian brain. There is other evidence that casts further doubt on these studies. Decamethonium is a very good inhibitor of the nicotinic receptor of mammalian muscle (48), but it is very poor on toxin binding to autonomic ganglia (39,40,41). In this respect, the α-bungarotoxin binding sites in house fly heads (13), Aplysia neurons (14) and rat brain (4) all resemble the questionable ganglionic site rather than the established neuromuscular receptor. The binding of α-bungarotoxin seems to be more nearly irreversible to muscle (48) and electroplax.

What we have just said implies a need for information about the physiological aspects of the insect receptor. Unfortunately, very little is known about the physiology of neurotransmitter receptors in any invertebrate, especially when compared to the extensive knowledge of vertebrate neuromuscular acetylcholine receptors. We will emphasize results on two preparations, ganglia of the mollusc Aplysia, and the synapse between the cercal nerve and giant fibers in cockroach ganglia.

Kehoe (49,50) has reported three different acetylcholine receptors on *Aplysia* neurons. One is very similar to the nicotinic neuromuscular junction acetylcholine receptor of vertebrates, and is blocked by α-bungarotoxin; one to the autonomic ganglion nicotinic receptor, not blocked by toxin; and one has a very different pharmacology. The third one is perhaps most similar to a muscarinic receptor, but the similarity is not close--atropine and several other classic muscarinic drugs are quite ineffective. Receptors similar to this have been reported elsewhere, notably in the hearts of molluscs (51,52,53). As it has not been reported in vertebrates, it may be worth examining as a site of action for drugs acting against invertebrates. It has not been reported yet in insects either, however.

The cockroach ganglion is a difficult preparation to study, as it is well covered by connective tissue. Desheathing is necessary to obtain good sensitivity to acetylcholine or other drugs. The pharmacology of this preparation has not been very extensively studied, but it is interesting to note that it is sensitive to both curare and atropine, classical nicotinic and muscarinic antagonists, at reasonable concentrations (54,55). There is some, even less extensive, information suggesting a similar "mixed" pharmacology acetylcholine receptor in other invertebrate preparations, including clam (56,57)& lobster muscle (58) and arthropod heart ganglia (59,60). In these cases there is no clear evidence that there is one mixed receptor, as opposed to separate nicotinic and muscarinic receptors.

Returning to the case of house flies, for which more biochemical data exists, it is plausible that there is at least one acetylcholine receptor in the head, especially bearing in mind that acetylcholine and acetylcholinesterase are present in high concentrations. It is not yet clear whether any or all of the three macromolecules revealed by binding studies are true acetylcholine receptors. Fortunately, the one present in largest amount, which is the mixed-type receptor, has a pharmacology rather different from that of vertebrates, encouraging the hope that one can develop inhibitors which are specific for it, as a lead to new selective insecticides.

ACKNOWLEDGEMENTS

This work was supported in part by Grant ES 00901 from the National Institutes of Health. The gifts of hycanthone from Dr. F. C. Nachod of Sterling-Winthrop and *cis*-2-methyl-6-pentadecyl piperidine from Dr. Murray Blum are gratefully acknowledged.

LITERATURE CITED

1. Sakai, M. (1969) Rev. Plant Protec. Res. 2, 17.
2. Farrow, J. T., and O'Brien, R. D. (1973) Molec. Pharmacol. 9, 33.
3. Schleifer, L. S., and Eldefrawi, M. E. (1974) Neuropharmacol. 13, 53.
4. Salvaterra, P. M., and Mahler, H. R. (1976) J. Biol. Chem. 251, 6327.
5. Yamamura, H. I., and Snyder, S. H. (1974) Proc. Nat. Acad. Sci. USA 71, 1725.
6. McCance, I., Phillis, J. W., Tebecis, A. K., and Westerman, R. A. (1968) Br. J. Pharmacol. Chemother. 32, 652.
7. Bradley, P. B., and Wolstencroft, J. H. (1965) Br. Med. Bull. 21, 15.
8. Anderson, P., and Curtis, D. R. (1964) Acta Physiol. Scand. 61, 100.
9. Dudai, Y., and Amsterdam, A. (1977) Brain Res. 130, 551.
10. Schmidt-Nielsen, B. K., Gepner, J. I., Teng, N. N. H., and Hall, L. M. (1977) J. Neurochem. 29, 1013.
11. Dudai, Y. (1978) Biochim. Biophys. Acta (in press.).
12. Sanes, J. R., Prescott, D. J., and Hildebrand, J. G. (1977) Brain Res. 119, 389.
13. Dudai, Y. (1977) FEBS Lett. 76, 211.
14. Shain, W., Greene, L. A., Carpenter, D. O., Sytkowski, A. J., and Vogel, Z. (1974) Brain Res. 72, 225.
15. Dudai, Y., and Ben-Barak, J. (1977) FEBS Lett. 81, 134.
16. Eldefrawi, A. T., and O'Brien, R. D. (1970) J. Neurochem. 17, 1287.
17. Eldefrawi, M. E., Eldefrawi, A. T., and O'Brien, R. D. (1971) Molec. Pharmacol. 7, 104.
18. Eldefrawi, M. E., Eldefrawi, A. T., and O'Brien, R. D. (1970) J. Agr. Food Chem. 18, 1113.
19. Jewess, P. J., Clarke, B. S., and Donnellan, J. F. (1975) Croat. Chem. Acta 47, 459.
20. Mansour, N. A., Eldefrawi, M. E., and Eldefrawi, A. T. (1977) Biochemistry 16, 4126.
21. Donnellan, J. F., Jewess, P. J., and Cattell, K. J. (1975) J. Neurochem. 25, 623.
22. Tripathi, R. K., and O'Brien, R. D. (1977) Biochem. Biophys. Acta 480, 382.
23. Clarke, B. S., and Donnellan, J. F. (1975) Biochem. Soc. Trans. 2, 1373.
24. Yeh, J. Z., Narahashi, T., and Almon, R. R. (1975) J. Pharmacol. Exp. Ther. 194, 373.
25. Daly, J. W., Karle, I., Myers, C. W., Tokuyama, T., Waters, J. A., and Witkop, B. (1971) Proc. Nat. Acad. Sci. USA 68, 1870.
26. Mansour, N. A., Eldefrawi, A. T., and Eldefrawi, M. E. (1978) Life Sci. (in press).

27. Albuquerque, E. X., Eldefrawi, A. T., Eldefrawi, M. E., Mansour, N. A. and Tsai, M.-C. (1978) Science 199, 788.
28. Eldefrawi, M. E., Britten, A. G., and O'Brien, R. D. (1971) Pestic. Biochem. Physiol. 1, 101.
29. Narahashi, T. (1974) Physiol. Rev. 54, 813.
30. Rollo, I. M. (1975) in "The Pharmacological Basis of Therapeutics", Goodman, L. S., and Gilman, A. G., Eds., Fifth Edition, Macmillan, New York, pp. 1018-1044.
31. Hillman, G. R., and Senft, A. W. (1975) Am. J. Trop. Med. Hyg. 24, 827.
32. Norton, S., and de Beer, E. J. (1957) Am. J. Trop. Med. Hyg. 6, 848.
33. Del Castillo, J., Morales, T. A., and Sanchez, V. (1963) Nature 200, 706.
34. Aziz, S. A. and Eldefrawi, M. E. (1973) Pestic. Biochem. Physiol. 3, 168.
35. Fertuck, H. C., and Salpeter, M. M. (1976) J. Cell Biol. 69, 144.
36. Denburg, J. L., Eldefrawi, M. E., and O'Brien, R. D. (1972) Proc. Nat. Acad. Sci. USA 69, 177.
37. Jones, S. W., Galasso, R. T., and O'Brien, R. D. (1977) J. Neurochem. 29, 803.
38. Marquis, J. K., Hilt, D. C., and Mautner, H. G. (1977) Biochem. Biophys. Res. Commun. 78, 476.
39. Greene, L. A., Sytkowski, A. J., Vogel, Z., and Nirenberg, M. W. (1973) Nature 243, 163.
40. Fumagalli, L., DeRenzis, G., and Miani, N. (1976) J. Neurochem. 27, 47.
41. Wilson, S. P., and Kirshner, N. (1977) J. Neurochem. 28, 687.
42. Chou, T. C., and Lee, C. Y. (1969) Eur. J. Pharmacol. 8, 326.
43. Obata, K. (1974) Brain Res. 73, 71.
44. Brown, D. A., and Fumagalli, L. (1977) Brain Res. 129, 165.
45. Bursztajn, S., and Gershon, M. D. (1977) J. Physiol. 269, 17.
46. Patrick, J. and Stallcup, B. (1977) J. Biol. Chem. 252, 8629.
47. Carbonetto, S., Fambrough, D. M., and Muller, K. J. (1977) Neurosci. Abstracts 3, 454.
48. Colquhoun, D., and Rang, H. P. (1976) Molec. Pharmacol. 12, 519.
49. Kehoe, J. S. (1972) J. Physiol. 225, 115.
50. Kehoe, J. S., Sealock, R., and Bon, C. (1976) Brain Res. 107, 527.
51. Kehoe, J. S., and Marder, E. (1976) Ann. Rev. Pharmacol. Toxicol. 16, 245.
52. Luduena, F. P., and Brown, T. G. (1952) J. Pharmacol. Exp. Ther. 105, 232.
53. Florey, E. (1967) Comp. Biochem. Physiol. 20, 365.
54. Shankland, D. L., Rose, J. A., Donniger, C. (1971) J. Neurobio. 2, 247.
55. Sattelle, D. B., McClay, A. S., Dowson R. J., and Callec, J. J. (1976) J. Exp. Biol. 64, 13.

56. Twarog, B. M. (1954) J. Cell. Comp. Physiol. 44, 141.
57. Cambridge, C. W., Holgate, J. A., Sharp, J. A. (1959) J. Physiol. 148, 451.
58. Marder, E. (1976) J. Physiol. 257, 63.
59. Carlson, A. J. (1907) Am. J. Physiol. 17, 177.
60. Naidu, M. B. (1955) Bull. Ent. Res. 46, 205.
61. Eldefrawi, M. E., Eldefrawi, A. T., Seifert, S., and O'Brien, R. D. (1972) Arch. Biochem. Biophys. 150, 210.

EFFECT OF STRUCTURE ON THE INTERACTION OF ORGANOPHOSPHORUS AND CARBAMATE ESTERS WITH ACETYLCHOLINESTERASE

T. Roy Fukuto

University of California

Riverside, CA 92521

INTRODUCTION

The toxicity of organophosphorus and carbamate esters to insects, mammals, and other organisms is attributable to the ability of these materials to inactivate acetylcholinesterase (AChE), the enzyme responsible for the hydrolysis of the neurotransmitting agent acetylcholine (ACh) into choline (Ch) and acetic acid. The neurohormone ACh is the synaptic mediator of nerve impulses in the central nervous system of insects as well as the autonomic nervous system in mammals.

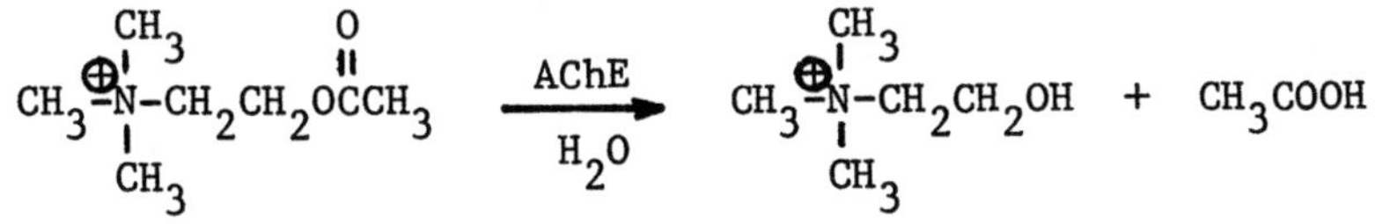

(ACh) (Ch)

Further, ACh is the agent responsible for innervation of muscles and glands associated with the neuromuscular junction of the parasympathetic nervous system.

ACETYLCHOLINESTERASE-ACETYLCHOLINE

In order to attain an appreciation of the AChE-ACh system let us attempt to picture (Figure 1) in rather simplistic terms the events which take place at the neuromuscular junction when a muscle is innervated (1). The same picture applies to the synapse but in this case innervation of the postsynaptic neuron occurs.

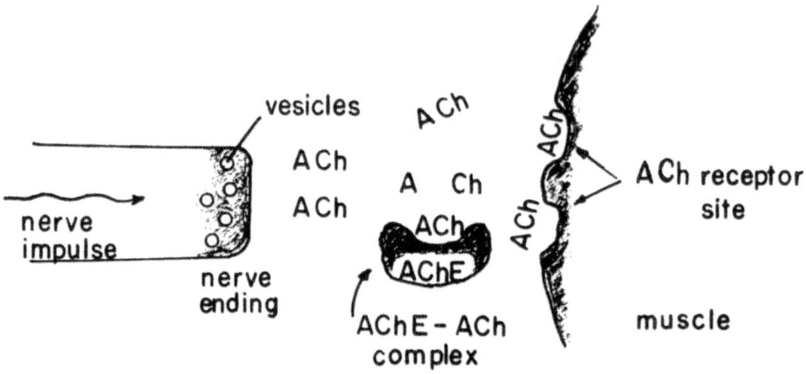

Figure 1. Pictorial presentation of events occurring at the neuromuscular junction.

Imagine a nerve impulse moving down a parasympathetic neuron and eventually reaching the nerve ending. With the arrival of the impulse, ACh, stored in vesicles at the nerve ending, is released into the junction and within 2-3 msec alights on the ACh receptor site of the muscle, causing contraction of the muscle. AChE hydrolytically converts ACh into Ch and acetic acid, neither of which is able to activate the muscle, resulting in decrease in the concentration of ACh in the junction and cessation of muscle contraction. When AChE is inactivated, e.g., by an organophosphorus or carbamate ester, it is no longer able to hydrolyze ACh; consequently, ACh concentration in the junctions remains high and continuous stimulation of the muscle occurs to the point of tetany and exhaustion. Thus, the maintenance of proper titers of AChE is of obvious importance.

Before proceeding on to the discussion concerning the interaction of organophosphorus and carbamate esters with AChE it is necessary first to describe briefly the interaction of AChE with ACh, the natural substrate. For a comprehensive review of the active site of AChE and its interaction with substrates, the reader is referred to an excellent chapter by Cohen and Oosterbaan (2). The hydrolysis of ACh by AChE undoubtedly occurs *via* an acid-base catalyzed reaction. Evidence for this is found in the bell-shaped relationship obtained in studies relating the effect of change in pH on ACh hydrolysis, i.e., low rates of hydrolysis at low (<6) and high (>10) pH and maximum hydrolysis rate at the isoelectric point of pH 8.25. Mechanistically, the hydrolysis of ACh may be depicted according to the scheme in Figure 2 (3). In this scheme, ACh is brought into close proximity to the active site by electrostatic attraction between the positively charged ammonium nitrogen atom and negative charge in the anionic site (structure E + S) resulting in the enzyme-substrate complex (ES). Acetylation of a serine hydroxyl (OH) in the esteratic site is catalyzed by the basic imidazole moiety B (histidine) and acidic moiety AH (tyrosine hydroxyl), leading to the acetylated enzyme EA. Deacetylation then takes place very fast, resulting in the free enzyme within milliseconds.

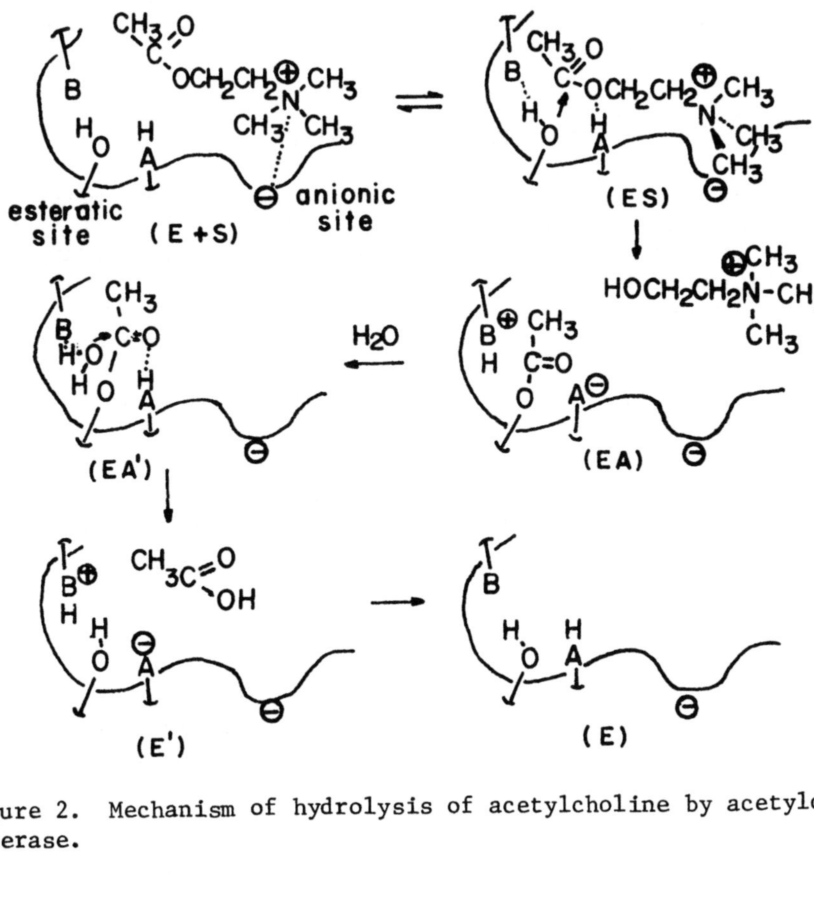

Figure 2. Mechanism of hydrolysis of acetylcholine by acetylcholinesterase.

The negative charge in the anionic site has been attributed to the carboxylate anion of aspartic or glutamic acid. As shown in structure EA', deacetylation also is catalyzed by the acidic and basic moieties AH and B with consequent expulsion of acetic acid (E') and reversion to the free enzyme (E). The overall process takes place extremely rapidly, and estimation of turnover of ACh as high as 750 mmol ACh/mg protein/hr has been reported for purified electric eel AChE (4). It should be added that the scheme presented above is a very elementary representation of a plausible mechanism for the hydrolysis of ACh. The enzyme is probably far more complex, e.g., it has been suggested that there are 2 basic centers in the esteratic site (5) as well as the presence of multiple binding zones which interact with different compounds in different ways (6, 7).

Evidently, there are at least two kinds of cholinesterases, (1) acetylcholinesterase or true cholinesterase and (2) pseudocholinesterase. AChE has the highest specificity for ACh while butyrylcholine appears to be the best substrate for pseudo-cholinesterase. Both cholinesterases are inhibited by organophosphorus and carbamate insecticides.

ORGANOPHOSPHORUS ESTERS

The inhibition of AChE by organophosphorus and carbamate esters is the result of an actual chemical reaction between the enzyme and inhibitor. The overall inhibition process, leading to phosphorylation of the serine hydroxyl group, is analogous to the acetylation process involving ACh (Figure 2). However, the phosphorylated enzyme, in contrast to the acetylated enzyme is highly stable and in some cases is irreversibly phosphorylated. The serine hydroxyl, thus blocked is no longer able to participate in the hydrolysis of ACh.

$$\text{AChE} + (\text{RO})_2\overset{\overset{\text{O}}{\|}}{\text{P}}\text{X} \underset{}{\overset{K_d}{\rightleftharpoons}} \underset{\text{complex}}{[\text{AChE}\cdot(\text{RO})_2\overset{\overset{\text{O}}{\|}}{\text{P}}\text{X}]} \xrightarrow{k_p} \underset{\text{inhibited enzyme}}{\text{AChE}-\overset{\overset{\text{O}}{\|}}{\text{P}}(\text{OR})_2} + X^-$$

$$\text{(overall: } k_i\text{)}$$

In the above equation X is an appropriate leaving group in the organophosphorus ester, K_d is the dissociation constant between complex and reactants, k_p is the phosphorylation constant, k_i is the bimolecular inhibition constant and is equal to k_p/K_d (8). Since K_d is a measure of the tendency for the enzyme-inhibitor complex to dissociate, it is regarded as a measure for binding and should be dependent on the structural and steric features of the molecule. In contrast, the effect of change in the reactivity of the ester has been associated with the phosphorylation constant, k_p (9). The overall rate constant, k_i, for the bimolecular inhibition reaction is dependent on the values of K_d and k_p.

Reactivity

Extensive studies on the relation between chemical structure and inhibition of AChE by organophosphorus esters have shown that anticholinesterase activity depends largely on the reactivity of the ester. For example, the inhibition of erythrocyte AChE by paraoxon and related substituted phenyl diethyl phosphates takes place in a bimolecular manner and there is a direct relationship between the rate constants for inhibition (k_i) and solvolysis in pH 8 phosphate buffer (10). Further, in a larger series of substituted phenyl diethyl phosphates, inhibition of house fly-head AChE was correlated with the effect induced by the ring substituent on the reactivity of the phosphorus center as estimated by Hammett's σ constants, shifts in P-O-phenyl stretching frequencies in the infrared region, and hydrolysis rates (11).

Steric Effects

While there is little doubt that appropriate reactivity of the phosphorus center is of major importance for high anticholinesterase activity, steric effects often strongly influence inhibitory activity of an organophosphorus ester. Steric effects generated by the structure of the organophosphorus ester may influence inhibitory potency in at least two different ways. Binding or fit on the enzyme active site may be hindered or aided, depending on the structure involved. This type of steric effect may be exemplified by the data in Table 1 (9).

As the size of the alkyl group R is increased from H to Me to *i*-Pr, anticholinesterase activity (k_i) against fly-head AChE increases about 8-fold while activity against bovine erythrocyte AChE decreases about 24-fold. Since reactivity, as estimated by solvolysis rates (k_b), decreases with substitution in the 3-position, the order of activity against fly-head AChE (3>2>1) indicates that steric effects are operating to increase anticholinesterase activity. Support for this belief is found in the change in K_d values with increasing alkyl size. Since K_d is an estimate of the tendency for the inhibitor-enzyme complex to dissociate, the K_d values in Table 1 suggest that enzyme-inhibitor complex formation is aided by interaction of the 3-alkyl substituent with a hydrophobic site in house fly AChE but is hindered in the case of bovine erythrocyte AChE. Binding of the 3-alkyl substituent probably occurs by hydrophobic interaction and interaction of 3 with the active site of fly-head AChE may be visualized

Table 1. Anticholinesterase data for fenitrothion oxon and related esters.

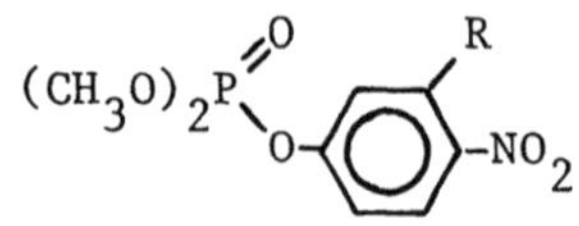

Comp'd	R	$k_i \times 10^{-5}$ ($M^{-1}min^{-1}$)	$K_d \times 10^5$ (M)	k_p (min^{-1})	k_b[a] (min^{-1})
		House fly-head AChE			
1	H	2.9	3.7	10.6	5.25
2	Me	7.6	1.1	8.3	3.50
3	*i*-Pr	22.6	0.33	7.5	2.89
		Bovine erythrocyte AChE			
1	H	5.2	1.3	6.5	-
2	Me	0.73	6.7	5.0	-
3	*i*-Pr	0.22	15.8	3.5	-

a/ First-order solvolysis constant in pH 8.0 phosphate buffer.

as follows:

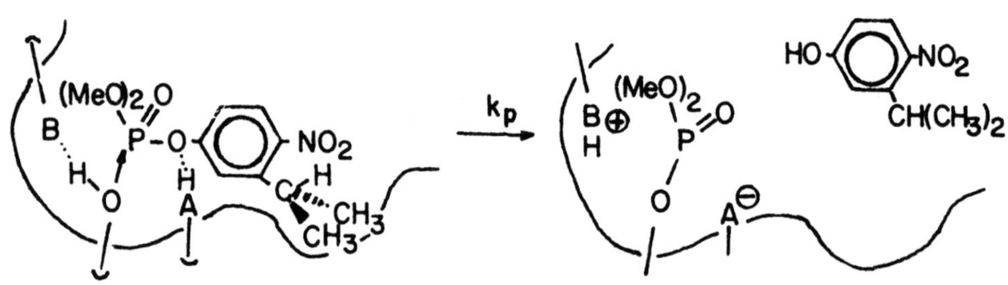

There are numerous other examples in which high anticholinesterase may be attributed to interactions of this nature. For example, 3-*t*-butyl and 3-dimethylaminophenyl diethyl phosphates are more than 10^3-fold stronger inhibitors than predicted from their intrinsic reactivity and this may be attributed to interaction of the 3-substituent with an anionic or hydrophobic site. The unusually potent anticholinesterase activity of organophosphorus esters containing a positive charge, typified by structures presented below, may be attributed in part to interaction of the charged moiety with the anionic site (12, 13, 14).

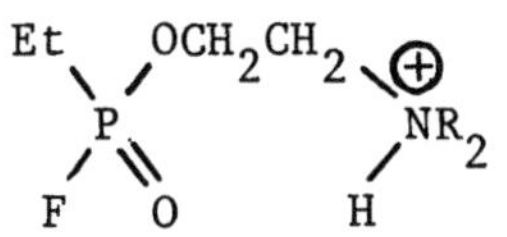

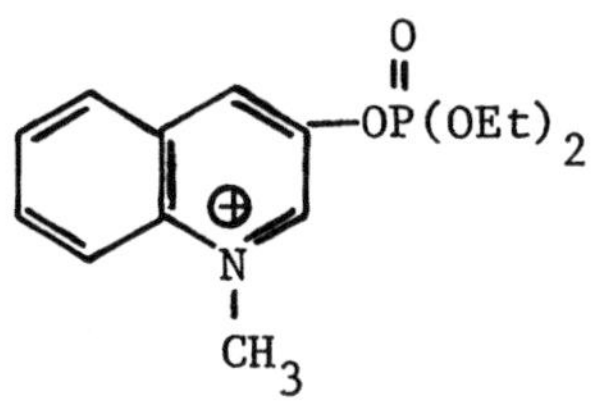

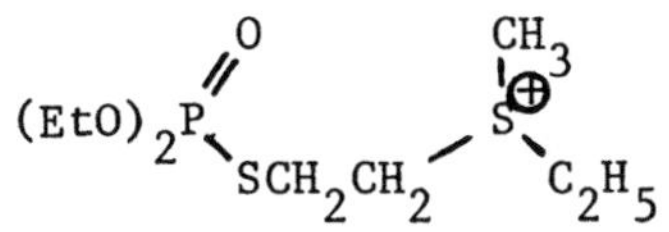

Bulky groups attached to the phosphorus atom also may interfere sterically with the reaction between the organophosphorus ester and the active site of the enzyme. This type of steric effect is evident from study of a series of phosphonate and phosphoramidate esters. For example, for a series of 13 ethyl ρ-nitrophenyl alkylphosphonate esters of the structure below, anticholinesterase

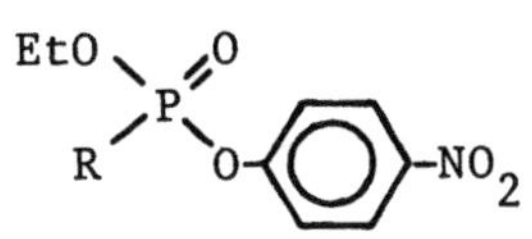

R = Me, Et, Pr, *i*-Pr, Bu, *i*-Bu, amyl, *i*-amyl, hexyl, cyclohexyl, phenyl, etc.

activity against house fly-head AChE was related to Taft's polar (σ*) and steric (E_s) substituent constants according to equations 1-3 (15). Since σ* provides an estimate of polar or electronic

	n	*s*	*r*	
$\log k_i = 5.68\ \sigma^* + 6.31$	13	1.574	0.407	(1)
$\log k_i = 2.58\ E_s + 7.94$	13	0.648	0.927	(2)
$\log k_i = 2.89\ E_s + 2.72\ \sigma^* + 7.92$	13	0.616	0.940	(3)

effects and E_s a measure of steric effects imposed on the reactivity of the phosphorus atom, the larger correlation coefficients (*r*) and smaller standard errors (*s*) associated with E_s (equations 2 and 3) clearly indicate that steric effects attributable to R are more important than polar effects in the inhibition of AChE by these esters.

Another example of this type is found in the analysis of fly-head anticholinesterase data obtained from a study of a series of methyl 2,4,5-trichlorophenyl *N*-alkylphosphoramidates of the structure below (16).

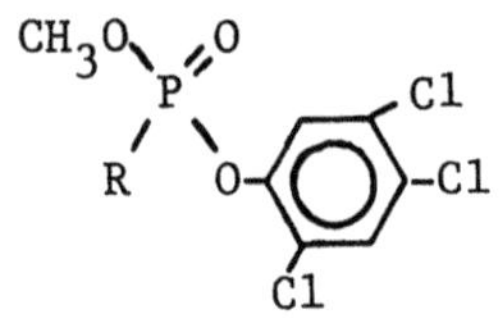

R = NH_2, NHMe, NHEt, NHPr, NH-*i*-Pr, NHBu, NH-*t*-Bu, NMe_2

Analysis of the data with different free energy parameters gave equations 4-6.

	n	*s*	*r*	
$\log k_i = 2.709\ \sigma^* + 4.490$	8	0.816	0.712	(4)
$\log k_i = 1.119\ E_s + 4.541$	8	0.563	0.875	(5)
$\log k_i = 2.359\ E_s - 3.913\ \sigma^* + 4.948$	8	0.438	0.939	(6)

Again, the regression equations reveal the importance of steric influence of the amido moiety on phosphorylation of AChE by organophosphorus esters. In the case of these phosphoramidates, k_i values for the esters containing the bulky NH-*i*-Pr and NH-*t*-Bu moieties were substantially poorer as anticholinesterases than predictable from reactivity considerations. The steric parameter E_s served the purpose of bringing these two compounds in line with the other phosphoramidates.

Overall, the results with the alkylphosphonates and alkylphosphoramidates reveal that the esteratic site where phosphorylation takes place is restricted in size and large groups attached to the phosphorus atom are poorly accommodated by the dimensions of the site.

Stereospecificity

Chirality at the phosphorus center often has a profound effect on the anticholinesterase activity of an organophosphorus ester. This probably is attributable to stereospecific accommodation by the esteratic site of the different groups attached to the phosphorus atom. Large differences in anticholinesterase activities between chiral isomers have been observed and the magnitude of the difference appears to depend on the differences in the size of the groups attached to phosphorus. For example, the rate constant for the inhibition of bovine erythrocyte AChE (BAChE) by (-)-*O*-isopropyl

S-2-(trimethylammonium)ethyl methylphosphonothioate is about 3000-fold greater than that observed for the (+) enantiomer (17). The (-)-isomer of soman (3,3-dimethyl-2-butyl methylphosphonofluoridate), assigned the R_P configuration, is reported to inhibit BAChE 12,000 times faster than its (+)-enantiomer (18).

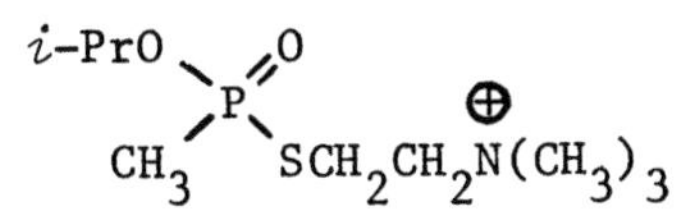

O-isopropyl *S*-2-(trimethylammonium)ethyl methylphosphonothioate

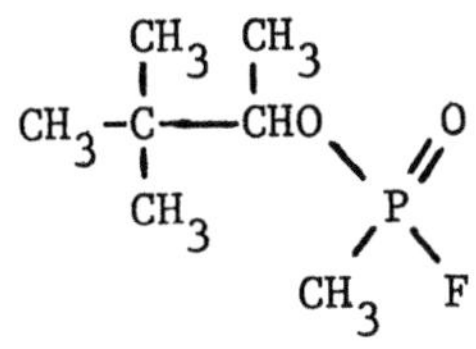

soman

In contrast to the above examples, chiral isomers which contain alkyl and alkoxy moieties of similar size show smaller differences in their ability to inhibit AChE. The rate of inhibition of fly-head AChE by (-)-*O*-ethyl *S*-2-(ethylthio)ethyl ethylphosphonothioate was 11-fold greater than that of the (+)-enantiomer (19). The (-)-isomer of fonofos oxon (established unequivocally to be S_P) was 9-, 9.3-, and 62.5-fold more effective in inhibiting BAChE, electric

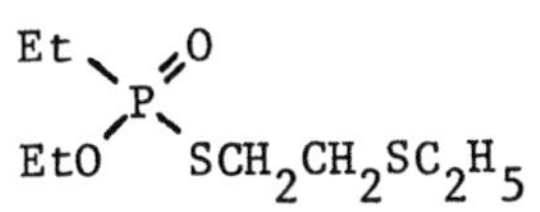

O-ethyl *S*-2-(ethylthio)ethyl ethylphosphonothioate

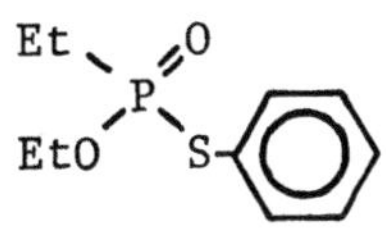

fonofos oxon

eel AChE, and house fly-head AChE, respectively, than the (+)-enantiomer (20). In a study of a series of *S*-alkyl *O*-4-nitrophenyl methylphosphonothioates, where the alkyl moiety was progressively increased from methyl to *n*-pentyl, differences in antiesterase activities between the chiral isomers generally increased with increasing alkyl size. Overall, the (+)-enantiomers were stronger inhibitors than the (-)-enantiomers (21, 22).

The absolute configurations of chiral organophosphorus anticholinesterases have been established by x-ray diffraction analysis and by chemical correlation and this has allowed greater insight into the nature of the interactions involved during the inhibition reaction. Table 2 presents values of K_d, k_p, and k_i for the inhibition of BAChE by the chiral isomers of *O*-2-butyl *S*-2-(dimethylammonium)ethyl ethylphosphonothioate (structure below) in which chirality resides at the 2-butyl carbon and phosphorus atom (23).

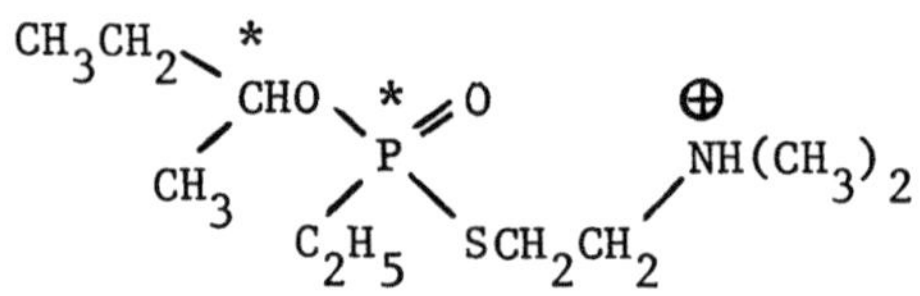

Table 2. Dissociation (K_d), phosphorylation (k_p) and bimolecular inhibition (k_i) constants for the inhibition of bovine erythrocyte AChE by the chiral isomers of *O*-2-butyl *S*-2-(dimethylammonium)ethyl ethylphosphonothioate.

Configuration	$K_d \times 10^5$ (M^{-1})	k_p (min^{-1})	$k_i \times 10^{-5}$ ($M^{-1}min^{-1}$)
S_c S_p	0.944	87.4	92.6
S_c R_p	234	15.5	0.0648
R_c S_p	2.00	56.7	28.4
R_c R_p	298	5.02	0.0174

The symbols S_c or R_c and S_p or R_p refer to the absolute configuration of the 2-butyl carbon and phosphorus centers. The values for k_i show that the most potent isomer (S_c S_p) is about 5300-fold stronger as an anticholinesterase than the weakest isomer (R_c R_p). Further, difference in anticholinesterase activity evidently is attributable mainly to differences in K_d, pointing out the importance of steric fit of the phosphonyl moiety in the esteratic site. The results also show that inhibitory potency is affected to a much greater extent by chirality at the phosphorus atom. A stereochemical drawing of the more active S_c S_p isomer is given below:

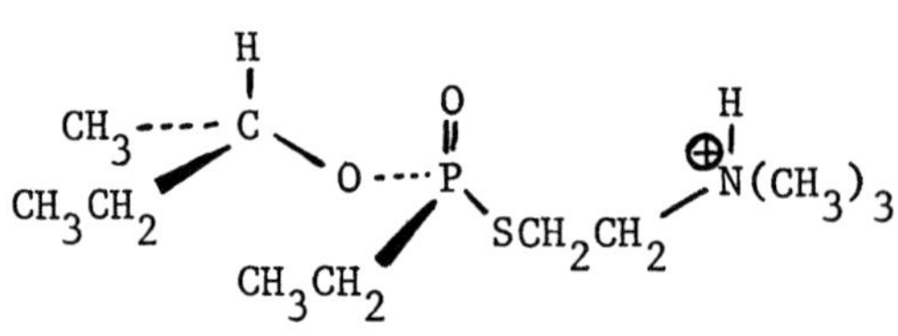

If it is assumed that the dimethylammonium moiety binds to the anionic site of AChE, then the approach of this compound to the active site may be visualized as shown in Figure 3.

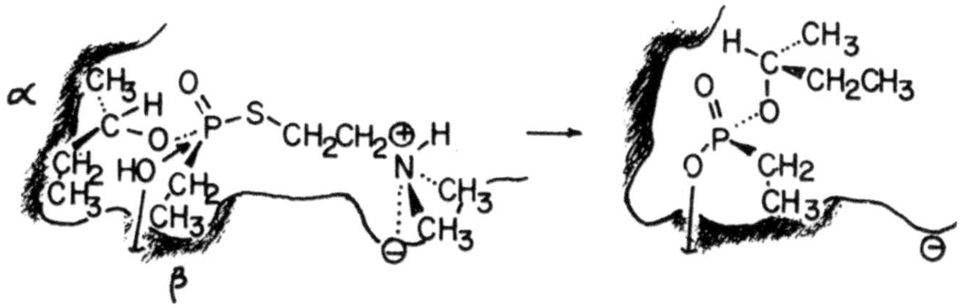

Figure 3. Possible interaction of S_cS_p *O*-2-butyl *S*-2-(dimethylammonium) ethyl ethylphosphonothioate with AChE active site. The acidic (HA) and basic (B) groups in the active site are not shown. α and β are hydrophobic areas.

In the figure, zones α and β are hydrophobic binding areas of appropriate dimension in which the *s*-butoxy and ethyl moieties fit or interact hydrophobically. These zones may correspond to the hydrophobic sites postulated by Kabachnik et al. (24) and O'Brien (6, 25). With the *s*-butoxy and ethyl groups in the most favorable position, attack of the serine hydroxyl on the phosphorus center occurs. The serine hydroxyl should approach the phosphorus atom from the direction opposite the sulfur atom, displacing the 2-(dimethylammonium)ethylthiolate moiety in an S_N2P type mechanism (26). The configuration of the phosphorus atom in the phosphorylated enzyme is thus inverted. Recent results with R_p and S_p fonofos oxon (*O*-ethyl *S*-phenyl ethylphosphonothioate) reveal that AChE inhibited by S_p-oxon is reactivated 2-fold faster by 2-PAM (2-pyridylaldoxime methiodide) than AChE inhibited by the R_p-oxon (20). Further, S_p-oxon was 62.5-fold more effective in inhibiting AChE than the R_p-oxon. These results suggest that the phosphonylation reaction occurs with some degree of stereospecificity and that the position of the ethoxy and ethyl groups in the phosphonylated enzyme is less critical to regeneration than to binding and eventual phosphonylation.

CARBAMATE ESTERS

In gross aspects, the inhibition of AChE by carbamate esters is mechanistically similar to the hydrolysis of ACh by AChE and the inhibition of AChE by organophosphorus esters. The inhibition process may be described by the following equation.

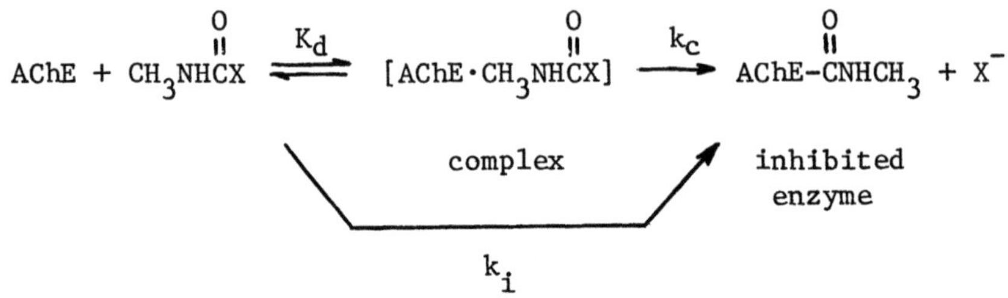

A methylcarbamate ester is used as an example here where X is an appropriate leaving group. K_d, k_c, and k_i are the dissociation, carbamylation, and bimolecular inhibition constants, respectively, analogous to the constants described earlier for the inhibition of AChE by organophosphorus esters.

In contrast to phosphorylated AChE, the carbamylated enzyme is less stable and spontaneous regeneration to the free enzyme occurs with a half-life velocity ($t_{1/2}$) of 30-60 minutes (27, 28).

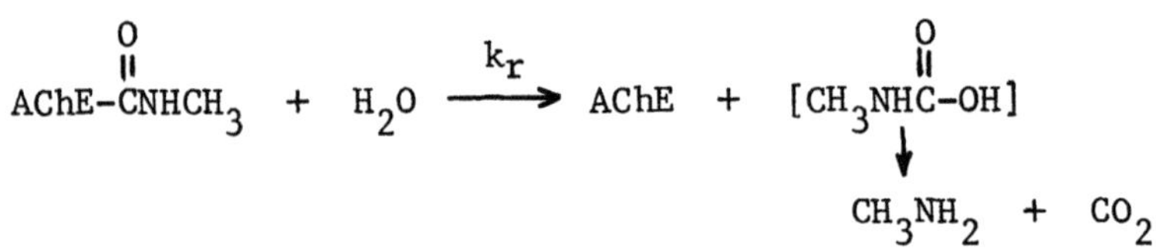

Thus, the order of spontaneous recovery of the acylated enzyme is:

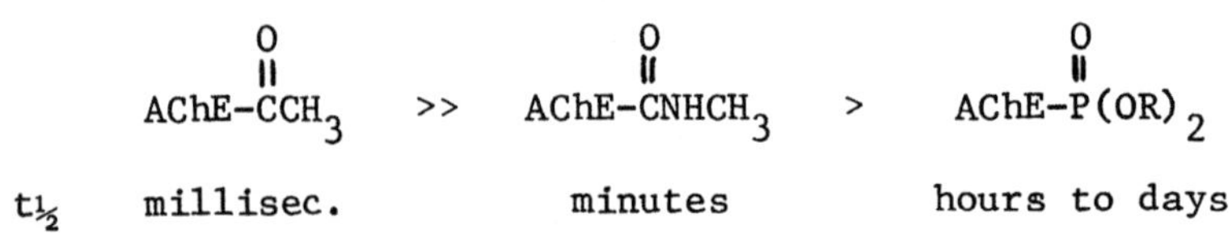

As in the case of the acetylation of AChE by ACh, carbamylation and decarbamylation doubtlessly are acid-base catalyzed (27). In many respects, carbamate esters may be regarded as poor substrates of AChE by virtue of the relatively high stability of the carbamylated enzyme compared to the transitory acetylated intermediate. The value for the first-order rate constant for deacetylation of the acetylated intermediate of electric eel AChE has been estimated to be at least $2.2 \times 10^{-7} min^{-1}$ (29). Decarbamylation of electric eel AChE inhibited by a methylcarbamate ester is reported to be about $1 \times 10^{-2} min^{-1}$ (30, 31) and, therefore, the methylcarbamylated enzyme is approximately 10^9-fold more stable to hydrolysis than the acetylated enzyme.

Although the reaction between AChE with carbamate and organophosphorus inhibitors appears to be mechanistically similar, structural requirements for anticholinesterase activity in carbamate esters differ markedly from those of organophosphorus esters. As indicated earlier, the most essential requirement in an organophosphorus ester for high anticholinesterase activity is proper chemical reactivity of the phosphorus center. In the case of carbamate esters, "fit" or molecular complementarity with the active site is essential for high anticholinesterase activity. The most potent carbamate anticholinesterases are generally those which structurally resemble the natural substrate ACh. The bulk of the evidence indicates that it is the binding step leading to AChE-carbamate complex formation which determines the magnitude of anticholinesterase activity.

Structures of several potent carbamate anticholinesterases are given below alongside that of ACh.

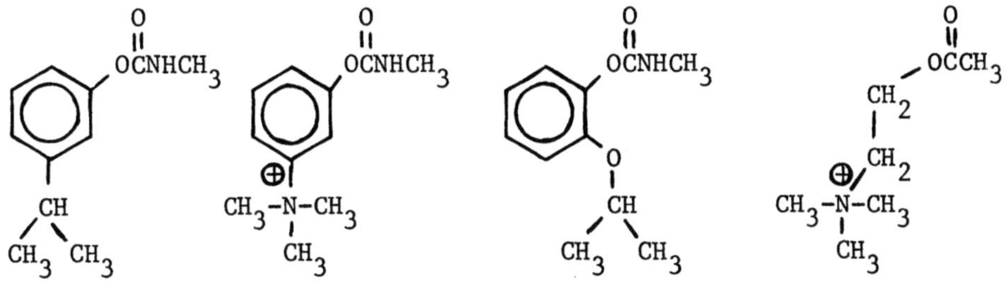

Substituent Effects

Most of the more effective carbamate anticholinesterases are aryl methylcarbamates (32). Evidently aryl methylcarbamates are intrinsically reactive enough so that they are able to carbamylate AChE. The second-order rate constant for the base catalyzed hydrolysis of phenyl methylcarbamate is reported to be 1.2×10^2 $M^{-1}min^{-1}$ at 25° (33). Thus, phenyl methylcarbamate is more reactive toward base than is paraoxon (34).

Although the carbamate moiety in phenyl methylcarbamate is more susceptible to attack by hydroxide ion than is the diethyl phosphate moiety in paraoxon, phenyl methylcarbamate, with a k_i value of 5.4 $\times 10^2$ $M^{-1}min^{-1}$ for the inhibition of BAChE, is a relatively poor anticholinesterase (32). However, introduction of substituents in the phenyl ring, particularly in the *ortho* and *meta* positions greatly increases anticholinesterase activity. For example, k_i values against BAChE for *m*-isopropylphenyl, *m*-trimethylammonium-phenyl, and *o*-isopropoxyphenyl methylcarbamate (structures given above) are 4.6×10^5, 2.8×10^7, and 1.1×10^5 $M^{-1}min^{-1}$, respectively. Therefore, introduction of the *m*-isopropyl, *m*-trimethyl-ammonium, and *o*-isopropoxy groups increased anticholinestrase activity 1642-, 55640-, and 391-fold over the unsubstituted carbamate.

The most potent aryl methylcarbamate anticholinesterases are those in which a positive formal charge resides in the ring at an appropriate distance from the carbamate moiety. This is evident in the extremely high activity of the *m*-trimethylammoniumphenyl methylcarbamate cited above. The corresponding uncharged isostere, *m*-*t*-butylphenyl methylcarbamate, is about 80-fold less effective as an anticholinesterase. The corresponding charged sulfur analog, *m*-dimethylsulfoniumphenyl methylcarbamate (structure below) is even more effective in inhibiting house fly-head AChE than the trimethylammonium analog (32).

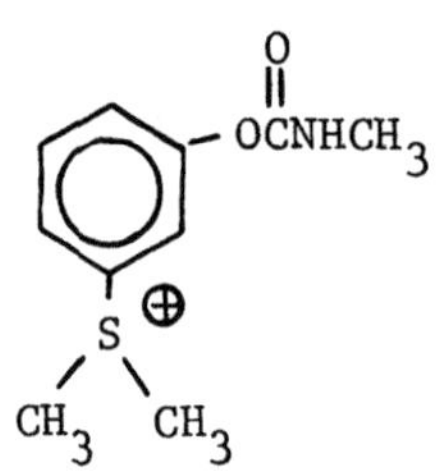

The outstanding anticholinesterase activity of aryl methyl-carbamates containing a positive charge suggests that the anionic site is involved in the interaction of these compounds with AChE, as depicted in Figure 4.

As depicted, carbamylation of the serine hydroxyl takes place in the same manner in which it is acetylated by ACh (Figure 2).

For the substituted alkylphenyl methylcarbamates, the effect of ring position on inhibitory potency against AChE is in the order *meta* > *ortho* > *para*. Further, for simple alkyl substituents inhibitory potency increases in the order $CH_3 < C_2H_5 < i\text{-}C_3H_7 = t\text{-}C_4H_9$, with $i\text{-}C_3H_7$ being 14-fold more effective than C_2H_5, which, in turn, is 3-fold more effective than CH_3. The increase in activity is almost identical to the increase in the binding of the various 2-aminoethanols to AChE upon sequential addition of methyl groups to the amino nitrogen atom (35). This suggests that van der Waals' forces or a similar type interaction between

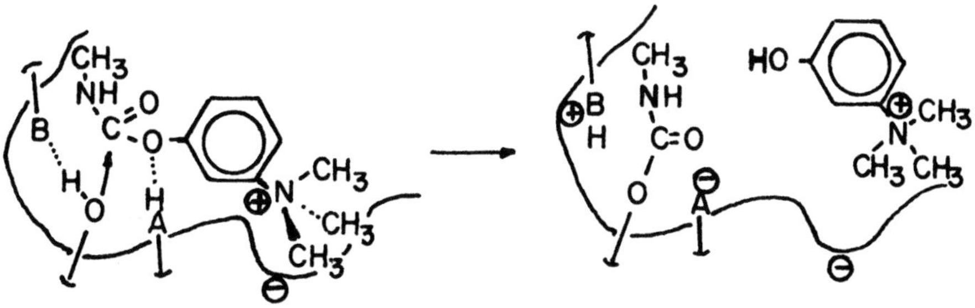

Figure 4. Possible interaction of *m*-trimethylammoniumphenyl methylcarbamate with the AChE active site.

each methyl group and a hydrophobic area at or near the anionic site is responsible for increased activity.

For halogen substituted phenyl methylcarbamates, the order of ring position or anticholinesterase activity is *ortho* > *meta* > *para* (32), with I > Br > Cl > F. In addition, there is almost a linear relationship between inhibitory potency and van der Waals' radius of the halogen substituent.

The importance of carbamate-AChE complex formation for high anticholinesterase activity, i.e., binding of the carbamate to the active site prior to carbamylation, is indicated by values obtained for K_d and k_c, the dissociation and carbamylation constants (36, 37). In general, k_c remains fairly constant for a wide variety of substituted phenyl methylcarbamates while K_d more or less parallels anticholinesterase activity. Values for K_d, k_c, and k_i for the 2-halophenyl and 3-alkylphenyl methylcarbamates are presented in Table 3 (37). The results clearly reveal the relationship between K_d and the bimolecular inhibition constant k_i.

Nishioka et al. (37) have examined K_d values obtained for a series of 53 substituted phenyl methylcarbamates by multiple regression analysis and have related $1/K_d$ with hydrophobic, electronic, hydrogen bonding and *ortho*-proximity effects. In assessing electronic effects ($\rho\sigma$) the substituents were divided into negative ρ substituents and positive ρ substituents. Based on the resulting regression equation, two possible mechanisms were suggested for the carbamylation reaction. In each case, nucleophilic attack of the serine hydroxyl on the carbonyl carbon and protonation of the carbonyl oxygen is postulated. In the positive ρ mechanism, serine hydroxyl-carbonyl carbon bond formation is more advanced in the transition state while catalytic protonation of the carbonyl oxygen is more important in the negative ρ mechanism.

Table 3. K_d, k_c, and k_i constants for the inhibition of bovine erythrocyte AChE by 2-halophenyl and 3-alkylphenyl methylcarbamates.

Substituent	K_d (M)	k_c (min^{-1})	k_i ($M^{-1}min^{-1}$)
H	3.02×10^{-3}	0.86	2.84×10^{2}
2-F	9.26×10^{-4}	1.10	1.19×10^{3}
2-Cl	2.53×10^{-4}	2.29	9.05×10^{3}
2-Br	6.81×10^{-5}	1.38	2.03×10^{4}
2-I	5.58×10^{-5}	1.82	3.22×10^{4}
3-Me	3.76×10^{-4}	1.40	3.72×10^{3}
3-Et	6.45×10^{-5}	2.22	3.43×10^{4}
3-*i*-Pr	5.55×10^{-6}	3.13	5.64×10^{5}
3-*t*-Bu	3.20×10^{-6}	1.42	4.43×10^{5}

Charge-transfer complex formation also may be involved in the binding of aryl methylcarbamates to the AChE active site. Hetnarski and O'Brien (38) have determined substituent constants for charge-transfer complex (C_T) by estimating equilibrium constants for complex formation between tetracyanoethylene and 23 substituted phenyl methylcarbamates. Multiple regression analysis between K_d (BAChE), C_T, and Hansch's π constant gave equation 7 for *para*-substituted carbamates and equation 8 for *meta*-substituted carbamates.

	n	*s*	*r*	
$K_d = -3.476\, C_T - 2.645\, \pi + 5.880$	11	0.367	0.989	(7)
$K_d = -1.228\, C_T - 1.960\, \pi + 2.444$	9	0.322	0.958	(8)

K_d is excellently correlated with C_T and π in a semilogarithmic relationship, thus implicating charge-transfer complex formation in binding.

Recently, a number of aryl *N*-methoxy and *N*-hydroxy-*N*-methylcarbamates were revealed to be reversible, competitive inhibitors of AChE (39). These compounds differ from methylcarbamates by the substitution of a methoxy or hydroxy moiety in place of the nitrogen proton and doubtlessly inhibit AChE by complexing with the active site. The equilibrium constant for enzyme-carbamate complex (K_i) therefore should provide a useful measure of the effect of ring substituents on binding of carbamate esters to the AChE active site. Regression analysis of different free energy related parameters (σ, E_s, π) and K_i for the inhibition of BAChE by a series of 20 substituted phenyl *N*-methoxy-*N*-methylcarbamates gave the following equation which gave best fit to the data (40)

$$-\log K_i = 4.323 + 0.663\,\pi + 0.455 d_1 - 0.275 d_2 \qquad (9)$$

$$n = 20 \qquad s = 0.30 \qquad r = 0.93$$

In the equation, d_1 and d_2 are ring position terms and are respectively equal to 1 and -1 for *ortho*-substituted compounds, both equal to 1 for *meta*-substituted compounds, and -2 and 0 for *para*-substituted compounds. Correlation was poor when π was used alone ($r = 0.31$). Based on t values d_1 was of greatest significance, followed by π and d_2. The values for d_1 and d_2 for the three ring positions gave an order for the inhibition of BAChE of *ortho* > *meta* > *para*. The heavy dependence of the correlation on d_1 (and d_2) suggests that distance between ring substituent and carbamate moiety is critical for maximum interaction of the carbamate ester with the active site, i.e., interaction with the esteratic site and a liphophilic or hydrophobic site situated approximately 0.4 nm from the esteratic site. Since π is often taken as an estimate of hydrophobic bonding by substituents, the presence of π in the equation suggests the involvement of hydrophobic interaction between the substituent and hydrophobic site.

REFERENCES

1. B. C. Saunders, *Some Aspects of the Toxic Action of Organic Compounds Containing phosphorus and Fluorine*, Cambridge University Press, Cambridge, 1957, p. 28.

2. J. A. Cohen and R. A. Oosterbaan, in *Handbuch der Experimentellen Pharmakologie*, 1963, Vol. 15, pp. 300-373, Springer, Berlin.

3. R. M. Krupka, *Canad. J. Biochem.* 42, 677 (1964).

4. W. Leuzinger and A. L. Baker, *Proc. Nat. Acad. Sci.* 57, 446 (1967).

5. R. M. Krupka, *Biochem. Biophys. Res. Commun.* 19, 531 (1965).

6. R. D. O'Brien, in *Insecticide Biochemistry and Physiology*, C. F. Wilkinson, Ed., Plenum Press, N.Y., 1976, p. 271.

7. W. N. Aldridge, *Croatica Chem. Acta* 47, 215 (1975).

8. A. R. Main, *Science* 144, 992 (1964).

9. R. M. Hollingworth, T. R. Fukuto, and R. L. Metcalf, *J. Agric. Food Chem.* 15, 235 (1967).

10. W. N. Aldridge and A. N. Davison, *Biochem. J.* 51, 62 (1952).

11. T. R. Fukuto and R. L. Metcalf, *J. Agric. Food Chem.* 4, 930 (1956).

12. K. J. M. Andrews, F. R. Atherton, F. Bergel and A. L. Morrison, *J. Chem. Soc.* 1638 (1954).

13. L.-E. Tammelin, *Svensk Kemisk Tidskrift* 70, 4 (1958).

14. T. R. Fukuto and R. L. Metcalf, *J. Amer. Chem. Soc.* 77, 3670 (1955).

15. C. Hansch, in *Biochemical Toxicology of Insecticides*, R. D. O'Brien and I. Yamamoto, Eds. Academic Press, New York, 1970, p. 33.

16. C. Hansch and E. W. Deutsch, *Biochim. Biophys. Acta* 126, 117 (1966).

17. L. D. A. DeJong and C. Van Dijk, *Biochim. Biophys Acta* 268, 680 (1972).

18. J. H. Keijer and G. Z. Wolring, *Biochim. Biophys. Acta* 185, 465 (1969).

19. T. R. Fukuto and R. L. Metcalf, *J. Econ. Entomol.* 52, 739 (1959).

20. P. W. Lee, R. Allahyari and T. R. Fukuto, *Pestic. Biochem. Physiol.* (in press).

21. A. J. J. Ooms and H. L. Boter, *Biochem. Pharmacol.* 14, 1839 (1965).

22. H. L. Boter and A. J. J. Ooms, *Biochem. Pharmacol.* 16, 1563 (1967).

23. D. A. Wustner and T. R. Fukuto, *Pestic. Biochem. Physiol.* 4, 365 (1974).

24. M. I. Kabachnik, A. P. Brestkin, N. N. Godovikov, M. J. Michelson, E. V. Rosengart and V. I. Rosengart, *Pharmacological Rev.* 22, 355 (1970).

25. R. D. O'Brien, in *Drug Design*, E. J. Ariens, Ed., Academic Press, 1971, p. 162.

26. R. Allahyari, P. W. Lee, G. H. Y. Lin, R. M. Wing, and T. R. Fukuto, *J. Agr. Food Chem.* 25, 471 (1977).

27. E. Reiner and W. N. Aldridge, *Biochem. J.* 105, 171 (1967).

28. W. D. Reed and T. R. Fukuto, *Pestic. Biochem. Physiol.* 3, 120 (1973).

29. I. B. Wilson and E. Cahib, *J. Amer. Chem. Soc.* 78, 202 (1956).

30. I. B. Wilson, M. A. Hatch and S. Ginsburg, *J. Biol. Chem.* 235, 2313 (1960).

31. I. B. Wilson, M. A. Harrison and S. Ginsburg, *J. Biol. Chem.* 236, 1498 (1961).

32. R. L. Metcalf, *Bull. Wld. Hlth. Organ.* 44, 43 (1971).

33. L. W. Dittert and T. Higuchi, *J. Pharm. Sci.* 52, 852 (1963).

34. D. F. Heath, *Organophosphorus Poisons*, Pergamon Press, New York, 1961, pp. 72-100.

35. J. B. Wilson, *J. Biol. Chem.* 197, 215 (1952).

36. R. D. O'Brien, B. D. Hilton and L. Gilmour, *Molec. Pharmacol.* 2, 593 (1966).

37. T. Nishioka, T. Fujita, K. Kamoshita, and M. Nakajima, *Pestic. Biochem. Physiol.* 7, 107 (1977).

38. B. Hetnarski and R. D. O'Brien, *J. Agr. Food Chem.* 23, 709 (1975).

39. Y. C. Chiu, M. A. H. Fahmy and T. R. Fukuto, *Pestic. Biochem. Physiol.* 3, 1 (1973).

40. D. A. Wustner, C. Smith and T. R. Fukuto, *Pestic. Biochem. Physiol.* (in press).

Contributors

James R. Chernobieff
Departments of Animal Physiology and Environmental Toxicology
University of California
Davis, California 95616

T. Roy Fukuto
Division of Toxicology and Physiology
Department of Entomology
University of California, Riverside
Riverside, California 92521

Derek W. Gammon
Department of Zoology
University of Cambridge
Cambridge, England 002-3 58717

Syed M. Ghiasuddin
Pesticide Research Center
Michigan State University
East Lansing, Michigan 44824

Robert M. Hollingworth
Department of Entomology
Purdue University
West Lafayette, Indiana 47907

Stephen W. Jones
Section of Neurobiology and Behavior
Cornell University
Ithaca, New York 14853

Albert E. Lund
Department of Pharmacology
Northwestern University Medical School
303 East Chicago Avenue
Chicago, Illinois 60611

Fumio Matsumura
Pesticide Research Center
Michigan State University
East Lansing, Michigan 44824

Thomas A. Miller
Division of Toxicology and Physiology
Department of Entomology
University of California
Riverside, California 92521

Toshio Narahashi
Department of Pharmacology
Northwestern University Medical School
303 East Chicago Avenue
Chicago, Illinois 60611

Dale M. Norris
642 Russell Laboratories
University of Wisconsin
Madison, Wisconsin 53706

Richard D. O'Brien
Office of Provost
University of Rochester
Rochester, New York 14627

Thomas L. Payne
Department of Entomology
Texas A&M University
College Station, Texas 77843

Puppala Sudershan
Section of Neurobiology and Behavior
Cornell University
Ithaca, New York 14853

Lawrence W. Reiter
National Environmental Research Center
Neurophysiology and Behavioral Branch
Environmental Protection Agency Technical Center
Research Triangle Park, North Carolina 27711

Wendell L. Roelofs
New York Agricultural Experiment Station
Department of Entomology
Entomology-Plant Pathology Laboratory
Geneva, New York 14456

Daniel L. Shankland
Department of Entomology
Mississippi State University
Mississippi State, Mississippi 39662

Joep van den Bercken
Institute of Veterinary Pharmacology and Toxicology
University of Utrecht
Utrecht, the Netherlands

Dorothy E. Woolley
Departments of Animal Physiology and Environmental Toxicology
University of California
Davis, California 95616

George K.W. Yim
Department of Entomology
Purdue University
West Lafayette, Indiana 47907

INDEX

GPSR Compliance

The European Union's (EU) General Product Safety Regulation (GPSR) is a set of rules that requires consumer products to be safe and our obligations to ensure this.

If you have any concerns about our products, you can contact us on ProductSafety@springernature.com

In case Publisher is established outside the EU, the EU authorized representative is:

Springer Nature Customer Service Center GmbH
Europaplatz 3
69115 Heidelberg, Germany

Batch number: 10684222

Printed by Printforce, the Netherlands